Advances in Geographic Information Science

For further volumes:
http://www.springer.com/series/7712

Daniel A. Griffith · Jean H.P. Paelinck

Non-standard Spatial Statistics and Spatial Econometrics

Springer

Prof. Daniel A. Griffith
University of Texas, Dallas
School of Economic, Political &
Policy Sciences
800 W. Campbell Road
75080 Richardson Texas
USA
dagriffith@utdallas.edu

Prof. Dr. Jean H.P. Paelinck
George Mason University, School
of Public Policy
Oranjelaan 36
3062 BT Rotterdam
Netherlands
j.paelinck@planet.nl

ISSN 1867-2434 e-ISSN 1867-2442
ISBN 978-3-642-16042-4 e-ISBN 978-3-642-16043-1
DOI 10.1007/978-3-642-16043-1
Springer Heidelberg Dordrecht London New York

Cover design: deblik, Berlin

Printed on acid-free paper

Springer is part of Springer Science+Business Media (www.springer.com)

Preface

Despite spatial statistics and spatial econometrics both being recent sprouts of the general tree "spatial analysis with measurement"—some may remember the debate after WWII about "theory without measurement" versus "measurement without theory"—several general themes have emerged in the pertaining literature.

But exploring selected other fields of possible interest is tantalizing, and this is what the authors intend to report here, hoping that they will suscitate interest in the methodologies exposed and possible further applications of these methodologies. The authors hope that reactions about their publication will ensue, and they would be grateful to reader(s) motivated by some of the research efforts exposed hereafter letting them know about these experiences.

Lectori salutem ...

Richardson, Texas — Daniel A. Griffith
Dallas, Texas — Jean H.P. Paelinck

Prologue

We began this work by writing the joint paper entitled "An equation by any other name is still the same: on spatial statistics and spatial econometrics," which appears in a very abridged form in *Annals of Regional Science* (2007, 41: 209–227). Here we present the unabridged version of this paper in its entirety, so that readers can appreciate the full set of discussions we were forced to dramatically condense to shorten the length of the published paper. They introduce the main theme of the present book: non-standard spatial statistical and econometric analysis.

Abstract

Statistics is a branch of mathematics concerned with the collection, quantification, analysis, interpretation, and presentation of real-world data, and the use of probability theory to estimate population parameters with these data. *Spatial statistics* is a subset of statistics that is concerned with handling the special problems associated with geographically distributed data, which include spatial point patterns, regional and lattice measurement aggregations, and irregularly spaced site-specific measurements on a surface. Meanwhile, econometrics is concerned with the application of statistical methods to the study of economic data and problems. When coining the term *spatial econometrics* in 1979, Paelinck and Klaassen characterized it as a subset of econometrics that is concerned with the role of spatial dependence in regional economic model response and explanatory variables, asymmetries in spatial relationships, the specification of geographic structure governing spatial interactions, and the explicit modeling of space. We outline and discuss similarities (e.g., testing for the presence of spatial autocorrelation) and differences (e.g., map generalization) between spatial statistics and spatial econometrics. In doing so, our goal is to help clarify past, present, and future relationships between these two subfields.

A reader should note that the table, figure, and equation numbering in the prologue is particular to it, and is not part of the number sequencings of these items in the book.

I Introduction

Spatial statistics (e.g., see Cressie, 1991) addresses the patterns and stochastic variation in attribute data across their geographic locations, given that all data have implicit, and hopefully explicit, geographic tags (i.e., georeferencing, such as longitude and latitude coordinates). The primary pattern being exploited is that data nearby in space tend to be more alike than those farther apart. This distance-based inter-values correlation complicates statistical inference—the assumption of independent observations is violated—whose principal form here often is prediction of values at unobserved locations from those at observed locations, and estimation of unknown parameters of models whose specifications incorporate this spatial correlation and, in some analyses, other forms of distance relationships. Spatial statistics primarily covers the topics of centrographic measures, statistics for spatial data (spatial autoregression and geostatistics), point pattern analysis, and image analysis. Most applications of spatial statistics to date have been concerned with modelling quantitative attribute measurements on an interval/ratio scale with a normal probability model—the auto-normal specification. More recently, binary, percentage, and counts data have been addressed through generalized linear modelling with the binomial and Poisson probability models–the auto-logistic, auto-binomial, and Winsorized (i.e., truncated) auto-Poisson specifications.

Spatial econometrics (e.g., see Anselin, 1988) addresses, inter alia (see Sect. III), two complications that arise when the locational tagging of sample economic data is explicitly recognized: (1) again, the spatial correlation (or dependence) that exists between observed values; and, (2) spatial heterogeneity, the place-to-place nonconstant variance of georeferenced data. These two features of geographic data are at odds with two Gauss-Markov Theorem assumptions used in regression modeling. With regard to spatial correlation, the Gauss-Markov view of sample data, say Y, is that geographic variation across locations can be accounted for with a battery of judiciously selected covariates, say **X**, where the values of each X are fixed at each location, and regression parameters are constant from place to place. Repeated sampling results in place-to-place variation in Y being attributed only to a stochastically varying error term. The critical assumption is that this error term has constant variance across locations, and zero covariance amongst its location-specific errors—the presence of non-zero spatial autocorrelation violates this second property. Meanwhile, spatial heterogeneity violates the Gauss-Markov assumption that a single linear relationship exists across locations.

In practice, spatial statistics and spatial econometrics reflect traditions of their respective parent disciplines, namely statistics and econometrics. In other words, they share much in common, with some notably differing emphases. "Statistics is the science of gaining information from numerical data" (Moore, 1995, p. 2). It provides data-interrogative tools and conceptual frameworks for gaining understanding through empirical-based induction, and involves data acquisition, data analysis, and statistical inference. Econometrics, which literally means economic measurement, is the "setting up of mathematical models describing economic relationships (such as that the quantity demanded of a good is dependent positively on income and negatively on price), testing the validity of such hypotheses and estimating the parameters in order to obtain a measure of the strengths of the influences of the different independent variables" (Bannock et al., 2003). It is a collection of quantitative techniques, both statistical and mathematical (e.g., operations research), that supports economic theory testing and decision-making.

The main objective of this paper is to describe conspicuous similarities and differences between the two subdisciplines of spatial statistics and spatial econometrics, especially as they are practiced in the social sciences. Of particular interest is the question asking where these two subdisciplines diverge.

II Similarities and Differences Between Spatial Statistics and Spatial Econometrics: An Overview

A simple comparison between spatial statistics and spatial econometrics may be derived from a tabulating of their focal problems, which are summarized in Table 2.1.

Both subdisciplines contain methodology, such as resampling techniques, and a focus on model diagnostics supporting model-based (i.e., the essential tool for describing a map is a model, and inferences are to a superpopulation), rather than simply design-based (i.e., locations have unique fixed but unknown values that can be estimated with a proper sampling design), inference (Brus and de Gruijter,

Table 2.1 Focal problems in spatial statistics and spatial econometrics

Spatial statistics	Spatial econometrics
Super population perspective (i.e., realizations from a theoretical population): model-based inference	
Properties of estimators	
Specification of geographic neighbourhood structure	
Modifiable areal unit problem (MAUP)	
Quantifying spatial autocorrelation	
Variable transformations: Box-Cox, Box-Tidwell	
Spatially adjusted statistical techniques	
Cluster detection: hot and cold spots; LISA statistics	
Distance as a covariate	
Bayesian hierarchical models	
Exploratory spatial data analysis	
Space-time modelling	
Sampling network structure: design-based inference	Constrained parameter estimation
Ecological fallacy	Optimization models
Map generalization: spatial interpolation	Endogenous versus exogenous variables
Missing spatial data imputation	Spatial complexity and spatial regimes
Auto- model specification: normal, Poisson, binomial	
Spatial structure as a covariate (spatial filtering)	
Bayesian smoothing of map values	
Error propagation	

1993). Until recently, the auto-normal model has been the specification of necessity, accompanied by a focus on Box-Cox variance stabilizing and Box-Tidwell linearizing transformations to strengthen auto-normal approximations for non-normal data (e.g., Griffith et al., 1998). A natural outcome of this work has been the development of cluster detection techniques, such as the LISA statistic (Anselin, 1995). Distributional properties of parameter estimates are common to both subdisciplines, too. Unbiasedness was the first property to be assessed (Cliff and Ord, 1981; Anselin, 1988), followed by efficiency (e.g., Cordy and Griffith, 1993) and consistency (Mardia and Marshall, 1984) assessments. And, more recent Markov chain Monte Carlo (MCMC) work has focused attention on the property of sufficiency (e.g., Graham, 1994) in spatial statistics, and Bayesian analysis in spatial econometrics (e.g., Le Sage 1997, 2000). Meanwhile, articulation of geographic neighbourhoods for constructing a geographic weights matrix was one of the first specification issues addressed (Florax and Rey, 1995; Griffith and Lagona, 1998). Both subdisciplines address the modifiable areal unit problem (MAUP)–the sensitivity of findings to the repartitioning of a landscape into a different set of n zones, and/or the reaggregation of locations into a different number of zones (e.g., Amrhein and Wong, 1996; Amrhein and Reynolds, 1996, 1997). Spatial statistics and spatial econometrics began with the problems of incorporating various distance effects into data analyses, and quantifying spatial correlation. Following these efforts, both subdisciplines moved on to hypothesis testing, and then to modelling issues (Cliff and Ord, 1973; Paelinck and Klaassen, 1979). One consequence of these efforts is that currently a wide range of autoregressive-based spatially adjusted statistical techniques is available to spatial scientists. These techniques also have been extended to space-time data analyses. Of note is that time, which is one-dimensional and unidirectional, can house stronger covariations than can space, whose two-dimensional and multidirectional nature can dilute covariations. Current fashionable analyses address regional convergence (see Arbia, 2004). Meanwhile, all real world data are noisy (are characterized by uncertainty/variability/stochastic error), dirty (are incomplete and/or include outliers/anomalies), and messy (permeated with observational dependencies and/or nonlinear relationships). Noisy data contain obscured/masked trends of various degrees; dirty data contain corruptions from inaccuracies and/or inconsistencies; messy data can motivate inscrutable model specifications. Exploratory spatial data analysis (ESDA) seeks to better understand georeferenced data in an attempt to neutralize effects of these three real world data properties. Geographic information systems (GIS) support some of the visualization critical to ESDA, helping to uncover potentially inexplicable data patterns. Spatial statistics furnishes some of the tools for geographic pattern detection, some of the inputs to spatial data generated hypothesis formulation, and part of the crucial perspective for spatial model specification. (See Haining et al., 1998; Anselin, 1998) And, today both subdisciplines are expanding the frontiers of Bayesian hierarchical modelling to include georeferenced data analysis (e.g., Haining, 2003; Le Sage, 1997, 2000).

The more catholic view of spatial statistics has produced a more diversified set of research problems. The data collection tradition of statistics results in spatial

Table 2.2 Thematic results obtained from four volumes[a] containing 36 papers about spatial econometrics

Topic	%
Spatial interaction and test	53
Model specification and test	14
Data analysis.	14
Estimation.	9

statistics addressing the problem of designing sampling networks to ensure appropriate geographic as well as inter-point distance coverage (e.g., Stehman and Overton, 1996; Diggle and Lophaven, 2004). Because considerable spatial data result from aggregating georeferenced individuals into regions, the ecological fallacy (i.e., an inference about some individual event based upon the aggregate group data to which it belongs being the observational unit in an analysis) has received much attention in spatial statistics (e.g., see Richardson, 1992; King, 1997; Freedman, 2001). Geographers have constructed contour maps for centuries (see Meijering, 2002); this spatial interpolation focus has promoted the development of kriging techniques, yielding best linear unbiased predictor map generalizations (e.g., Stein, 1999). This tradition also has spawned research addressing the data imputation problem of filling holes in maps: missing data estimation, as well as Bayesian smoothing to bolster grouped data based upon insufficiently small sample sizes (see Pascutto et al., 2000; Griffith and Layne, 1999). Meanwhile, the MCMC procedures enabling implementation of Bayesian analysis also support maximum likelihood estimation of auto- model parameters other than those for the auto-normal probability model (e.g., Kaiser and Cressie, 1997; Gotway and Stroup, 1997; Huffer and Wu, 1998). Consequently, generalized linear spatial modelling now is feasible, from both a frequentist and a Bayesian point of view. The theory and development of spatial filtering models, which have some parallels with impulse-response time-series modelling, is beginning to unfold (see Getis, 1995; Griffith, 2000a, 2002, 2004; Getis and Griffith, 2002; Borcard and Legendre, 2002). Finally, error propagation–both locational and attribute—especially through GIS operations and its impacts on spatial statistical inference, has been topical for about two decades.

Meanwhile, a scanning of four sources produces the tabulation for recent spatial econometrics work appearing in Table 2.2. Of course, papers may treat different aspects at the same time, but the central interest of a study determined its classification in the listing. The absolute majority of these papers treat spatial interaction. The model specification tests and estimation topics are *enfants pauvres*, constituting a mere 23% of the total. It is precisely on these two topics that our non-standard view about spatial econometrics concentrates.

III Toward Non-standard Spatial Econometrics

The following are selected non-conventional problems encountered in the practice of spatial econometrics:

(1) the fundamental bias of regional statistical data, resulting from spatial aggregation;
(2) the specification of spatial models is very often "classical," but other alternatives exist and should be explored;
(3) spatial characteristics continue to pose econometric problems, alluding to the need to develop more appropriate estimators; and,
(4) complexity, especially spatial complexity, is the feature that covers these three preceding points, often in terms of uncovering latent spatial regimes.

Possible solutions to these specific problems are addressed in the ensuing discussion.

3.1 Spatial Bias

A problem studied in particular by spatial statisticians is the "Modifiable Areal Unit Problem" (MAUP), the possible use of territorial units of different sizes. In a genuine econometric spirit, this can be treated as a spatial aggregation problem, producing some disturbing consequences for a spatial econometrician. One of these consequences may be summarized as follows: "The important result is that in general econometric aggregation, if only one macro-aggregate is considered, *just one* parameter bias is present in the macro-model; in meso-aggregation, as it took place here, *every meso-area* has its own specific aggregation bias, which leads to parameter variability between meso-areas, and this might result, in econometric estimation, in some sort of (biased) average value, depending on the characteristics of the sample being investigated and the particular spatial aggregation specification" (Paelinck, 2000).

In larger models the implicit bias will be even more complex; moreover, the stochastic terms of a model will reveal heteroscedasticity and spatial autocorrelation under very general conditions. Of note here is that resulting conclusions impose the use of appropriate specifications adapted to each problem at hand (see Sect. 3.2); a possible technique for achieving this end is that of composite parameters—at least when the number of degrees of freedom permits—in order to take account of the *specific* bias inherent in each meso-economic spatial unit included in a cross-section analysis. But then, what is spatial heterogeneity, and what is spatial bias? Recently filtering data for observational errors, and then for spatial aggregation bias, was proposed by Paelinck (2003). The method was applied to a series with maximal spatial complexity (see Sect. 3.4), after which complexity was reduced by two thirds, and a simple linear model could be fitted to the filtered data.

3.2 Specification

The specification of spatial models—either regional or urban in nature—should obligatorily reproduce the workings of spatial economies. Consider, for example, the problem of multi-regional convergence in terms of per capita incomes. A possible specification here could be the so-called Lotka-Volterra model, which allows spatial interdependencies to be introduced; this specification is non-linear in form. If column-vector **y** represents regional per capita incomes, this model can be written as:

$$\mathbf{\Delta'}\ln\left(\mathbf{y_t}\right) = \mathbf{A}\,\mathbf{y_{t-1}} + \mathbf{a}, \tag{3.1}$$

where **Δ'** is the backward difference operator, **A** is a transition matrix, and **a** is a column vector of autonomous growth rates that take into account all factors not covered by the transition matrix **A** (e.g., regional policy, foreign trade). The interesting point, with respect to pure tendency models, is that one can verify the presence of convergence (via the eigenvalues of matrix **A;** Paelinck, 1992)—i.e., whether a stable singular point exists—the latter being defined by

$$\mathbf{y^o} = -\mathbf{A^{-1}a}, \tag{3.2}$$

if $\mathbf{A^{-1}}$ exists. Vector $\mathbf{y^o}$ should not differ significantly from the unit vector when convergence is attained. An appropriate estimation method for the parameters of **A** and **a** is mentioned in the subsequent discussions. The model for 119 European regions converged mathematically (119 eigenvalues have the correct signs and values), but diverged in an economic sense; those eigenvalues are of course stochastic variables, but if the condition were not to be satisfied, the model would even be mathematically divergent, and certainly no economic convergence could be present.

An even more important problem is that of the algebraic structure to be given to the model under construction. Paelinck (2002) investigates the possibilities of model specification based on a so-called *min-algebra*, which he uses to generalize the specification of the European FLEUR-model (Ancot and Paelinck, 1983), the latter being based on the idea of a *growth threshold.* In min-algebra, one or several explanatory terms (variables with their reaction coefficients) *of minimal value* determine the value of the endogenous variable(s). This perspective is reminiscent of the minimum requirements approach found in the urban geography literature (see Ullman and Dacey, 1960). Thus, instead of considering a (linear or non-linear) combination of endogenous, exogenous or predetermined variables, only one (or a limited number of) explanatory variable(s) appears in each equation. For instance, the development of a region could be hampered by the absence of a strategic factor, such as technologically highly trained manpower. In mathematical terms, an equation of the model presents itself as:

$$\mathrm{y_i} = \min\left(\mathrm{a_{i1}y_1} + \mathrm{ci_1},\ ..,\ \mathrm{a_{in}y_n} + \mathrm{c_{in}};\ \mathrm{b_{i1}x_1} + \mathrm{d_{i1}},\ ..,\ \mathrm{b_{im}x_m} + \mathrm{d_{im}}\right), \tag{3.3}$$

where the y_i are endogenous variables, and the x_js are exogenous variables or economic policy instruments (i.e., control variables). Such a specification can be illustrated in terms of the efficiency of instruments problem of regional policy or physical planning. Often such instruments reveal themselves as inefficient, one reason being that a development process is blocked by an absence of the needed minimal value of one or another of the driving terms. This reasoning sheds new light on the theory of endogenous regional growth. It is indeed possible that one of the key-factors of regional development (e.g., entrepreneurial initiatives, appropriate manpower, even cultural factors) is not sufficiently present in a region. To promote its growth in this context, a region has to favour factor expansion, rather than outcomes from applying *classical* recipes, such as financial stimuli or the creation of *technopoles*.

A bad specification of regional models becomes really dramatic when they are used to derive *baskets* of regional policy measures. For example, borrowing an *optimal* regional policy from the aforementioned study (Paelinck, 2003), assume the following objective function:

$$y = \min\,(65 + x_1,\ 13 + x_2,\ 2 + x_3,\ 60 + x_4)\,, \tag{3.4}$$

where the x_i, $i=1, ..,4$ represent restrictive factors. Equation (3.4) has to be maximised under the condition

$$4\,x_1 + 3\,x_2 + 2\,x_3 + x_4 \le 50, \tag{3.5}$$

together with the usual non-negativity conditions.

One can show that the following is the solution to this problem:

Variables	x_1	x_2	x_3	x_4	y
Values	0	5.6	16.6	0	18.6

The logic of the chosen algebra produces a solution with *two* non-zero decision variables, whereas a linear program (LP) under a *classical* algebra would, in general, produce only one non-zero decision variable; the solution could of course be obtained by LP, but given the algebra used, extra side conditions would have to be introduced.

Finite automaton is still another specification (for a formal definition, see Linz 1996, p. 2) to be systematically investigated. This specification can be viewed as an "if"-specification; in symbolic terms,

$$y:\ \mathrm{if}(\alpha x_i + \beta < \gamma z_i + \delta; \alpha x_i + \beta; \gamma z_i + \delta), \tag{3.6}$$

which reads as follows: if $\alpha x_i + \beta < \gamma z_i + \delta$, then the values of the left-hand side hold; otherwise, those on the right-hand side hold. Comparing expressions (3.5) and (3.6) reveals that both specifications are in fact isomorphic. Therefore, problems of the types just listed can be treated by either method.

In order to subject a finite automaton model to a well-documented empirical test, gross regional product figures for the Netherlands were divided into two macro-regional sets, one for the western provinces (Noord-Holland, Zuid-Holland and Utrecht, the so-called "Rimcity"), and the other comprising data for the remaining provinces. Both a binary and a fuzzy version of the automaton model were developed. One conspicuous, and rather curious, insight gleaned from the obtained results is the behavior of the growth rate values for the non-Rimcity provinces: whatever the state of the location factors' attractiveness, they follow the ups and downs of the Rimcity growth rates. This finding is consistent with the Rimcity being the *motor* of the Dutch economy (Paelinck, 1973, pp. 25–40, especially pp. 37–40), imposing its evolutionary rhythm on the other regions, and corresponding to a sort of non-*Fick* diffusion in thermodynamics.

Of note is that the specifications presented here can be readily generalized to three or more alternatives (e.g., regions, test specifications). For the finite automaton version, for example, the following illustration shows how *AND* and *OR* statements can be added:

$$\begin{aligned} &y_i : \text{if}((cz_i + d < ax_i + b) \\ &\mathit{AND}\,(eu_i + f < ax_i + b);\ (cz_i + d)\,\mathit{OR}(eu_i + f);\ ax_i + b). \end{aligned} \tag{3.7}$$

All of these specifications should be tested against each other; examples of this pairwise testing can be found in Griffith and Paelinck (2009).

Finally, the specification of spatial lags, in the endogenous and/or exogenous variables, may be referred to as the *W*-matrix problem. Several suggestions for dealing with this problem appear in the literature, some of them being purely *mechanical*. As spatial econometrics is about economics in pre-geographical space, some economic background for a solution is desirable. One possibility is the inspection of the residuals (the "doggy-bag principle"); relatively high and/or low, positive and/or negative values should be inspected, in an attempt to generate assumptions (e.g., competition and/or cooperation could be present at short or long distances: distance can be *hampering* or *protecting*). Examples are known where, for instance, mapped locations of residuals have led to identification of the correct complementary variable missing from a model specification (e.g., see Thomas, 1968).

A more recent solution has been proposed in Chapter 11, to wit the use of a bivariate Poisson distribution to jointly estimate the parameters of space- *and* time-lags; a first application to regional products of Belgian regions has shown the method to be operational.

3.3 Estimators

The fundamental structure of spatial models invites development of different types of estimators adapted to the special situations encountered. Indeed, established software does not always fit the specific estimation problem encountered.

Now Chapter 17 proposes various estimators especially appropriate for handling particular circumstances. One is a simultaneous least squares estimator, perfectly well adapted for use with the Lotka-Volterra-type models already mentioned—i.e., models with simultaneous spatial and temporal interdependencies. The computation can proceed by iterative ordinary least squares (OLS) employing the expression

$$\hat{\boldsymbol{\beta}} = (\mathbf{X'\hat{X}})^{-1}\mathbf{X'Y}, \tag{3.8}$$

where **X** endogenous variables take on the values computed by the model itself, the method integrating the computation of optimal—spatial and temporal—*starting points* for endogenous simulations. This estimator is consistent, and the probability limit of its variance-covariance matrix is known (in fact, it is the usual OLS matrix). Meanwhile, in recent research (Paelinck, 2006), the estimating procedure has been endogenized, resulting in the parameters and the estimated—endogenous—variables being computed in the same iteration. This method can be applied to static and dynamic spatial models—with their typical spatial lags.

Also of note is that most spatial models are inherently non-linear, so that after appropriate specification, estimation methods other than OLS need to be used. For example, a recently devised estimation method combines Box-Cox and Box-Tidwell transformations (Paelinck and van Gastel, 1995; Griffith, Paelinck and van Gastel, 1998), and proceeds from the (partial) elasticities of the transformed function.

Here estimation may be obtained with some semi-parametric method. Consider the following second order differential expansion (derived from a second order MacLaurin expansion):

$$\mathrm{df}(x, y) = f_x dx + f_y dy + f_{xx} x\, dx + f_{yy} y\, dy + f_{xy} (y\, dx + x\, dy), \tag{3.9}$$

where the coefficients of the linear terms are changed by adding periodically the coefficients of the quadratic terms, which, in production functions, for example, express the changes in marginal productivities. Now the question of interest asks how to apply this technique, originally developed for time series, to a problem in *spatial* econometrics? The difficulty originates from the difference between *time's arrow* and the non-oriented multi-directional and reciprocal dependencies latent in spatial data.

Suppose regions 1, 2 and 3 coexist on a line, such that the degree of contiguity between 1 and 2, and between 2 and 3 is one; we will limit ourselves to that degree and skip further problems of spatial autocorrelation. Let the regressand be *y*, and the unique regressor *x*. The model, adapted from Eq. (3.9), can be specified as follows.

$$\Delta' y_{12} = a_{12} \Delta' x_{12} + b x_1 \Delta' x_{12} \tag{3.10}$$

$$\Delta' \acute{y}_{23} = a_{23} \Delta' x_{23} + b x_2 \Delta' x_{23} \tag{3.11}$$

$$\Delta' y_{21} = a_{21} \Delta' x_{21} + b x_2 \Delta' x_{21} \tag{3.12}$$

$$\Delta' y_{32} = a_{32} \Delta' x_{32} + b x_3 \Delta' x_{32} \tag{3.13}$$

$$a_{12} = a_{21} \tag{3.14}$$

$$a_{23} = a_{32} \tag{3.15}$$

$$a_{23} = a_{12} + (d_{12} - e_{12})b \tag{3.16}$$

$$a_{21} = a_{32} + (d_{32} - e_{32})b \tag{3.17}$$

$$d_{12} + e_{12} = 1 \tag{3.18}$$

$$d_{32} + e_{32} = 1 \tag{3.19}$$

$$d_{12} = e_{32} \tag{3.20}$$

$$d_{32} = e_{12} \tag{3.21}$$

The variables in equations (3.10)–(3.21) are binary 0–1. This specification does not privilege any direction in space, and allows for increases or decreases of the reaction parameter between pairs of regions with the same separating distance. Finally, min-algebraic and finite automata parameters can be estimated and the specification tested (see Paelinck, 2003).

3.4 Complexity, Estimation and Testing

The problem studied in this section arose when spatial data were tested to determine whether or not they belonged to one or more possible regimes. In particular, a classical linear model and a min-algebraic one (Paelinck, 2003) were considered.

One idea about the relevance of one or another specification is to look into the *computational* complexity of a problem (Chaitin, 1975; Wolfram, 2002, pp. 557–559). This complexity, which we will call *conditional complexity*—due to the presence of exogenous variables—can be expressed as a function of the number of parameters necessary to fit a polynomial to endogenous variable. Consider the following index on [0,1] (Getis and Paelinck, 2004):

$$c = (n_p - 1)/(n_{pm} - 1)\,, \tag{3.22}$$

where n_p is the number of non-zero parameters, and n_{pm} is their maximum number (equal to the length of the series of endogenous variables; i.e., the size of a sample). Suppose especially the endogenous variables are void of measurement errors; after all, the observed values are the only ones available for analysis.

In Paelinck (2004), Eq. (3.22) has been applied to a series of test data. It first resulted in $c = 1$, meaning that the number of parameters equal to the sample size was necessary to satisfy the cubic equation:

$$y_i = a_i + \mathbf{a}'\mathbf{u}_i + \mathbf{u}_i{}'\mathbf{A}\mathbf{u}_i + \mathbf{u}_i{}'\mathbf{B}'\,\hat{\mathbf{u}}_i\,\mathbf{B}\mathbf{u}_i, \tag{3.23}$$

where $\mathbf{u}_i$ is observation i's vector of exogenous variables. When this test was applied to figures generated by $y_i = x_i + 2z_i$, only two parameters were necessary, rendering c = 0 .11. This finding gives a clue to a more complex specification for the first series than would be the case for the second one.

The following model was formulated to test the first series according to this aforementioned clue:

$$y_i = \theta\,(ax_i + bz_i + c) \;+\; (1 - \theta)\min(\alpha x_i + \beta; \gamma z_i + \delta) \;+ \varepsilon_i, \tag{3.24}$$

where θ is binary, and for which $\varphi = min\ \Sigma_i \varepsilon_i^2$ was chosen as a selection criterion—in fact, a minimal variance one (Theil, 1971, pp. 543–545; Aznar Grasa, 1989, p. 133). The second term on the right-hand side of Eq. (3.32) represents a min-algebraic specification (Paelinck, 2003).

The computation based upon Eq. (3.24) results in $\theta = 0$. Quite logically, the min-algebraic model was selected over the classical linear combination one. This procedure can be generalized to more than two competing model specifications. Equation (3.32) was set up in its present form because this form naturally leads to a *fuzzy* generalisation, by first relaxing the binary condition on θ to $0 \le \theta \le 1$, and to the split between min-regimes. In both cases, the min-algebraic regime remained dominant. Finally, if one applies this method to the *exact* case, then $\theta = 1$ and the exact equation is selected with $\varphi = 1.96x10^{-13}$.

In conclusion, the observed series may contain errors, and moreover (see Sect. 3.1), spatial bias is always present. Thus, the question arises as to whether corrections for these two features could be devised. This problem has been addressed in Paelinck (2004; see Sect. 3.1), where a solution to this twofold complication has been proposed. Applied to the data mentioned earlier in this section, the procedure resulted in a significant decline of the complexity coefficient given by Eq. (3.22).

IV On the Frontiers of Applied Spatial Statistics

Methodological advances are underway in the eight thematic areas of spatial statistics in which spatial econometricians tend to have less interest (see Table 2.1)–sampling network design, the ecological fallacy, map generalization, missing spatial data imputation, non-normal auto- model specification, spatial filtering, Bayesian smoothing of map values, and error propagation.

4.1 Non-normal Georeferenced Data Analysis

The description and explanation of map patterns of objects or events has been a continuing interest in geography and regional science for more than half a century. The normal probability model describes a bell-shaped curve, a statistical frequency distribution that adequately characterizes many attributes, especially interval-ratio measurement scale ones. The Poisson probability model has played an important role in quantitative geographic work involving counts and rare events. And, the binomial probability model describes situations in which responses are categorical or limited to a particular range, resulting in binary indicator variables or percentages. In this first binomial situation a response is, say, presence/absence; in this second situation a response is a constrained event count. Historically, because of difficulties associated with analyzing Poisson or binomial georeferenced data, Box-Cox types of transformations often were employed in an attempt to devise good bell-shaped or normal curve approximations for non-normal data. The development of generalized linear modelling techniques was a first step supporting the direct analysis of non-normal data. Formulation of MCMC procedures supplied a necessary second step for their auto- versions. Now, autoregressive versions of a wide range of probability models can be estimated. Consequently, the use of normal approximations, with or without spatial model elements, should be a practice of the past.

Beginning with the work of Nelder and McCullagh (1983), applied statisticians increasingly have been successful in devising user-friendly implementations of probability models beyond that for the normal curve. Beginning with Wrigley (1985), spatial auto- versions of these model implementations have been developed in parallel, but with a time lag. "The central role of the Poisson distribution with respect to the analysis of counts is analogous to the position of the normal distribution in the context of models for continuous data" (Upton and Fingleton, p. 71). But development of an auto-Poisson model proved to be a failure, with this particular model specification being unable to capture the more commonly encountered case of positive spatial autocorrelation (Besag, 1974). Circumventing this restriction has been achieved in several different ways: Kaiser and Cressie (1997) propose an approach based upon Winsorizing counts (i.e., systematically replacing extremely high counts with the value of some cut-off criterion; after Barnett and Lewis, 1978); and, Griffith (2002) proposes an approach based upon spatial filtering (i.e., transforming a variable containing spatial dependence into one free of spatial dependence by partitioning the original georeferenced attribute variable into two synthetic variates, a spatial filter variate capturing latent spatial dependency that otherwise would remain in the response residuals, and a nonspatial variate that is free of spatial dependence; e.g., Griffith, 2000a). Meanwhile, statistical techniques for analyzing correlated binary variables are not as plentiful as those for analyzing correlated continuous data. When binary spatial data are of interest, the first model that should come to mind for describing these data and their latent spatial autocorrelation is the auto-logistic/binomial specification, whose development has been successful, although estimation of its parameters is rather daunting. Again, Griffith (2004) proposes an approach based upon spatial filtering.

Parameters of both the Winsorized auto-Poisson and the auto-logistic/binomial models became estimable with the advent of MCMC techniques. A Markov chain is a process consisting of a finite number of *states* and known probabilities, p_{ij}, of moving from state i to state j. Markov chain theory is based on the *Ergodicity Theorem*: the transition matrix of state-to-state probabilities must be irreducible, recurrent non-null, and aperiodic. If a Markov chain is ergodic, then a unique steady state distribution exists, independent of the initial geographic distribution:

for transition matrix $\mathbf{M}$, $\lim_{k \to \infty} \mathbf{M}^k = \mathbf{M}^*$. Meanwhile, the Monte Carlo method provides approximate solutions to a variety of mathematical problems by performing statistical sampling experiments with a computer using pseudo-random numbers. Consequently, MCMC provides a mechanism for taking dependent samples in situations where regular sampling is difficult, if not completely impossible. The standard situation is where the normalizing constant for a joint or a posterior probability distribution is either too difficult to calculate or analytically intractable. Accordingly, MCMC begins with conditional (or marginal) distributions, and MCMC sampling outputs a sample of parameters drawn from their joint (or posterior) distribution.

For auto-Poisson and auto-logistic/binomial models, MCMC is implemented with Gibbs sampling, a recipe for producing a Markov chain that yields simulated data that have the correct unconditional model properties, given the conditional frequency distributions of those variables under study (see Casella and George, 1992). MCMC exploits sufficient statistics, makes use of marginal probabilities, and frequently utilizes pseudo-likelihood results. Beginning with pseudo-likelihood estimates, and a set of initial random numbers geographically distributed across a map, Gibbs sampling involves visiting each location, in turn, and updating its value by computing a new value with the auto- model specification of interest, using pseudo-likelihood parameter estimates obtained from observed data. A single MCMC iteration is completed when all n locations have had their values updated. Iterations are repeated until the sufficient statistics converge, which often involves tens of thousands of iterations. Finally, MCMC maximum likelihood estimates (MCMC-MLEs) are calculated by constructing a ratio of two likelihood functions, one with the unknown parameters and a reference one based upon the observed data (e.g., the pseudo-likelihood specification). The formulae for this, as well as for the accompanying asymptotic standard errors, appear in Huffer and Wu (1998).

For illustrative purposes, consider results of the empirical example appearing in Table 4.1; two additional examples can be found in Griffith (2005b, 2006b). As with auto-normal models, these examples illustrate that one of the principal impacts of spatial autocorrelation is on the estimates of standard errors. This findings holds even when the numbers involved are quite large (e.g., for Wales, 1,112,912 of 2,218,850 voters cast ballots, yielding very small standard errors).

MCMC techniques also have made Bayesian analysis based upon hierarchical generalized linear models feasible. This category of model is called "hierarchical" because it has two levels. At the *higher level*, hyper-parameter distributions are described by multivariate priors. Such distributions are characterized

Table 4.1 Auto-binomial description of % Welsh voter turnout: 1997 referendum

	Maximum pseudo-likelihood		MCMC-MLE	
Parameter	Estimate	Standard error	Estimate	Asymptotic standard error
α	–0.1463	0.003	–0.1296	0.0004
ρ	0.0519	0.001	0.0179	0.0003

by vectors of means and covariance matrices; spatial autocorrelation is captured here. At the *lower level*, individuals' behavior is described by probabilities of achieving some outcome that are governed by a particular model specification. This type of analysis can be implemented with software such as GeoBUGS (see Casella, 1985; Casella and George, 1992), the spatial statistical module add-on to WinBUGS (see Cowles, 2004), an MSWindows-based program supporting **B**ayesian inference **U**sing **G**ibbs **S**ampling (BUGS; see http://www.mrc-bsu.cam.ac.uk/bugs/winbugs/contents.shtml). Bayesian inference is used to spatially smooth georeferenced data values using MCMC methods. GeoBUGS implements models for data that are collected within discrete regions (not at the individual level), and smoothing is done based on Markov random field models for the neighborhood structure of the regions relative to each other, specified in terms of a conditional autoregressive model and the queen's definition of geographic adjacency as the default. Random effects are included that are either spatially structured (i.e., contain spatial autocorrelation) or spatially unstructured (i.e., contain no spatial autocorrelation). The perspective here is similar to that for spatial filtering. Rather than having observed values directly correlated, spatial autocorrelation is accounted for through the mean response parameter capturing spatial dependency effects. In fact, spatial filter eigenvectors can be used as covariates to account for spatial autocorrelation in GeoBUGS.

For illustrative purposes, consider the 2000 geographic distribution of percentage urban population by municipio across the island of Puerto Rico (see Fig. 4.1a) employing a binomial probability model. The MCMC Bayesian analysis involved 50,000 iterations, of which 10,000 were discarded for the burn-in period, after which only ever 5th map was retained for analysis. The iteration time-series plot and correlogram for the spatial autocorrelation parameter, $\hat{\rho}_{\text{CAR}}$, appear to be well behaved. Grouping the MCMC results for the spatial autocorrelation parameter into 100 groups of 100 consecutive estimates produced the following ANOVA results: for variance homogeneity, Levene = 1.14 (p = 0.156), and for difference of means, F = 1.30 (p = 0.023)—suggesting, perhaps, the need for more iterations or designating a larger burin-in period. Given a multivariate normal conditional autoregressive (CAR) model specification, a uniform prior for the logistic intercept term, and a gamma prior for the spatial autocorrelation parameter, the 8,000 retained maps rendered $\hat{\rho}_{\text{CAR}} = 0.9378(S_{\hat{\rho}} = 0.0526)$–where $\left[< 1/\sqrt{\sum_{j=1}^{n} c_{ij}} >_{\text{diag}} (\text{I} - \rho_{\text{CAR}}\text{C}) < 1/\sqrt{\sum_{j=1}^{n} c_{ij}} >_{\text{diag}} \right]^{-1}$ is the covariance structure matrix, with $\left[< 1/\sqrt{\sum_{j=1}^{n} c_{ij}} >_{\text{diag}} \right]$ being a diagonal matrix. For positive spatial autocorrelation, this specification of the covariance matrix allows positive values of $\hat{\rho}_{\text{CAR}}$ to fall between 0 and 1, permitting a more intuitive interpretation for it. The geographic distribution of random effects appears in Fig. 4.1b, and reflects

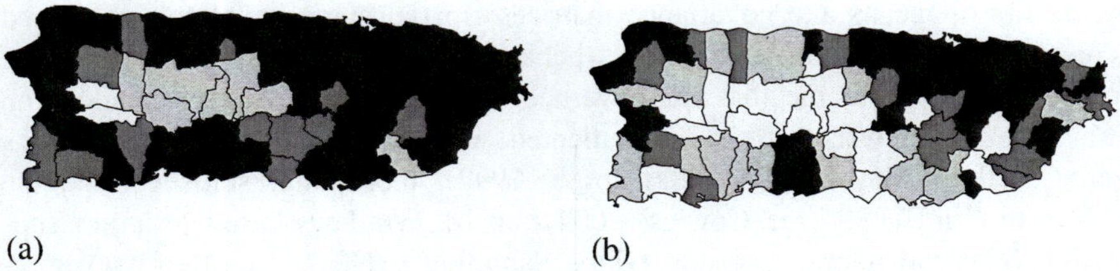

Fig. 4.1 Quantile classifications. *Left*: (**a**) geographic distribution of the 2000 percentage of urban population: white – 34–86%; light grey – 87–92%; medium grey – 93–96%; dark grey – 97–99%; and, black – 100%. *Right*: (**b**) geographic distribution of the mean random effect: white – –0.6 – 1.8; light grey – 1.8 – 2.5; medium grey – 2.5 – 3.1; dark grey – 3.1 – 4.8; and, black – 4.8 – 13.3

much of the geographic pattern visible in Fig. 4.1a. Spatial autocorrelation indices for this random effects map are: Moran Coefficient = 0.49836, Geary Ratio = 0.44382.

Clearly MCMC furnishes a powerful implementation tool for spatial statistics dealing with non-normal data and especially mixed effects modelling.

4.2 Sampling in Geographic Space

As spatial autocorrelation latent in georeferenced data increases, the amount of duplicate information contained in these data also increases. This property suggests the research question asking what is the number of independent observations, say n^*, that is equivalent to the sample size, n, of a geographic data set (Griffith, 2005a; also see Cressie, 1991, p. 15). This is the notion of effective sample size. Intuitively speaking, for a univariate situation, when zero spatial autocorrelation prevails, $n^* = n$; when perfect positive spatial autocorrelation prevails, $n^* = 1$. Equations may be derived for estimating n^* based upon the sampling distribution of a sample mean with the goal of obtaining some predetermined level of precision, using the following spatial statistical model specifications: (1) simultaneous autoregressive; (2) geostatistical semivariogram; and, (3) spatial filter. In this first case, for a given simultaneous autoregressive (SAR) model estimate of spatial autocorrelation, $\hat{\rho}$

$$\hat{n}^* \approx n \times \left[1 - \frac{1}{1 - e^{-1.92369}} \frac{n - 1}{n} \left(1 - e^{-2.12373\hat{\rho} + 0.20024\sqrt{\hat{\rho}}}\right)\right]. \quad (4.2.1)$$

In this second case,

$$n^* = \frac{n}{n - \sum_{i=1}^{n} \sum_{\substack{j=1 \\ j \neq 1}}^{n} \gamma(d_{ij})/(C_0 + C_1)/n}, \quad (4.2.2)$$

where $\gamma(d_{ij})$ denotes a particular semivariogram model with respective nugget and slope parameters of C_0 and C_1. And, in this third case,

$$n^* = \left(1 - R^2\right) n, \qquad (4.2.3)$$

where R^2 is the squared multiple correlation for the corresponding spatial filter regression model.

These preceding three formulae enable the computation of appropriate sample sizes for quantitative studies when nonzero spatial autocorrelation is present in georeferenced data. In order to do so, a pilot study must be carried out to obtain initial estimates of spatial autocorrelation and variable variance (also see Flores et al., 2003). If a spatial scientist chooses to obtain a variance estimate from the literature, then assuming moderate, positive spatial autocorrelation for most variables, and extremely strong, positive spatial autocorrelation for remotely sensed images, would be reasonable, too. Although this model-informed sampling design approach is somewhat sensitive to the way in which spatial autocorrelation is modelled, all three alternative model specifications indicate that geographic studies require substantially larger sample sizes than are suggested by conventional statistical theory.

4.3 Error Propagation in a GIS Environment

Source errors in georeferenced data give rise to further errors when data analysis operations (e.g., overlay, addition and ratioing) are performed with them present, resulting in error propagation. The errors that are present in maps are modified by data transformations in ways that undermine the purpose of analysis and lead to uncertainty in the validity of inferences/conclusions that are drawn. Arbia et al. (1998, 1999) analyze, for a univariate context, how source map error propagates as a result of map operations; extend this evaluation to the multivariate case of linear combinations of georeferenced variables. Conceptualizing this problem of error propagation requires recognition that attribute values on maps are spatially correlated, attribute measurement error also may be spatially correlated, and the location of attribute values may contain errors. There also may be inter-map correlation.

One preliminary set of findings includes an importance ranking of sources of error, established with quantitative analyses of simulation experimental results, ordering their relative magnitudes in terms of error propagation. For various overlay operations, error propagation primarily is attributable to the underlying attribute spatial correlation in a source map, the signal-to-noise ratio, location error, spatial correlation in the error process, and interaction effects among selected pairs of these components. For addition and ratioing, error process variance tends to play the single most important role. For addition, location error is the next most important factor, followed by location error interacting with different scales of source-map spatial correlation. For ratioing, factors other than error process variance play a far less important role; nevertheless, error-process spatial correlation and source-map spatial correlation are important contributors to error propagation. Meanwhile, inter-map correlation impacts upon error properties of map operations, but apparently without altering the relative importance of the sources of error.

In a study of positional error attributable to geocoding via automated address matching of individual observations, Griffith et al. (2004) found that positional error also matters in terms of error propagation in sophisticated spatial statistical analysis. In a study of pediatric lead poisoning in Syracuse, NY, location error (the difference between TIGER line file-based and cadastral parcel-based geocoding) caused model parameter estimation results to be noticeably different. But for all models estimated, essentially every parameter estimate obtained with TIGER-based geocoding falls within the 95% confidence interval of its corresponding cadastral parcel-based geocoding estimate, across the 10% to 50% range of location error found in the sample data. In other words, on average, positional error may well produce conspicuous, but not shockingly dramatic, differences in spatial statistical analysis results.

4.4 The Ecological Fallacy Revisited

Two serious weaknesses of the ecological approach commonly employed in spatial statistical analysis are an inability to: draw proper inferences from areally aggregated entities to the individual entities constituting these aggregates; and, disentangle the impacts of overlooked confounders. Richardson and Monfort (2000, pp. 218–19) emphasize that such ecological correlation studies assessing the health effects of environmental exposure, for example, are in increasing demand because individual-level assessments in these types of situations frequently are too difficult. Emphasizing advantages of ecological analysis, they also point out (p. 206) that geographically aggregated data: tend to be straightforward to obtain, dampen measurement error by averaging, lead to increased power by increasing the range of a response over that for individuals, and furnish "natural experiments" when a response variable links to contextual physical geography features of a landscape.

Recognizing that migration introduces a serious complication to spatial analysis studies, Elliott and Wakefield (2000, p. 71) underscore that some georeferenced phenomena have less opportunity for bias to occur because of cause-effect corruptions attributable to migration factors. Furthermore, individual-level inferences can be drawn from ecological analyses when the predictor-response relationship is linear (Elliott & Wakefield, 200, p. 77; Richardson & Monfort, 2000, p. 207):

$$\mathrm{Y} = \alpha\mathbf{1} + \mathbf{X}\boldsymbol{\beta}, \tag{4.4.1}$$

where α denotes a conditional mean, $\mathbf{1}$ is an n-by-1 vector of ones,[1] $\mathbf{X}$ is an n-by-p vector of p mean aggregate predictors, and $\beta$is a p-by-1 vector of regression coefficients. But in practice this additive specification often is unlikely. Rather, predictors frequently act synergistically, resulting in a disproportionate increase in a response outcome as they individually increase, due to amplifications by their interactions. This nonlinear, exponential functional form of the predictor-response relationship, which is multiplicative in nature, is the norm in many spatial analyses. Accordingly, the mean response for geographic aggregates is a function of within-area individual-level means and variances/covariances:

$$\mathrm{Y} = \mathrm{EXP}[\alpha + \mathbf{X}\boldsymbol{\beta} + 0.5\boldsymbol{\beta}'\boldsymbol{\Omega}\boldsymbol{\beta}], \qquad (4.4.2)$$

where Ω is the within-area predictor covariances matrix. If some power function of an individual-level response within a geographic aggregate is normally distributed, a third functional form can be established that represents an increase in predictor values accompanied by a decline in their impacts on the response outcome:

$$\mathrm{Y} = [\alpha + \mathbf{X}\boldsymbol{\beta} + 0.0370(-1+\gamma)\boldsymbol{\beta}'\boldsymbol{\Omega}\boldsymbol{\beta}]^{\gamma},\ 0 < \gamma < 1. \qquad (4.4.3)$$

In other words, as the level of one predictor increases while all other predictors remain the same, beyond some point the resulting increase in the response outcome becomes smaller and smaller. This decline may arise because of diminishing returns, and/or because the suite of predictors is multicollinear. A fourth functional form pertains to solutions where interactions amplify but are not purely multiplicative:

$$\mathrm{Y} = (\alpha + \mathbf{X}\boldsymbol{\beta})^{\gamma} + \sum_{\mathrm{j}=1}^{\mathrm{k}} \mathrm{c_j} < \beta^{\frac{\gamma}{2}-\mathrm{j}} >' \boldsymbol{\Omega}^{\mathrm{j}} >, \mathrm{j} \le \frac{\gamma}{2}, \qquad (4.4.4)$$

where <> denotes a vector, $\mathrm{c_j} = \prod_{\mathrm{h}=1}^{[[\gamma]]} (\frac{1}{2\mathrm{h}})[-\frac{1}{4} + (\gamma - 2\mathrm{h} + \frac{3}{2})^2], \gamma \ge 1$, and [[]] denotes the integer value.[2] This last specification reduces to the additive case when the exponent is $\gamma = 1$, and in a limiting sense relates to the multiplicative case as the exponent γ increases. Additional research needs to confirm these theoretical specifications, which expand the family of models that allows a minimization of ecological bias due to nonlinearity, in terms of correlations between the mean response and covariances, based upon integration of the response relationship at the individual level in order to approximate the ecological aggregate mean response. Most databases fail to contain enough information to compute an estimate of the within-area covariance matrix Ω. Of note is that Richardson and Monfort (2000, p.

[1]This vector of ones often is denoted by the n-by-1 vector **i** in the spatial econometrics literature.

[2]This equation was established with bootstrap analyses of selected empirical data using 25,000 replications.

209) find that approximating this matrix by assuming independence (hence, matrix Ω is diagonal) is an improvement over completely neglecting it.

Meanwhile, ecological bias can result from confounding both within and across areal units. Socio-economic/demographic variables often furnish useful surrogates for unknown/unmeasured confounders, and tend to dominate geographic variation (Elliott and Wakefield, 2000, pp. 78–79). In addition, accounting for spatial autocorrelation when analyzing a georeferenced variable further accounts for the aggregate effect of unobserved confounding predictors (Richardson and Monfort, 2000, p. 210), as does including random effects terms in a hierarchical Bayesian analysis (see Sect. 4.1).

4.5 Filling Maps with Gaps: Imputation of Georeferenced Data Values

The *E*stimation-*M*aximization (EM) algorithm (Dempster, Laird, and Rubin 1977), an iterative procedure for computing MLEs when datasets are incomplete, is a useful device for helping to solve model-based small area—especially small geographic area—estimation problems, which currently are receiving considerable attention (see, for example, Datta et al., 1999, and Rao, 1999), with even more attention being argued for (see Citro, 1998).

Descriptions of the EM algorithm may be found in Flury and Zoppè (2000), Meng (1997), and McLachlan and Krishnan (1997), among others. Because the EM procedure requires imputation of the complete-data sufficient statistics, rather than just simply the individual missing observed values, the equivalency discussed here derives from an assumption of normality, for which the means and covariances constitute the sufficient statistics. Of note is that this generally is not true of non-normal populations, although it is for the Poisson and binomial probability models. An assumption of normality links OLS and MLE regression results; application of the Rao-Blackwell factorization theorem of mathematical statistics verifies that the means and covariances are sufficient statistics in this situation.

Yates (1933) shows for analysis of variance (ANOVA) that if each missing observation is replaced by a parameter to be estimated (i.e., the conditional expectation for a missing value), the resulting modified analysis becomes straightforward by treating the estimated missing value as an observation (i.e., an imputation). Rewriting the ANOVA as an OLS regression would involve introducing a binary indicator variable for each missing value—the value of –1 denoting the missing value observation in question, and 0 otherwise—with the estimated regression coefficients for these indicator variables being the missing value estimates. Generalizing this regression formulation allows missing values to be estimated with an analysis of covariance (ANCOVA) regression specification, one in fact suggested by Bartlett (1937) and by Rubin (1972).

Consider a bivariate set of n observed values, each pair denoted by (y_i, x_i), i=1, 2, . . ., n. Suppose only the response variable, Y, contains incomplete data. First, the

n_m missing values need to be replaced by 0. Second, n_m 0/–1 indicator variables, I_m (m = 1, 2, . . ., n_m), need to be constructed; I_m contains (n–1) 0 s and a single –1 corresponding to the mth missing value observation. Regressing Y on X together with the set of indictor variables constitutes the ANCOVA.

Suppose $\mathbf{Y}_o$ denotes the n_o-by-1 ($n_o = n - n_m$) vector of observed response values, and $\mathbf{Y}_m$ denotes the n_m–by-1 vector of missing response values. Let $\mathbf{X}_o$ denote the vector of predictor values for the set of observed response values, and $\mathbf{X}_m$ denote the vector of predictor values for the set of missing response values. Further, let $\mathbf{1}$ denote an n-by-1 vector of ones that can be partitioned into $\mathbf{1}_o$, denoting the vector of ones for the set of observed response values, and $\mathbf{1}_m$, denoting the vector of ones for the set of missing response values. Then the ANCOVA specification of the regression model may be written as

$$\begin{pmatrix} \mathbf{Y}_o \\ \mathbf{0}_m \end{pmatrix} = \begin{pmatrix} \mathbf{1}_o & \mathbf{X}_o \\ \mathbf{1}_m & \mathbf{X}_m \end{pmatrix} \begin{pmatrix} \alpha \\ \beta \end{pmatrix} + \begin{pmatrix} \mathbf{0}_{o,m} \\ -\mathbf{I}_{m,n} \end{pmatrix} (\boldsymbol{\beta}_m) + \begin{pmatrix} \boldsymbol{\varepsilon}_o \\ \mathbf{0}_m \end{pmatrix} \Rightarrow \mathbf{Y} = \alpha \mathbf{1} + \mathbf{X}\boldsymbol{\beta} - \sum_{m=1}^{n_m} y_m \mathbf{I}_m + \boldsymbol{\varepsilon}, \tag{4.5.1}$$

where $\mathbf{0}_m$ is an n_m-by-1 vector of zeroes, $\mathbf{0}_{o,m}$ is an n_o-by- n_m matrix of zeroes, α and β respectively are the bivariate intercept and slope regression parameters, $\boldsymbol{\beta}_m$ is an n_m-by-1 vector of conditional expectation regression parameters, $\mathbf{I}_{m,m}$ is an n_m-by- n_m identity matrix, and ε_0 is an n_o-by-1 vector of random error terms. The bivariate OLS regression coefficients, a and b, for this ANCOVA specification are the regression results for the observed data only [e.g., $\mathbf{b} = (\mathbf{X}_o^T\mathbf{X}_o)^{-1}\mathbf{X}_o^T\mathbf{Y}_o$, where T denotes matrix transpose]. In addition, the regression coefficients, $\mathbf{b}_m$, for the indicator variables are given by

$$\mathbf{b}_m = a\mathbf{1}_m + b\mathbf{X}_m = \hat{\mathbf{Y}}_m, \tag{4.5.2}$$

which is the vector of point estimates for additional observations (i.e., the prediction of new observations) that should have X values within the interval defined by the extreme values contained in the vector $\mathbf{X}_o$. This is a standard OLS regression result, as is the prediction error that can be attached to it (see, for example, Montgomery and Peck, 1982, pp. 31–33). Dodge (1985, p. 159) cautions that the OLS equivalency highlighted here rests on the existence of estimable equations, which in some instances means that the ANCOVA solution is appropriate only when the number of missing values is not excessive. If enough observations are missing, the number of degrees of freedom can become zero or negative, the matrix $\begin{pmatrix} n_o & \mathbf{1}_o^T\mathbf{X}_o \\ \mathbf{X}_o^T\mathbf{1}_o & \mathbf{X}_o^T\mathbf{X}_o \end{pmatrix}$ can become singular, and as such not all of the parametric functions would be estimable.

For the SAR model specification, Eq. (4.5.3) is replaced with

$$\mathbf{Y} = \rho\mathbf{W}\mathbf{Y} + (\mathbf{I} - \rho\mathbf{W})\mathbf{X}\boldsymbol{\beta} + \sum_{m=1}^{M} y_m(-\mathbf{I}_m + \rho\mathbf{W}_{om}) + \boldsymbol{\varepsilon}, \tag{4.5.3}$$

where matrix $\mathbf{W}$ often is the row-standardized version of the conventional binary geographic connectivity matrix. Estimation of this equation is discussed in Martin

Table 4.2 Imputation of turnip production in 3 vandalized field plots

Field plot	Conventional EM estimate	Spatial SAR-EM estimate $\hat{\rho} = 0.443$	Spatial filter: 3 selected eigenvectors
(6,5)	28.9	29.99	24.31
(5,6)	18.8	17.66	13.62
(6,6)	27.8	28.26	23.93

(1984) and Haining et al. (1989). For the spatial filter model, equation (4.5.3) is replace with

$$\mathbf{Y} = \mathbf{X}\boldsymbol{\beta}_\mathrm{X} - \sum_{m - 1}^{M} \mathbf{y_m I_m} + \mathbf{E}_k\boldsymbol{\beta}_{\mathrm{E}_k} + \boldsymbol{\varepsilon}, \quad (4.5.4)$$

where $\mathbf{E}_k$ are the set of k selected eigenvectors extracted from the Moran Coefficient numerator matrix $(\mathbf{I} - \mathbf{11}^\mathrm{T}/n)\mathbf{C}(\mathbf{I} - \mathbf{11}^\mathrm{T}/n)$ (see Griffith, 2000a); the projection matrix $(\mathbf{I} - \mathbf{11}^\mathrm{T}/n)$ centers a variable and is found throughout the multivariate statistics literature.

For illustrative purposes, consider the agricultural field plot turnip production data reported by Rayner (1969). The originally devised latin square experimental design was destroyed by a cluster of three plots in one corner of the field being vandalized. Imputation results from Eqs. (4.5.2), (4.5.3), and (4.5.4) are reported in Table 4.2. Equation (4.5.2) simply yields the mean of observed harvest values for the variety type that is missing. Equation (4.5.3) yields modifications of these imputations that exploit redundant locational information in the values, increasing the two preceding imputations immediately adjacent to observed values, and decreasing the preceding imputation that is adjacent only to missing values. And, Eq. (4.5.4) shrinks all three spatial autocorrelation adjusted imputations toward the field-wide mean.

Substantially more sophisticated versions of these estimators, involving both missing X and Y values and constrained estimation, appear in Griffith (1999, 2000b).

V Concluding Comments: Commonalities and Distinctions Between Spatial Statistics and Spatial Econometrics

Not surprisingly, spatial econometrics and spatial statistics share much subject matter. But they also embrace a number of thematic differences. The preceding discussion describes conspicuous similarities and differences between these two subdisciplines, especially as they are practiced in the social sciences, and particularly highlighting where they diverge.

5.1 Lessons from Spatial Econometrics

As to spatial econometrics, its practitioners should have to learn how to work systematically with spatially biased data, that bias being responsible not only for spatial heterogeneity and asymmetry, but also for complex spatial patterns.

A second conclusion outlined above refers to the importance of pre-econometric specifications, especially if models are going to be used for future exploratory purposes, such as forecasts or regional policy computations. Almost certainly future spatial econometric work will remain fundamentally theory-laden (Aznar Grasa, 1989, p.10), and accordingly theoretical spatial economics will continue to be an indispensable guide to spatial econometric modelling.

The same remark applies to estimators employed in spatial econometrics; these, too, need to be decided on in terms of the uses to which a model will be put. As a last word of caution here: attention should be given to identifiability, a topic rarely encountered in spatial econometric work.

Finally, complexity of spatial patterns has to be addressed explicitly and more fully. An appropriate complexity analysis should be the firsts step in every spatial econometric exercise.

5.2 Lessons from Spatial Statistics

The analysis of non-normal data and missing value imputation are two topics for which considerable progress has been made to date. MCMC allows spatial scientists to have a much richer data analytic toolbox, one that includes Bayesian modelling. A spatial EM algorithm allows both redundant attribute and locational information to be exploited for imputation purposes; algebraically, the resulting equations are the same as those used for kriging in geostatistics. At this point in time, dissemination of these techniques has become an important goal for spatial statistics.

As soon as individuals must be aggregated in order to establish sound incidence rates, or because their attributes are to be related to areal-extensive environmental exposure, the ecological approach is the only sensible option for data analysis. The ecological fallacy problem, which now is much closer to being resolved, must be solved in order to bolster ecological inferences/conclusion.

Finally, much more research work is needed before a sound understanding is attained about how to devise appropriate spatial sampling designs, and how to control error propagation in georeferenced data analyses.

Contents

Part I
Non-standard Spatial Statistics

[Georeferenced v]ariables … tend to have distributions with high temporal and spatial autocorrelation, very high variance due to outliers and nonrandom missing value[s]. In such cases, the departures from normality are so great that even robust techniques such as the Gauss-Markov linear model may behave unpredictably.

Philip A. Schrodt, *Patterns, Rules and Learning* (2nd ed.), 2004, p. 19.

Chapter 1
Introduction: Spatial Statistics

A wide array of topics in spatial statistics introduce methodological controversy: aggregate versus disaggregated data inference (e.g., the ecological fallacy), modelling the spatial covariance versus the spatial inverse covariance matrix, including fixed and/or random effects terms in a model specification, spatial autocorrelation specified as part of the mean response versus part of the variance parameter, and methods for simulating spatially autocorrelated random variables.

A spatial statistician often pursues a data-driven, rather than a model-specification-driven, analysis. This perspective reflects the sampling design origins of statistical inference. A critical issue in this approach is accounting for all trends in a data set, in turn allowing residual values to be reduced to ones that mimic independent and identically distributed (iid) random variables. These trends may be related to covariate as well as autoregressive relationships. Complexity associated with these trends often is a function of noisy (e.g., considerable dispersion), dirty (e.g., nonlinear relationships), and/or messy (e.g., unbalanced factors) data. Experience with the normal-linear statistical model (especially with regard to variable transformations) has taught that flexibility is needed in order to properly address this complexity.

In this part, a certain number of working papers are brought together, most of which have been presented and commented on in departmental colloquia and/or in special sessions devoted to spatial statistics and its applications at national and international conferences.

Although this set of papers appears to be articulated with rather loose couplings, these papers share the common thread of dealing with spatial autocorrelation and its associated problems in an advanced way. Hopefully they will stimulate fresh thinking about some of the more complicated problems in spatial statistics.

D.A. Griffith, J.H.P. Paelinck, *Non-standard Spatial Statistics and Spatial Econometrics*, Advances in Geographic Information Science 1,
DOI 10.1007/978-3-642-16043-1_1, © Springer-Verlag Berlin Heidelberg 2011

Chapter 2
Individual Versus Ecological Analyses

2.1 Introduction

Analyses of disease maps frequently require the use of an ecological approach, partially because aggregates of cases allow such measures as rates to be computed. In addition, group averages of individual measures often are more readily available, tend to reduce impacts of measurement error, and help to preserve the confidentiality of individuals in each aggregation group. Given this context, the resulting problematic issue concerns drawing sound inferences about individuals from such grouped data. The general drawback to this type of inference is known as the ecological fallacy: most often a difference exists between an ecological regression and the regression based upon individuals underlying it (i.e., aggregate-level relationships do not necessarily hold at the individual level). Well-recognized impacts corrupting inference are aggregation bias (i.e., distortions of the information content of data attributable to loss of variability through observation aggregation), confounding variables (i.e., hidden or unknown variables lurking about in a study that cause distortions through their correlations with the response variable), and nonlinearity. One interesting exchange about this topic appears in the *Annals of the Association of American Geographers* (2000).

In this chapter, results of experiments with Syracuse, NY pediatric lead poisoning data demonstrate selected nonstandard spatial statistical analyses concerning individual versus ecological inference.

2.2 Spatial Autocorrelation Effects

Frequently georeferenced data comprise geographic aggregates, with geographic variability constituting part of the focus of a study. Accordingly, analyses of disease maps are further complicated by the presence of spatial autocorrelation (SA) effects associated with georeferenced data, especially because less is known about impacts of these effects on binomial or Poisson random variables. Generally speaking, variance inflation is the principal impact of positive SA in linear statistical analyses.

D.A. Griffith, J.H.P. Paelinck, *Non-standard Spatial Statistics and Spatial Econometrics*, Advances in Geographic Information Science 1,
DOI 10.1007/978-3-642-16043-1_2,

This holds for binomial and Poisson variables, too, where it operates as a source of overdispersion.

Consider a P-by-Q regular square tessellation network of locations. Simple binomial models were estimated for $P = 91$ and $Q=92$ (i.e., $n = 8{,}372$), and the Syracuse pediatric blood lead level (BLL) data parameter estimates based upon the three current threshold values of concern: 5 micrograms/deciliter (μg/dl; the detection level), 10 μg/dl (the concern threshold), and 20 μg/dl (the intervention threshold); these data contain 8,343 child-parcel matched locations, with global parameter estimates reported in Table 2.1. Impacts of SA in this numerical example are illustrated in Fig. 2.1. As SA latent in the data increases from none, to a moderate level, and then to a marked level, variance indeed increases, with the principal impact being a noticeable decrease in kurtosis (i.e., peakedness; Fig. 2.1a). In other words, the distribution is being flattened, with more extreme counts becoming increasingly likely, and more central counts becoming increasingly less likely.

The moderate levels of positive SA (msa) employed to construct Fig. 2.1 are those more commonly encountered in the real world. These levels are accompanied by a noticeable, but not a dramatic, distortion of the affiliated histogram. The strong level of positive SA (ssa) employed to construct Fig. 2.1 is rarely encountered in the real world. Nevertheless, it distorts histograms in a way that makes them more closely resemble a uniform distribution, even when the sample size implies a bell-shaped curve should be expected. Figure 2.2 portrays the impact of near-perfect positive SA. It demonstrates that further increasing the level of positive SA results in additional squashing of the more central frequencies, essentially forcing all counts to be either of the two extremes of the range of counts. In other words, the frequency distribution now is sinusoidal in form.

2.3 Aggregation Impacts

For independent and identically distributed (iid) observations, the number of ways the total number of individuals (P) can be allocated to n aggregate groups is given by the following Stirling number of the second kind (Abramowitz and Stegun, 1964):

$$\frac{1}{n!}\sum_{k=0}^{n}(-1)^{n-k}\frac{n!}{k!(n-k)!}k^{P} \tag{2.1}$$

One reason to note SA impacts, beyond variance inflation, is that the clustering of similar values on a map means the actual number of geographic areal unit aggregates is constrained to be less than the quantity rendered by expression (1). Accordingly, positive SA reduces within areal unit variation, and hence accentuates between areal unit variation. For example, if all of an even number of observations were linked pairs (i.e., correlated), with the net effect being that P/2 is the total number of items for allocation, then for two groups and 10 observations, this constraint reduces the

Table 2.1 Logistic model comparisons when spatial autocorrelation is induced via the mean response: numerical results preserving $\hat{\alpha}_0$ and VAR($\hat{\alpha}_0$), based upon the entire Syracuse pediatric blood lead levels (BLLs) data

Feature	N	BLL > 5 μg/dl	BLL > 10 μg/dl	BLL > 20 μg/dl
$\hat{\mathbf{p}}$	***	5557/8343 = 0.66607	1700/8343 = 0.20376	122/8343 = 0.01462
$\hat{\boldsymbol{\alpha}}_0$	***	0.69045	–1.36294	–4.21043
		iid: zero spatial autocorrelation		
VAR($\hat{\alpha}_0$)	10	0.67052^2	0.78508^2	2.63439^2
	30	0.38712^2	0.45327^2	1.52097^2
	100	0.21204^2	0.24826^2	0.83307^2
		Moderate spatial autocorrelation		
$\hat{\alpha} = \hat{\boldsymbol{\alpha}}_0\ \mathbf{1} + 50(\mathbf{E}_{1,2} + \mathbf{E}_{2,1})/\sqrt{2}$	10	0.15890^2	0.12489^2	0.01401^2
	30	0.08086^2	0.07120^2	0.01293^2
	100	0.03001^2	0.02858^2	0.01018^2
		Marked spatial autocorrelation		
$\hat{\alpha} = \hat{\boldsymbol{\alpha}}_0\ \mathbf{1} + 100(\mathbf{E}_{1,2} + \mathbf{E}_{2,1})/\sqrt{2}$	10	0.06506^2	0.05867^2	0.01245^2
	30	0.02544^2	0.02441^2	0.00959^2
	100	0.00813^2	0.00803^2	0.00532^2

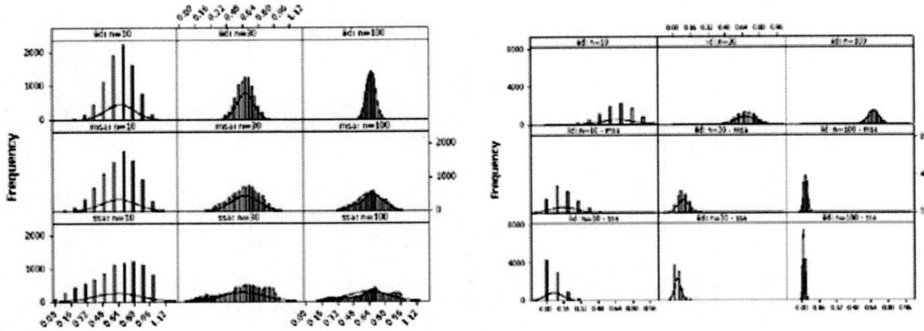

Fig. 2.1 Binomial distribution histograms for $n = 8{,}372$. *Left* (**a**): impacts of spatial autocorrelation. *Right* (**b**): comparable binomial histograms based upon the logistic regression intercept term variance

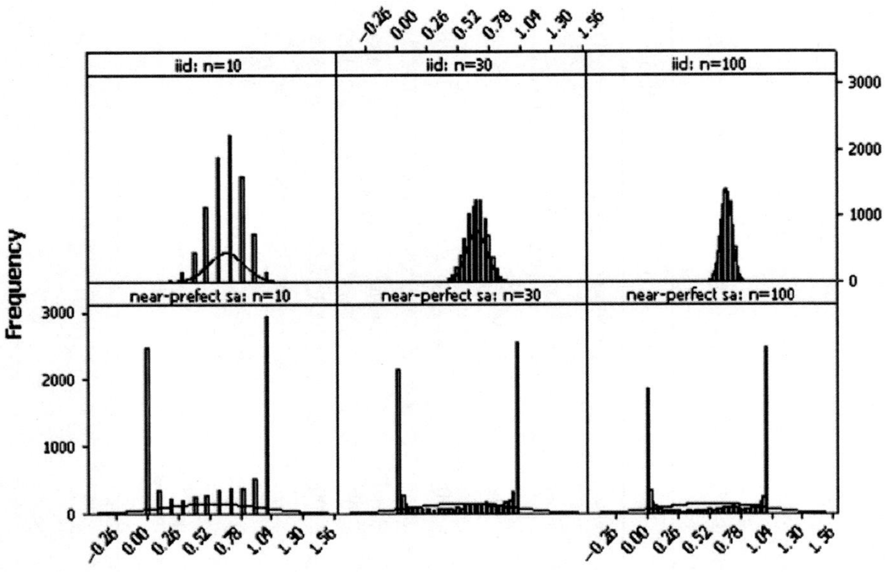

Fig. 2.2 Binomial distribution histograms for $n = 8{,}372$: impacts of near-perfect positive spatial autocorrelation

number of possible groups from 511 to 15. In other words, SA may well help data analysts contend with the ecological fallacy to some degree.

2.3.1 The Syracuse Data

BLL data were collected by the Onondaga County Health Department for children, ages 0–6, residing in the City of Syracuse during 1992–1996, and then made digitally available for scientific analysis, with confidentiality being maintained by masking names with unique identification numbers. These data have undergone considerable editing and cleaning, and have been geocoded using the 2002 cadastral property tax map, which contains 35,500 parcels (Griffith et al., 2008). This data set comprises a total of 16,383 BLL measurements, of which 37 fail to have addresses that matched any of the city parcel addresses (i.e., they are located outside of the city boundaries), and 73 final address matchings fail to have consistent block and block group allocations (which introduces a small amount of noise into some of the aggregate data analyses). Repeated measures for children are summarized by retaining only the maximum BLL for each child. These observations are geographically distributed across 8,208 parcel locations in the City (see Fig. 2.3), with three parcels failing to link to census tracts (of which there are 57) or census block groups (of which there are 147), and an additional two parcels failing to link to census blocks (of which there are 2,025).

The handful of cases available for a non-geographic analysis that had to be set aside for a geographical analysis introduce some, but not much, noise into the analysis. In all cases for BLL > 5 μg/dl, regardless of geographic aggregation, the simple constant mean logistic regression model yields an intercept estimate of 0.6965, with a standard error of 0.0234 (see Table 2.2). In other words, the geographic aggregation does not distort this parameter estimate or the inference that accompanies it. Rather, ecological distortion enters here in terms of the deviance statistic. Although somewhat meaningless for a binary variable, the individual data analysis is accompanied by a deviance statistic of 1.27. This value increases to 2.06 for census blocks, to 6.02 for census block groups, and to 19.81 for census tracts. Results for BLL >

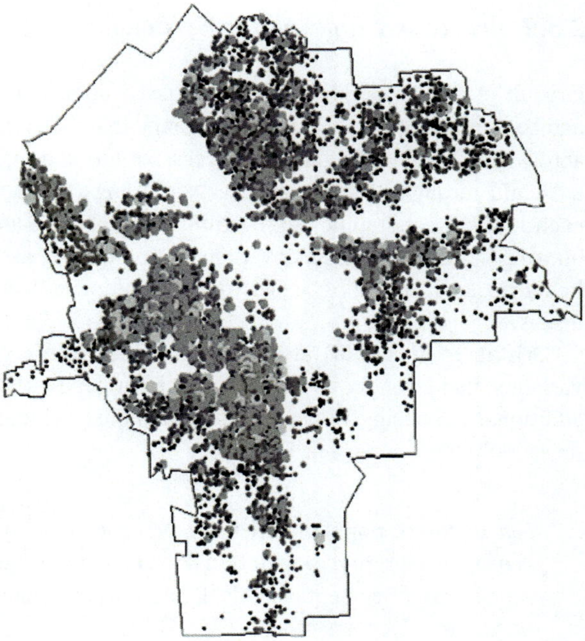

Fig. 2.3 The geographic distribution of individual BLLs across the City of Syracuse. *Black*: 0–5 μ g/dl; *dark gray*: 5–10 μg/dl; μ *medium gray*: 10–20 μ g/dl; and, *light gray*: 20–47 μ g/dl

Table 2.2 Logistic regression estimation results for a constant mean model specification, for threshold BLL values and the different levels of geographic aggregation

Statistic	Individual		Block		Block group		Tract	
	Estimate	SE	Estimate	SE	Estimate	SE	Estimate	SE
				BLL >5 μg/dl				
$\hat{\alpha}$	0.6965	0.0234	0.6965	0.0234	0.6965	0.0234	0.6965	0.0234
Deviance	1.27		2.06		6.02		19.81	
				BLL >5 μg/dl				
$\hat{\alpha}$	−1.3643	0.0274	−1.3643	0.0274	−1.3643	0.0274	−1.3643	0.0274
Deviance	1.01		1.49		4.17		12.99	
				BLL >20 μg/dl				
$\hat{\alpha}$	−4.2532	0.0939	−4.2532	0.0939	−4.2532	0.0939	−4.2532	0.0939
Deviance	0.15		0.34		0.76		1.74	

10 and >20 μg/dl (see Table 2.2) are consistent with these findings. Not only may the deviance statistic be detecting a mixture of heterogeneous Bernoulli random variables, but it also may be detecting the presence of SA.

In summary, for the simple intercept-only logistic regression model, ecological distortions appear to manifest themselves most noticeably through the deviance statistic, with aggregate data cross-tabulated by geographic areal units rendering the same inference as individual data.

2.3.2 Previous Findings for Syracuse

Griffith et al. (1998) report findings based upon a spatial analysis of part of the database employed here. Their study found that the general pattern of elevated BLLs across the City persists through successive levels of aggregation, from the individual child through 1990 census tract groupings. Conspicuous SA is identifiable at each level of geographic aggregation. On both substantive and empirical grounds, housing value is the single covariate that is strongly associated with elevated BLLs. Pediatric lead poisoning tends to be a completely preventable inter-city/poverty disease.

Griffith et al. (1998) also report sets of socio-economic/demographic census variables that strongly covary with pediatric lead poisoning at aggregate levels. In additional to housing value (e.g., median house value, percentage renter occupied), these include:

census tracts: population density, percentage in cohort < 18 years of age
census block group: population density, percentage black, number of cases
census block: percentages black and Hispanic, number of cases, percentage in cohort < 18 years of age

Covariate surrogates for SA also appear in the models. In addition, the census block resolution is sufficiently fine that many geographic areas are non-residential, resulting in many areal units having zeroes; this is one problematic feature associated with using fine resolution census geographies or individual data for analysis purposes.

2.4 Spatial Autocorrelation in the Syracuse Data

Two sources of SA in the Syracuse BLL data are of particular interest. The first is latent in the BLL values themselves: children who are neighbors tend to have similar BLLs. The second is latent in the housing value covariate: neighboring houses tend to have a similar market value.

2.4.1 Spatial Autocorrelation in the Syracuse Data: LN(BLL + 1) Values

A Thiessen polygon partitioning of the Syracuse city surface based upon locations with children for which BLL values have been measured appears in Fig. 2.4. Below-detection-level BLL anomalies are conspicuous, whereas high BLL anomalies are not, according to a simple normal quantile plot of individual LN(BLL + 1) values, where one is the maximum likelihood translation parameter estimate for aligning the log-BLL values with a bell-shaped curve (see Fig. 2.5).

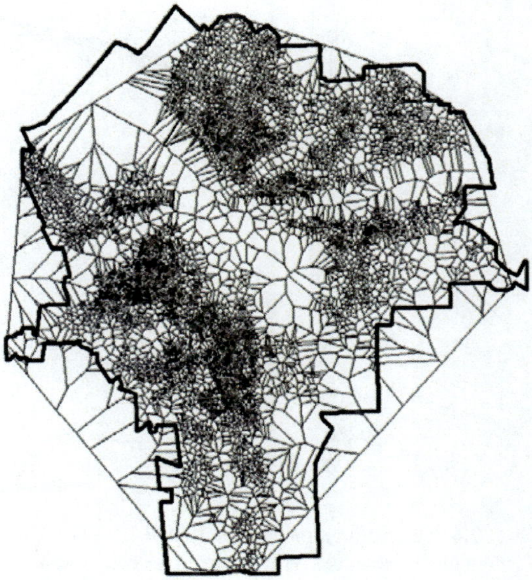

Fig. 2.4 Thiessen polygon surface partitioning of the City of Syracuse, for the locations of children for which BLL values were obtained during 1992–1996

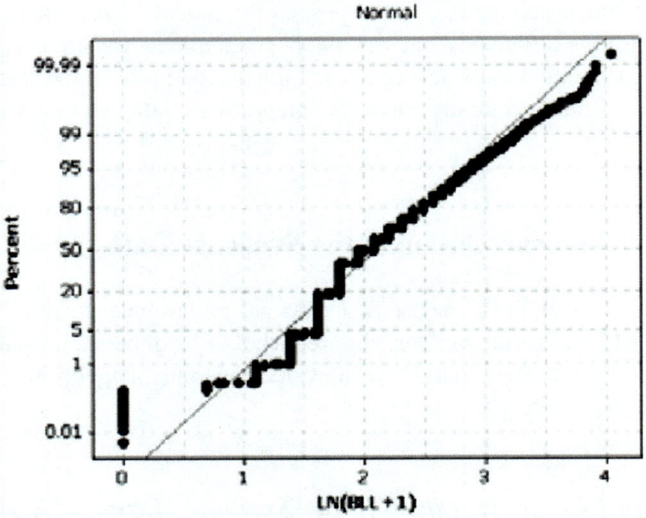

Fig. 2.5 Normal quantile plot for individual log-BLL values

SA for individual LN(BLL + 1) value locations (a total of 8,208 parcels), portrayed with a semivariogram plot (see Fig. 2.6) for distance not exceeding roughly a third of the span of the geographic landscape, is weak-to-moderate and positive. Based upon roughly 37.3 million distance pairs, where distance hasbeen

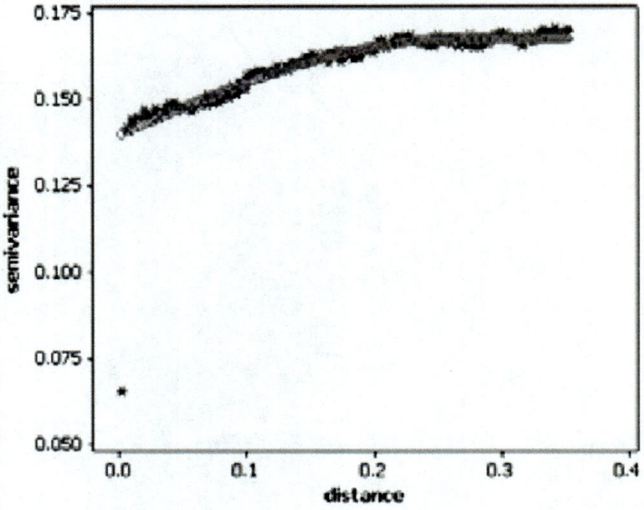

Fig. 2.6 Semivariogram plot for LN(BLL + 1) values, City of Syracuse, NY. Black asterisks denote observed values; gray open circles denote spherical model predicted values

standardized to the unit square, the following spherical and circular semivariogram models[1] (where $\hat{\gamma}_{ij}$ denotes semivariance) best describe these data:

$$\text{penta-spherical model: } \hat{\gamma}_{ij} = 0.14 + 0.03\left[\frac{15}{8}\frac{d_{ij}}{0.32} - \frac{5}{4}\left(\frac{d_{ij}}{0.32}\right)^3 + \frac{3}{8}\left(\frac{d_{ij}}{0.32}\right)^5\right],$$

$$d_{ij} \le 0.32;$$

$$\hat{\gamma}_{ij} = 0.14 + 0.03 = 0.17,\ d_{ij} > 0.32$$

$$\text{spherical model: } \hat{\gamma}_{ij} = 0.14 + 0.03\left[\frac{3}{2}\frac{d_{ij}}{0.26} - \frac{1}{2}\left(\frac{d_{ij}}{0.26}\right)^3\right],\ d_{ij} \le 0.26$$

$$\hat{\gamma}_{ij} = 0.14 + 0.03 = 0.17,\ d_{ij} > 0.26$$

These models respectively yield 0.074 and 0.075 relative error sums of squares. The scatterplot reveals very marked in situ variability of log-BLL values, and a well-defined geographic pattern to their covariation.

2.4.2 Spatial Autocorrelation in the Syracuse Data: Appraised House Value

The correlation between individual log-BLLs and 2002 appraised house values is –0.29 (see Fig. 2.7).

In general, house values tend to display strong positive SA. Indices for the City of Syracuse, calculated with median values for geographic aggregates, are as follows (also see Fig. 2.7):

These statistics are based upon 2002 assessed values, per \$10,000, for houses in which children were tested for pediatric lead poisoning (a total of 7,057 houses).

aggregation unit	*Moran Coefficient (MC)*	*Geary Ratio (GR)*	*n*
census tract	0.40902	0.62080	56 (#32 missing)
census block group	0.55331	0.45103	145 (#32.001 and #32.002 missing)
census block	0.66111	0.32304	1,485 (540 blocks missing)

[1] The semivariance is one half of the squared difference between the values of an attribute at two locations. A scatterplot is constructed between these values and the distance separating the two locations. A semivariogram model (e.g., penta-spherical, spherical, circular) describes the nonlinear trend line for this scatterplot.

Fig. 2.7 Scatterplot and trend line portraying the relationship between BLL and 2002 appraised house value

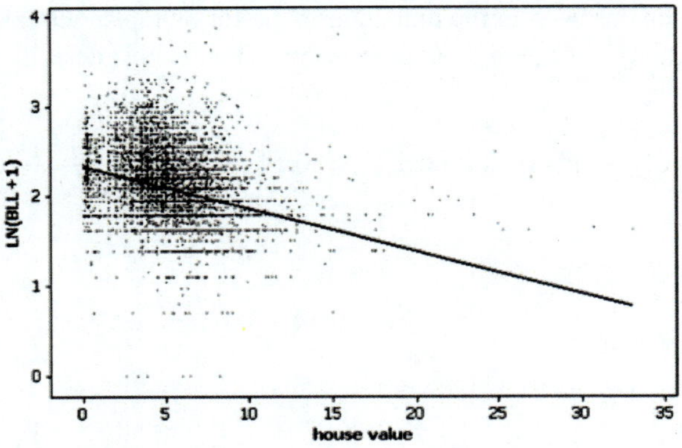

Areal units without residential properties were set aside during the SA index computations. These results simply indicate that latent SA in the geographic aggregations is moderate and positive, increasing with increasingly finer resolution.

SA for individual residential properties, portrayed with a semivariogram plot (see Fig. 2.8) for distance not exceeding a third of the span of the geographic landscape, is strong and positive. Based upon roughly 16.9 million distance pairs, where distance has been standardized to the unit square, the following spherical and circular semivariogram models (again where $\hat{\gamma}_{ij}$ denotes semivariance) best describe these data:

$$\text{circular model: } \hat{\gamma}_{ij} = 1.58 + 4.82\frac{2}{\pi}\left[\frac{d_{ij}}{0.18}\sqrt{1-\left(\frac{d_{ij}}{0.18}\right)^2} + \mathrm{SIN}^{-1}\left(\frac{d_{ij}}{0.18}\right)\right],$$

$$d_{ij} \leq 0.18\,;$$

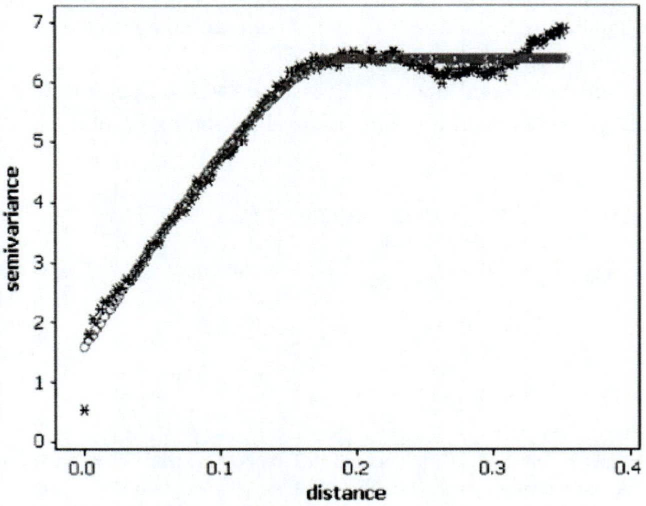

Fig. 2.8 Semivariogram plot for 2002 appraised house values, City of Syracuse, NY. Black asterisks denote observed values; gray circles denote circular model predicted values

$$\hat{\gamma}_{ij} = 1.58 + 4.82 = 6.40\,,\ d_{ij} > 0.18$$

$$\text{spherical model: } \hat{\gamma}_{ij} = 1.47 + 4.96\left[\frac{3}{2}\frac{d_{ij}}{0.21} - \frac{1}{2}\left(\frac{d_{ij}}{0.21}\right)^3\right],\ d_{ij} \leq 0.21$$

$$\hat{\gamma}_{ij} = 1.47 + 4.96 = 6.43,\ d_{ij} > 0.21$$

These models respectively yield 0.005 and 0.008 relative error sums of squares. The scatterplot reveals sizeable in situ variability of house values, a pronounced geographic pattern to their covariation, and a not surprising city-wide trend possibility.

Including house value in the logistic regression specification accounts for some of the SA in BLLs. Because appraised house values are not reported for apartment complexes, the values for these locations were set to 0, and then an indicator variable was created to differentiate these rental locations from the other residential locations (the numeral 1 denotes non-rental, and –1 denotes rental). Logistic regression estimation results for this situation appear in Table 2.3. As expected, house value is negatively related, whereas rental property is positively related, to elevated BLLs. Inclusion of the housing variables reduces overdispersion across the individual and ecological analyses (see Sect. 3.1). In addition, ecological bias now is detectable in all of the parameter estimates as well as their corresponding standard

Table 2.3 Logistic regression estimation results when house value is used as a covariate, for threshold BLL values and the different levels of geographic aggregation

Statistic	Individual		Block		Block group		Tract	
	Estimate	SE	Estimate	SE	Estimate	SE	Estimate	SE
				BLL > 5 μg/dl				
$\hat{\alpha}$	1.4552	0.0484	1.2539	0.0838	0.9241	0.1710	1.2624	0.2613
$\hat{\beta}_{\text{house value}}$	–0.2442	0.0118	–0.2669	0.0120	–0.3192	0.0135	–0.3540	0.0141
$\hat{\beta}_{I_{\text{house value}}}$	0.5686	0.0484	0.7491	0.0838	1.3071	0.1710	1.1288	0.2613
Deviance	1.21		1.71		4.60		8.89	
				BLL >10 μg/dl				
$\hat{\alpha}$	–0.8165	0.0539	–1.2567	0.1396	–1.1597	0.3038	–0.8042	0.3821
$\hat{\beta}_{\text{house value}}$	–0.2731	0.0142	–0.2787	0.0162	–0.3404	0.0186	–0.3773	0.0198
$\hat{\beta}_{I_{\text{house value}}}$	0.8462	0.0539	1.1290	0.1396	1.2680	0.3038	1.0676	0.3821
Deviance	0.96		1.26		3.50		6.21	
				BLL > 20 μg/dl				
$\hat{\alpha}$	–4.1665	0.2006	–3.5970	0.2306	–3.4234	0.2554	–3.2939	0.2693
$\hat{\beta}_{\text{house value}}$	–0.1384	0.0429	–0.1429	0.0505	–0.1897	0.0582	–0.2210	0.0622
$\hat{\beta}_{I_{\text{house value}}}$	0.6742	0.2006	***		***		***	
Deviance	0.14		0.38		1.09		1.67	

errors (Green, 1993; Wrigley, 1995; Holt et al., 1996). Although inferences tend not to be dramatically altered for BLL > 5 or 10 μg/dl, nevertheless they are altered. The case of BLL > 20 μg/dl illustrates how ecological analysis findings can deviate radically from individual-based findings. Furthermore, the rareness of BLLs > 20 creates numerical problems with estimation of the house value binary 0–1 indicator variable parameter, which had to be set aside for its aggregate analyses. This complication resulted in a loss of observations: 121 blocks, five block groups, and one census tract.

2.5 Spatial Autocorrelation in the Syracuse Data: Other Sources

Other sources of SA (e.g., geographic concentration of poverty, siblings)—which may well represent the presence of confounders—beyond house value can be captured in part by employing a spatial filter (SF) model specification. Spatial filtering involves regressing a disease map variable on a set of synthetic variates representing distinct map patterns that accounts for SA. Griffith (2003) develops one form of spatial filtering whose synthetic variates are the set of n eigenvectors extracted from matrix $(\mathbf{I} - \mathbf{ii}^{\mathrm{T}}/n)\mathbf{C}(\mathbf{I} - \mathbf{ii}^{\mathrm{T}}/n)$, the matrix appearing in the numerator of the MC index of SA, where **C** is a binary 0–1 n-by-n geographic weights matrix (i.e., $c_{ij} = 1$ if areal units i and j are neighbors, and 0 otherwise), and **i** is an n-by-1 vector of ones.[2] This procedure is similar to executing a principal components analysis in which the covariance matrix is given by $(\mathbf{I} - \mathbf{ii}^{\mathrm{T}}/n)\mathbf{C}(\mathbf{I} - \mathbf{ii}^{\mathrm{T}}/n)$. But rather than using the resulting eigenvectors to construct linear combinations of attribute variables, the eigenvectors themselves (instead of principal components scores) are the desired synthetic variates, each containing n elements, one for each areal unit. The extracted eigenvector $\frac{1}{\sqrt{n}}\mathbf{i}$ relates to the mean response, and the remaining (n–i) extracted eigenvectors relate to distinct map patterns characterizing latent SA—whose MCs are given by standardizing their corresponding eigenvalues (Tiefelsdorf and Boots, 1995)—that can materialize with matrix **C**. Furthermore, for a given geographic landscape surface partitioning, the eigenvectors represent a fixed effect in that matrix $(\mathbf{I} - \mathbf{ii}^{\mathrm{T}}/n)\mathbf{C}(\mathbf{I} - \mathbf{ii}^{\mathrm{T}}/n)$ does not, and hence they do not, change from one attribute variable to another.

Because this eigenfunction decomposition yields n eigenvectors, a spatial scientist needs to restrict attention to only those eigenvectors describing substantive positive/negative (e.g., MC > 0.25) SA, reducing the candidate set to a more manageable number for describing a given disease map. Supervised stepwise selection from this set of eigenvectors is a useful and effective approach to identifying the subset of eigenvectors that best describes latent SA in a particular disease map. This procedure begins with only the intercept included in a regression specification. Next, at each step an eigenvector is considered for addition to the model specification. For

[2]This vector almost always is denoted by **1** in the spatial statistics literature.

the stepwise generalized linear binomial model regression, the eigenvector that produces the greatest reduction in the log-likelihood function chi-square test statistic is selected, but only if it produces at least a prespecified minimum reduction; this is the criterion used to establish statistical importance of an eigenvector. At each step all eigenvectors previously entered into a SF equation are reassessed, with the possibility of removal of vectors added at an earlier step. The forward/backward stepwise procedure terminates automatically when some prespecified threshold chi-square statistic values are encountered for entry and removal of all candidate eigenvectors.

SFs were constructed for the three geographic aggregations from the 15 candidate eigenvectors for census tract, the 37 for block group, and the 483 for block surface partitionings. Spatial filtering results appear in Table 2.4. Although SA is being accounted for in the parameter estimations for these models, ecological bias still persists. The constructed SFs represent moderate-to-strong levels of positive SA:

Aggregation	BLL >5		BLL >10		BLL >20	
unit	MC	GR	MC	GR	MC	GR
census tract	0.52360	0.46773	0.57387	0.42043	0.82900	0.19180
census block group	0.78798	0.21419	0.80439	0.24604	0.89953	0.22550
census block	0.96443	0.28303	0.90625	0.29532	0.97343	0.31957

Individual results are not available here, since eigenvectors were not computed for the set of individual locations (see Fig. 2.4 for a possible surface partitioning supporting this purpose). Of note is that, as before, the rareness of BLLs > 20 continues to create numerical problems with estimation of the house value binary 0–1 indicator variable parameter, which has been removed from the model specification.

2.6 Bayesian Analysis Using Gibbs Sampling (BUGS) and Model Prediction Experiments

The parallel analyses of individual and ecological data in preceding sections reveal the presence of positive spatial dependence beyond house value, most likely attributable to other unmeasured cofounders with spatial structure, in elevated pediatric BLLs. These parallel analyses also document the presence of ecological biases. A previous ecological investigation of these data uncovers population density, an indicator of urban poverty that could not be detected with the individual-level data, as a covariate of elevated BLLs. This finding illustrates Darby et al.'s contention that "the ecological result [is not always the one] that is wrong" (2001, p. 202). But even findings reported here from ecological analyses conducted by changing geographic aggregation resolution do not agree. This ecological variation arises from a suppression of within-areal unit variability, a finding established in Sect. 2.3.1: "within-area information ... is vital for analysis and interpretation" (Wakefield and Salway, 2001, p. 136). Wakefield (2003) notes that this is particularly true for regression analyses, in which SA components potentially account for unmeasured confounders.

Table 2.4 Ecological logistic regression estimation results when house value is used as a covariate and spatial autocorrelation is accounted for, for threshold BLL values

Statistic	Block			Block group			Tract		
	Estimate	SE	Eigen-vectors	Estimate	SE	Eigen-vectors	Estimate	SE	Eigen-vectors
				BLL >5 μg/dl					
$\hat{\alpha}$	0.9796	0.0892	3–6, 9–12,	0.2982	0.1746	2–10, 13, 16,	1.1629	0.2614	1, 2, 5, 6,
$\hat{\beta}_{\text{house value}}$	–0.1618	0.0120	14–16, 18,	–0.1644	0.0147	17, 20, 22, 25,	–0.2296	0.0153	8–10, 12,
$\hat{\beta}_{I_{\text{house value}}}$	0.3438	0.0900	. . ., 478	1.0975	0.1725	31, 34, 36	0.5298	0.2634	14, 15
$\hat{\beta}_{\text{spatial filter}}$	1	0.0395	(total of 70)	1	0.0501		1	0.0592	
Deviance		1.19			1.65			3.46	
				BLL >10 μg/dl					
$\hat{\alpha}$	–1.8391	0.1480	2, 5, 9, 12,	–1.8370	0.3063	1–4, 6, 7, 9, 10,	–1.0875	0.3825	1, 5, 6, 9,
$\hat{\beta}_{\text{house value}}$	–0.1275	0.0174	14–16, 18,	–0.1803	0.0197	15–17, 23, 25,	–0.2483	0.0209	13, 14, 15
$\hat{\beta}_{I_{\text{house value}}}$	0.6953	0.1450	. . ., 475	0.9761	0.3043	30, 33	0.5802	0.3835	
$\hat{\beta}_{\text{house value}}$	1	0.0493	(total of 41)	1	0.0579		1	0.0676	
Deviance		0.92			1.45			2.00	
				BLL > 20 μg/dl					
$\hat{\alpha}$	–4.5827	0.2569	5, 6, 12, 27,	–3.4947	0.2589	1, 3, 5, 7, 9, 21	–3.5835	0.2794	1, 8
$\hat{\beta}_{\text{house value}}$	–0.0560	0.0483	37, . . ., 446	–0.2394	0.0598		–0.1865	0.0623	
$\hat{\beta}_{I_{\text{house value}}}$	1	0.1184	(total of 19)	1	0.1574		1	0.2719	
Deviance		0.30			0.76			1.42	

Accordingly, the question of interest now asks if this within-areal unit variation can be recovered. Richardson and Montfort (2000) argue that one method of recovery is to posit a parametric form for this variation in order to adjust the corresponding individual-level model, noting that even a parametric form that describes the variation poorly is better than none at all. Wakefield and Salway (2001) allude to the use of random effects, which is explored in this section.

The experiments conducted to explore the utility of random effects estimates as surrogates for within-areal unit variation include those ecological covariates found in the previous study (Griffith et al., 1998). Besag et al. (1991) suggest that these random effects could be spatially structured using a conditional autoregressive (CAR) covariance specification. Wakefield and Salway (2001) suggest that the simplest approach is to employ non-spatial random effects. As a compromise between these two specifications, a SF is employed here to specify spatially structured random effects; the SF becomes the mean of the effects. As is done in the tradition of principal components regression, this SF is computed exogenously, and then its coefficient—which subsequently is distributed across the linear combination of eigenvectors—is estimated; this procedure is analogous to introducing starting values in nonlinear regression estimation (e.g., logistic regression). Next, this analysis is repeated with a proper CAR specification for spatially structured random effects.

Various different completed analyses facilitate exploring relationships between individual- and ecological-based model predictions. One hypothesis evaluated here may be stated as follows:

> The variance of a spatially structured ecological random effects term is directly proportional to the within areal unit variability suppressed by undertaking an ecological analysis.

Preparatory work for assessing this hypothesis involved a Bayesian analysis of the pediatric BLL data. This analysis was executed with the WinBUGS software (the Windows version of BUGS; Thomas et al., 2004), employing a SF model specification, normal priors for the parameter estimates and the random effects term, a gamma prior for the inverse of the error variance, a 25,000-iteration burn-in period, and 500,000 subsequent Markov chain Monte Carlo (MCMC) iterations that then had only every hundredth one retained (weeding), yielding chains of length 5,000 for estimation purposes. With regard to diagnostics, accompanying temporal correlograms and time series plots suggest the generated chains are sound. A CAR comparison also is made, using a 5,000-iteration burn-in period, and 100,000 subsequent MCMC iterations that then had only every hundredth one retained, yielding chains of length 1,000 for estimation purposes.

A second hypothesis evaluated here may be stated as follows:

> Individual level prediction improves by adding to its model specification those neighborhood variables identified as important factors with ecological modeling.

The resulting model is labeled mixed here.

2.6.1 Results for the 2000 Census Tracts

Results of parameter estimation for both generalized linear and BUGS binomial regressions are reported in Table 2.5. For the most part, the BUGS results corroborate the frequentist generalized linear model results. The SFs capture strong positive SA. Maps for two eigenvectors common to all three SFs (i.e., $\mathbf{E}_3$ and $\mathbf{E}_9$) appear in Fig. 2.9. One conspicuous difference between these two sets of results is the standard errors for BLL > 5 μg/dl and BLL > 10 μg/dl: Bayesian-based standard errors tend to be noticeably larger in these two cases. Nevertheless, models for BLL > 5 μg/dl and BLL > 10 μg/dl appear to furnish respectable descriptions of the ecological data.

The suppressed variation induced by aggregation for ecological analysis is for appraised house values. The following battery of descriptive statistics for the 5,000 MCMC generated random error terms, aggregated by census tract, were calculated: mean, median, standard deviation, minimum value, maximum value, skewness, and kurtosis. Next, a stepwise regression was executed using these statistics as predictor variables, and the standard deviation of house value as the regressor variable. Kurtosis was the single statistic selected in the stepwise analysis for BLL > 5 μg/dl; it accounts for roughly 15% of the variability in the standard deviation of house values. The standard deviation was the single statistic selected in the stepwise analyses for BLL > 10 μg/dl and BLL > 20 μg/dl; it accounts for, respectively, roughly 6.6% and 4.6% of the variability in the standard deviation of house values. Meanwhile, replacing kurtosis with this standard deviation for BLL > 5 results in roughly 4.6% of the variability in the standard deviation of house values being accounted for. The ideal result would be for nearly 100% of the variability in the standard deviation of house values to be accounted for by the standard deviation in estimated random error terms. Therefore, the hypothesis positing direct proportionality between these two statistics is not supported here. Apparently the type of approach promoted by Richardson and Montfort (2000) can neither be recaptured nor receive empirical guidance from ecological Bayesian spatial modeling.

Of note is that random effects results from a proper CAR model also were generated for BLL > 5 μg/dl. Here the spatial autoregressive parameter estimate is 0.7870 (SE = 0.2063), indicating the presence of strong, positive SA; now the degrees of freedom are 13. These random effects failed to exhibit any covariation whatsoever with the suppressed variability.

A cross-tabulation of individual observed and prediction results for 0 (non-elevated BLL) and 1 (elevated BLL) appear in Table 2.6; predicted probabilities less than 0.5 have been classified as and rounded to 0, whereas those greater than 0.5 have been classified as and rounded to 1. As the ecological fallacy warns, applying an ecological model to individuals is unsuccessful here. Of note is that even the individual-level model predictions loose reliability as elevated BLL increasingly becomes a rare event. Nevertheless, as Darby et al. (2001) argue, enhanced model results are obtained by formulating a mixed individual-ecological model specification. Not only are covariates like population density detectable at the aggregate level, while not at the individual level, but adding these covariates to an individual-level

Table 2.5 Tract-level ecological logistic regression results when selected socio-economic/demographic variables are used as covariates and spatial autocorrelation is accounted for, for threshold BLL values

Statistic	BLL > 5 μg/dl			BLL > 10 μg/dl			BLL > 20 μg/dl		
Generalized linear binomial regression model									
	Esti-mate	SE	Eigen-vectors	Esti-mate	SE	Eigen-vectors	Esti-mate	SE	Eigen-vectors
$\hat{\alpha}$	0.4825	0.0277	3, 4, 8, 9, 14	−1.8507	0.0410	3, 4, 9, 10, 14	−4.7522	0.1529	3, 9, 11
$\hat{\beta}_{\text{population density}}$	0.2548	0.0292		0.1935	0.0368		0.0630	0.1272	
$\hat{\beta}_{<18\text{ years of age}}$	0.5064	0.0508		0.6703	0.0654		0.9467	0.2326	
$\hat{\beta}_{\text{house value}}$	−0.1985	0.0536		−0.0227	0.0732		0.3598	0.2368	
$\hat{\beta}_{\text{spatial filter}}$	1	0.0847		1	0.1022		1	0.2812	
$MC_{\text{spatial filter}}$		0.72673			0.71626			0.69890	
$GR_{\text{spatial filter}}$		0.27972			0.32717			0.33507	
Deviance		2.53			1.84			1.14	
Pseudo-R^2		0.745			0.781			0.195	
BUGS logistic regression model									
	Estimate		SE	Estimate		SE	Estimate		SE
$\hat{\alpha}$	0.4838		0.0428	−1.8556		0.0507	−4.7881		0.1560
$\hat{\beta}_{\text{population density}}$	0.2641		0.0454	0.2040		0.0513	0.0639		0.1316
$\hat{\beta}_{<18\text{ years of age}}$	0.4878		0.0781	0.6455		0.0909	0.9602		0.2362
$\hat{\beta}_{\text{house value}}$	−0.2179		0.0804	−0.0202		0.0946	0.3616		0.2394
$\hat{\beta}_{\text{spatial filter}}$	1.0164		0.1324	1.0228		0.1369	1.0175		0.2805
df		20			30			50	
Variance		0.0518			0.0355			0.0058	

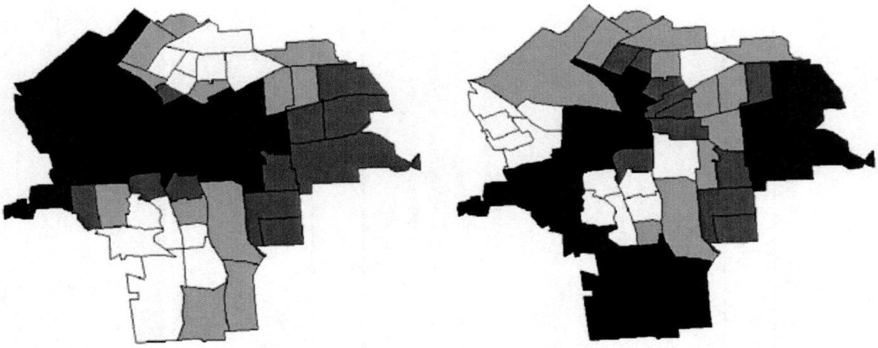

Fig. 2.9 Eigenvectors common to the spatial filters for the BLL >5 μg/dl, BLL > 10 μg/dl, and BLL > 20 μg/dl. *Left* (**a**): eigenvector $\mathbf{E}_3$. *Right* (**b**): eigenvector $\mathbf{E}_9$

Table 2.6 Cross-tabulations of observed and model predicted elevated BLLs, for threshold BLL values

Equation		BLL >5 μg/dl		BLL >10 μg/dl		BLL >20 μg/dl	
	Predicted observed	0	1	0	1	0	1
Ecological	0	2698	31	6535	0	8090	0
	1	5413	63	1670	10	115	0
		($\hat{\varphi} = 0.001$)		($\hat{\varphi} = 0$)		($\hat{\varphi} = 0$)	
Individual	0	367	2362	6516	19	8090	0
	1	291	5185	1663	7	115	0
		($\hat{\varphi} = 0.141$)		($\hat{\varphi} = 0.009$)		($\hat{\varphi} = 0$)	
Mixed	0	983	746	6522	13	8090	0
	1	660	4816	1659	11	115	0
		($\hat{\varphi} = 0.282$)		($\hat{\varphi} = 0.034$)		($\hat{\varphi} = 0$)	

model also improves predictability for BLL > 5 μg/dl, and very marginally for BLL > 10 μg/dl. Of note is that any individual-model gains by including these ecologically determined covariates is lost as these covariates become statistically nonsignificant in their ecological analyses.

Because the results here were so poor, analyses were not repeated for either the census block group or census block aggregations.

2.7 Discussion and Implications

The empirical case study explored here reveals that geographic aggregation combined with SA can cause diagnostic statistics to be misleading. Nevertheless, four general ecological inference conclusions can be drawn from findings summarized here. First, spatial filtering may furnish a blurred, but still unsatisfactory, glimpse of within-areal unit covariation by serving as the spatial structuring term for random effects. Second, the failure of estimated random effects to furnish a useful within-areal units variability surrogate implies that the Richardson-Montfort suggestion of specifying individual-level covariance structure *a priori* should be a more fruitful pursuit. But guidelines for undertaking this task remain to be established; the ultimate goal is to be able to draw the same statistical inferences from aggregate-level data that would be drawn from individual-level data, but without having the individual details. Third, a posited covariance structure should include prominent attributes identified via ecological analysis, resulting in a mixed formulation, as advocated by Darby et al. (2001). Prominent ecological covariates that remain invisible at an individual level of analysis offer the potential to dramatically improve statistical description. In addition, these ecologically-based attributes may at least partially account for SA that impacts upon individual data. Finally, the ability to develop far better ecological-level predictive models for rare events is a continuing need.

Chapter 3
Statistical Models for Spatial Data: Some Linkages and Communalities

3.1 Introduction

Introductory mathematical statistics textbooks discuss topics such as the sample variance by invoking the assumption of independent and identically distributed (*iid*). In other words, in terms of second moments, of the n^2 possible covariations for a set of n observations, the independence assumption posits that n(n – 1) of these covariations have an expected value of 0, leaving only the n individual observation variance terms for analysis. This independence assumption is for convenience, historically making mathematical statistical theory tractable. But it is an arcane specification that fails to provide an acceptable approximation to reality in many contexts. Moreover, the *iid* assumption so popular in theoretical statistics over the years "should not be taken for granted, particularly when there are good physical reasons to abandon it" (Cressie 1989, p. 197). Pursuing this thinking, theoretical and applied statisticians began to explore situations in which the independence part of the *iid* assumption is relaxed. Most notably is time series analysis; more recently is spatial statistics.

Spatial statistics includes spatial autoregression and geostatistics, two branches that evolved separately over a number of decades. In a very general sense spatial statistics is concerned with the statistical analysis of georeferenced data, or data for which observations may be ordered on a two-dimensional surface and tagged with Cartesian coordinates. These observations are correlated strictly due to their relative locational positions [spatial autocorrelation (SA)—data located relatively close together geographically tend to be correlated], resulting in information redundancies being present in georeferenced data values. Spatial autoregression links directly to the Moran Coefficient used to index SA, while geostatistics links directly to the Geary Ratio index. The purpose of this chapter is to focus attention on the linkages between and commonalities spanning these two branches of spatial statistics, and progress that has been made in merging them. The interested reader also should read the two seminal articles by Ord (1975) and Cressie (1989); the two classic treatises reviewing and extending spatial statistical theory are by Cliff and Ord (1981), who have motivated research involving spatial autoregression, and Cressie (1991), who has crystallized research involving geostatistics.

D.A. Griffith, J.H.P. Paelinck, *Non-standard Spatial Statistics and Spatial Econometrics*, Advances in Geographic Information Science 1,
DOI 10.1007/978-3-642-16043-1_3, © Springer-Verlag Berlin Heidelberg 2011

3.2 Background: Quantifying Spatial Autocorrelation

SA, the underlying statistical concept linking spatial autoregression and geostatistics, directly parallels that of correlation in traditional statistics (see Griffith, 1992b). Given a conventional statistical situation, if two variables, say X and Y, are positively correlated, then high values of X tend to be paired with the high values of Y, medium values of X with the medium values of Y, and low values of X with the low values of Y. The spatial statistical parallel involves a single variable, say Y. If SA is positive, then locations with high values of Y tend to be surrounded by nearby high values of Y, locations with medium values of Y tend to be surrounded by nearby medium values of Y, and locations with low values of Y tend to be surrounded by nearby low values of Y. Similar sets of statements can be composed for correlations and autocorrelations that are either negative or zero.

SA's literal definition is self-correlation. Extending the foregoing parallel between traditional correlation and SA, the Pearson product moment correlation coefficient (r_P) formula can be translated into the Moran Coefficient (MC):

$$r_P = \frac{\sum_{i=1}^{n} \frac{(x_i - \bar{x})(y_i - \bar{y})}{n}}{\sqrt{\sum_{i=1}^{n} \frac{(x_i - \bar{x})^2}{n}} \sqrt{\sum_{i=1}^{n} \frac{(y_i - \bar{y})^2}{n}}} \rightarrow MC = \frac{\sum_{i=1}^{n} \sum_{j=1}^{n} c_{ij} \frac{(y_i - \bar{y})(y_j - \bar{y})}{\sum_{i=1}^{n} \sum_{j=1}^{n} c_{ij}}}{\sqrt{\sum_{i=1}^{n} \frac{(y_i - \bar{y})^2}{n}} \sqrt{\sum_{i=1}^{n} \frac{(y_j - \bar{y})^2}{n}}}. \quad (3.1)$$

The left-hand term in the denominator of r_P accounts for the variation in variable X. This term is replaced with that for the variation in Y (the right-hand term in the denominator of r_P), in keeping with the meaning of the prefix auto-. The numerator of r_P is an average over the total number of pairings of X and Y, namely n. This term is replaced with an average over the total number of geographic pairings, namely $\sum_{i=1}^{n} \sum_{j=1}^{n} c_{ij}$, where c_{ij} is a binary (i.e., 0–1) indicator variable denoting whether or not locations i and j are nearby. In other words, variable X in the formula for r_P is replaced by variable Y. Unlike conventional correlation coefficients, however, MC is not restricted to the range [–1, 1]; rather, its range is dictated by what essentially are the second largest positive and most negative eigenvalues of matrix **C**, which is constructed from the n^2 c_{ij} binary values (de Jong et al., 1984), and consequently can go slightly beyond these two endpoint values.

SA also can be interpreted as quantifying pattern on a map (e.g., trends, gradients, mosaics, hot/cold spots). As such, positive SA indicates that similar values tend to cluster on the map, while negative SA indicates that dissimilar values tend to cluster. This particular interpretation may be translated into paired comparisons of nearby values, with positive SA rendering a near-zero average paired comparison value. This perspective produced the Geary Ratio (GR) index:

$$\mathrm{GR} = \frac{\sum_{i=1}^{n}\sum_{j=1}^{n} c_{ij}\frac{(y_i - y_j)^2}{2\sum_{i=1}^{n}\sum_{j=1}^{n} c_{ij}}}{\sqrt{\sum_{i=1}^{n}\frac{(y_i-\bar{y})^2}{n-1}}\sqrt{\sum_{i=1}^{n}\frac{(y_i-\bar{y})^2}{n-1}}}. \tag{3.2}$$

The denominator of this index employs the unbiased variance estimate (division by n–1 rather than n). The numerator has a squared differences term, so that the sign of the difference is neutralized. And, the numerator has division by 2 because $(y_i - y_j)^2 = (y_j - y_i)^2$, both of which appear in the numerator, producing $2{y_i}^2$ and $2{y_j}^2$. The numerator term is the basis of geostatistics, with c_{ij} being replaced with d_{ij}, the distance separating locations i and j. Division by 2 is retained, and the relationship between $(y_i - y_j)^2/2$, labelled the semivariance, and d_{ij} characterizes latent spatial dependency.

The Moran scatterplot, for spatial autoregression, and the semivariogram plot, for geostatistics, are two graphical tools, which are special cases of the conventional bivariate scatter diagram, that can be used to portray these relationships. This first scatter diagram plots the sum or the average of neighbouring values (vertical axis) against the values themselves (horizontal axis). The second scatter diagram plots half the squared difference between two values (vertical axis) against the distance separating these two values (horizontal axis); virtually always, in part because there are so many distance pairs, the set of paired comparisons is aggregated into distance groups, and then group averages are plotted.

3.2.1 The Moran Scatterplot

When undertaking a spatial autoregression analysis, a useful graphical tool for first gauging the spatial dependency structure latent in georeferenced data is the Moran scatterplot (Anselin, 1995). This scatterplot allows an investigator to examine the nature, degree and extent of SA latent in georeferenced data. A modified version of the scatterplot links the MC to regression by focusing on the numerator of Eq. (3.1). Because the MC is computed with deviations from the mean, the georeferenced data first needs to be centred; hence, the plot becomes $\mathbf{C}(\mathbf{Y} - \bar{y}\mathbf{1})$ on the vertical axis versus $(\mathbf{Y} - \bar{y}\mathbf{1})$ on the horizontal axis. Now the MC can be computed with results obtained from two simple linear regressions. In the first regression, the spatially lagged deviation vector $\mathbf{C}(\mathbf{Y} - \bar{y}\mathbf{1})$ is regressed on the deviation vector $(\mathbf{Y} - \bar{y}\mathbf{1})$, producing regression coefficient $b_{\mathbf{YCY}}$. In the second regression, the row sum of connections $\mathbf{C1}$ is regressed on the vector $\mathbf{1}$, producing regression coefficient $b_{\mathbf{1C1}}$. Combining these results produces $\mathrm{MC} = \frac{b_{\mathbf{YCY}}}{b_{\mathbf{1C1}}}$. The regression analyses should be executed with no-intercept equations. This result reveals that the MC actually is a rescaled slope of the best-fitting line traversing the Moran scatterplot.

Calculating the Geary Ratio requires the additional bivariate regression of $(y_i - \bar{y}) \sum_{j=1}^{n} c_{ij}$ on $(y_i - \bar{y})$, again executed with a no-intercept option, yielding the regression coefficient b_{GR}, and equals

$$\frac{n-1}{n}\left(\frac{b_{GR}}{b_{1C1}} - \frac{b_{YCY}}{b_{1C1}}\right).$$

These regressions emphasize that SA calculations are traditional regressions involving geographical weightings.

3.2.2 The Semivariogram Plot

SA in geostatistics is visualized using the semivariogram plot (Cressie, 1989). The semivariogram allows an investigator to examine the nature, degree and extent of SA latent in georeferenced data. It is denoted by $\gamma(d)$, and may be written as

$$\gamma(\bar{d}_g) = \frac{1}{2n_g} \sum_{j=1}^{n} \sum_{i=1}^{n} \delta_{ij}(y_i - y_j)^2, \tag{3.3}$$

where n_g denotes the number of location pairs contained in distance group g, d_{ij} is the distance separating locations i and j, δ_{ij} is a binary 0–1 variable denoting whether or not both locations I and j belong to group g, and $\bar{d}_g$ is the average distance for group g.

The semivariogram plot is constructed using the average interlocation distance for distance interval g, $\bar{d}_g$, along the horizontal axis, and the average semivariances, $\gamma(\bar{d}_g)$, along the vertical axis. Pairings of similar attribute values give smaller values of $\gamma(\bar{d}_g)$; if positive SA is present in georeferenced data, $\gamma(\bar{d}_g)$ goes toward zero as $\bar{d}_g$ decreases. For many data sets, $\gamma(\bar{d}_g)$ increases with increasing distance until it reaches a relatively constant value, a value that commonly is referred to as the sill. The scatter of points in the semivariogram plot can portray various functional forms, all of which give some description of SA across a surface. A high degree of positive SA results in a semivariogram plot displaying a relatively shallow slope, while near-zero SA results in a steep slope. Here weighted nonlinear regression analysis is required to analytically determine the best-fitting line describing the semivariogram plot.

3.3 Specifications of Spatial Autoregressive and Geostatistical Models

Rather than invoking the *iid* assumption, suppose a set of observed values contains correlation, with their covariation denoted by $\mathbf{V}^{-1}\sigma^2$, where σ^2 is the common variance across the set of values, and $\mathbf{V} = \mathbf{I}$ in classical statistics. A distinction between the two spatial statistics modelling approaches is that spatial autoregression parameterises matrix **V** whereas geostatistics parameterises matrix $\mathbf{V}^{-1}$. These respective perspectives are analogous to those in time series analysis, with spatial autoregression exploiting the partial correlogram (the partial SA function) and geostatistics exploiting the correlogram (the SA function). In their simplest forms, each is specified such that spatial covariation behaves the same regardless of position on a map surface or direction in which the covariation occurs (i.e., isotropy).

3.3.1 Spatial Autoregressive Models

The simplest way to view spatial autoregression is to consider a standard linear regression model in which the predictor matrix **X** is augmented by a vector whose element i is calculated from values for locations nearby to location i. In keeping with Eq. (3.1), one example of this specification may be written as

$$\mathbf{Y} = \rho\mathbf{CY} + \mathbf{X}\boldsymbol{\beta} + \boldsymbol{\epsilon}, \tag{3.4}$$

where ρ denotes the regression coefficient that captures SA, and **CY** is the vector containing sums of nearby values. This specification directly links spatial autoregression to classical regression techniques, and hence to the normal distribution. But Eq. (3.4) furnishes an example where ordinary least squares (OLS) results are not equivalent to maximum likelihood results.

Gaussianity has been central to spatial autoregressive analyses, in part because the accompanying probability density function yields tractable maximum likelihood estimators, and in part because of the versatility of regression. Three model specifications dominate the spatial autoregressive literature: the conditional autoregressive (CAR, a 1st-order model), the simultaneous autoregressive (SAR, a 2nd-order model), and the autoregressive response (AR, a second-order model) models. Descriptions of these models are presented in Anselin (1988), Cressie (1991), Griffith (1988), Haining (1990), Ripley (1988), and Upton and Fingleton (1985), among others. In general, though, the constant mean and constant variance spatial autoregressive log-likelihood function based upon any inverse-covariance matrix $\mathbf{V}\sigma^{-2}$ may be written, in terms of a standard probability density function expression found in introductory multivariate textbooks, as

$$\text{constant} - \frac{n}{2}\ln(\sigma^2) + \frac{1}{2}\ln[\det(\mathbf{V})] - (\mathbf{X} - \mu\mathbf{1})^{\mathrm{T}}\mathbf{V}(\mathbf{X} - \mu\mathbf{1})/(2\sigma^2), \tag{3.5}$$

where det denotes the matrix determinant operation, and matrix **V** is a function of the connectivity matrix **C** and the SA parameter ρ. The normalizing constant (technically speaking, the Jacobian of the transformation from a spatially autocorrelated to a spatially unautocorrelated mathematical space, a concept discussed both in calculus and in introductory mathematical statistics textbooks), $\frac{1}{2}\ln[\det(\mathbf{V})]$ in this case, complicates spatial autoregressive analyses. Fortunately, for the Gaussian distribution it can be easily rewritten in terms of the eigenvalues of matrix **V**. Griffith and Sone (1995) also propose a very good approximation for this normalizing constant, one that facilitates the analysis of extremely large georeferenced data sets.

For the CAR spatial autoregressive model, matrix **V** in expression (3.5) becomes $(\mathbf{I} - \rho\mathbf{C})$. For the AR or SAR spatial autoregressive models, matrix **V** becomes $(\mathbf{I} - \rho\mathbf{C})^{T}(\mathbf{I} - \rho\mathbf{C})$, where T denotes the operation of matrix transpose, or $(\mathbf{I} - \rho\mathbf{W})^{T}(\mathbf{I} - \rho\mathbf{W})$ if a row-standardized (or stochastic) version of matrix **C** (namely matrix **W**) is preferred. Estimation of the parameters μ, σ^2, and ρ is complicated because there is no closed form estimator for the spatial autoregressive parameter ρ. The simplest way to estimate these three parameters is to use a non-linear algorithm to simultaneously solve the following triplet of first derivative equations: $\frac{\partial \ln(L)}{\partial \mu} = 0, \frac{\partial \ln(L)}{\partial \sigma^2} = 0$, and $\frac{\partial \ln(L)}{\partial \rho} = 0$. Initial values for the parameter estimates can be obtained by: (1) shifting the data to a mean of 0 for μ; (2) using a descriptive statistics procedure to compute s^2 for σ^2; and (3) regressing vector **X** on vector **CX** for ρ. The non-linear estimate for μ then can be added to $\bar{x}$ in order to compute the estimated mean of the original data. A non-constant mean version can be obtained by replacing μ with a vector of regression coefficients, $\boldsymbol{\beta}$, and then including a differential equation in the non-linear analysis for each of the regression coefficients.

This estimation processes can be recast as a weighted least squares regression problem. Now a simple way to estimate the three parameters is to solve the pair of differential equations $\frac{\partial \ln(L)}{\partial \mu} = 0$ and $\frac{\partial \ln(L)}{\partial \sigma^2} = 0$, substitute the resulting two estimators into the likelihood function, and then employ a non-linear algorithm to optimise the reduced likelihood function. To illustrate this situation, this substitution results in the following non-linear regression model for the AR autoregression specification employing matrix **W**:

$$\mathbf{Y}/EXP\left[-\frac{1}{n}\sum_{i=1}^{n}\mathrm{LN}(1-\rho\lambda_i)\right] = [\rho\mathbf{W}+\mathbf{X}\boldsymbol{\beta}]/EXP\left[-\frac{1}{n}\sum_{i=1}^{n}\mathrm{LN}(1-\rho\lambda_i)\right] + \boldsymbol{\varepsilon}/EXP\left[-\frac{1}{n}\sum_{i=1}^{n}\mathrm{LN}(1-\rho\lambda_i)\right] \tag{3.6}$$

This non-linear regression formulation uncovers the interesting feature that a systematic set of values spanning the feasible parameter space for ρ, namely $[1/\lambda_{min}, 1/\lambda_{max}]$, can be sequentially substituted into Eq. (3.6), each time followed by the execution of a conventional linear regression routine. The estimated value of ρ is that value for which the accompanying linear regression has the smallest MSE. A non-linear algorithm merely automates this sequence of linear regressions, being able to converge more efficiently on a more precise estimate of ρ.

SAS and SPSS code for implementing the estimation of parameters for the CAR and SAR spatial autoregressive specifications appears in Griffith and Layne (1999); SAS code for the AR model appears in Griffith (1993a).

3.4 Geostatistical Models

Various valid semivariogram functions can be fitted to a semivariogram plot in order to more concisely summarize the spatial similarity that may be present in georeferenced data (Cressie 1991; Christensen 1991; Isaaks and Srivastava 1989). Then parameter estimates of these functions can be used to define coefficients for optimal linear predict of unknown attribute values for unsampled areas from the data collected for sampled locations; that is, kriging (Cressie 1989). Semivariogram models that have been shown to be valid and/or useful include the exponential, circular, spherical, Gaussian, Bessel, power and wave/hole; they routinely appear throughout the geostatistics literature. As an example, the equation for the exponential model may be written as

$$\gamma(d) = \begin{cases} 0, & \text{for } |d| = 0 \\ C_0 + C_1\,[(1 - e^{(d/r)})], & \text{for } |d| > 0 \end{cases}, \tag{3.7}$$

where C_0 is an intercept term, C_1 defines the slope of the semivariogram curve, and r is the range parameter for spatial dependency. Semivariogram models have only a few parameters that need to be estimated. Theoretically, $\gamma(d)$ at lag 0 should be 0; however, a discontinuity may exist, called the nugget effect, which is represented as C_0 in Eq. (3.7). Since the semivariogram is similar to a covariance plot, the other two parameters that need to be estimated can be defined in terms of a covariance. The range parameter (r) defines the distance at which the covariance effectively becomes 0. The sill is the value at which the covariance stabilizes with increasing distance. Because a semivariogram plot essentially is an inverted covariance plot, the range for a semivariogram plot is identified as the distance at which $\gamma(d)$ becomes approximately constant.

Another semivariogram model that shows considerable data analytic promise, but has not been extensively employed by practitioners, is the modified Bessel function (motivated by Whittle 1954); it is one model that appears to be quite relevant but has been little discussed and little used. Preliminary investigation by Griffith and Csillag (1993) found it to be an important geostatistical model that provides a link to the SAR model of spatial autoregression. This model is specified as

$$\gamma(d) = \begin{cases} 0, & \text{for } |d| = 0 \\ C_0 + C_1\left[\left(1 - \left(\frac{d}{r}\right) K_1\left(\frac{d}{r}\right)\right)\right], & \text{for } |d| > 0 \end{cases}, \tag{3.8}$$

where K_1 is the modified Bessel function of the first-order, second-kind.

SAS and SPSS code used for estimating parameters of semivariogram models, including the Bessel function, appears in Griffith and Layne (1999). ESRI's *Geostatistical Analysis* (Johnson et al., 2001) also offers this estimation capability.

3.5 Linkages Between Spatial Autoregression and Geostatistics

Two salient distinctions between spatial autoregression and geostatistics can be made. For the most part, spatial autoregressive models are used to describe discretized georeferenced data (i.e., attributes located at distinct points or of objects aggregated because they are contained in the same cell of a surface partitioning), while geostatistical models are used to describe continuous georeferenced data (e.g., air pollution). In practice a researcher can move between these two types of georeferenced data by using areal unit centroids to perform a geostatistical analysis on aggregates of locations, and using a Thiessen polygon (see Okabe et al., 1992) surface partitioning to perform a spatial autoregression on sampled continuously distributed data. Also, in contrast with spatial autoregression, the emphasis of geostatistics is that of description and prediction and not inferential testing, per se, although prediction/confidence intervals can be constructed for predicted values and model parameter estimates.

Given that matrix **V** is common to both spatial autoregression and geostatistics, natural linkages should exist between models each sub-field specifies. Consider a regular square tessellation surface partitioning like the one associated with a remotely sensed image. Theoretical spatial correlations for this surface can be calculated using the following spectral density function:

$$\rho_{h,k} = \frac{\int_0^{\pi}\int_0^{\pi} \frac{\cos(hu) \times \cos(kv)}{[1 - 2\rho[\cos(u) + \cos(v)]]^{\kappa}} dxdy}{\int_0^{\pi}\int_0^{\pi} \frac{1}{[1 - 2\rho[\cos(u) + \cos(v)]]^{\kappa}} dxdy} \tag{3.9}$$

(after Bartlett 1975), where u and v respectively are the horizontal and vertical axes, and h represents the easting lag distances and k the northing lag distances. The CAR model is specified when the exponent is $\kappa = 1$ (rendering it a first-order model) and the SAR specified when $\kappa = 2$ (rendering it a second-order model). The values of $\rho_{h,k}$ are the entries in matrix $\mathbf{V}^{-1}$.

Geostatistics specifies models to describe the quantity $(1 - \rho_{h,k})$. Griffith and Csillag (1993) show that the exponential semivariogram model provides a near-perfect fit to this quantity when $\kappa = 1$ (the case of the CAR spatial autoregressive model). Griffith et al. (1996) show that the Bessel function semivariogram model provides a near-perfect fit to this quantity when $\kappa = 2$ (the case of the SAR spatial autoregressive model). Griffith and Layne (1996) show that generation of the correlations entered into matrix $\mathbf{V}^{-1}$ with Eq. (3.7) or (3.8), respectively for the exponential and Bessel function semivariogram models, followed by inversion to obtain matrix **V**, the covariance matrix form used in spatial autoregression, performs well but fails to render as close a correspondence between these two model pairings. Because of the differences between discretized and continuously distributed phenomena, these linkages are not necessarily exact correspondences.

Empirical evidence is less supporting of these numerical linkages. In an analysis of 35 data sets, Griffith and Layne (1999) find that frequently there is no clearly

Table 3.1 Numerical links between spatial statistical models used in selected empirical analyses

	Spatial autoregressive model		
Geostatistical model	CAR	SAR	AR
Power	1	4	1
Spherical	0	17	2
Exponential	1	6	3

Table 3.2 ANOVA table for difference of mean levels of spatial autocorrelation

Model source	Sequential sum of squares	Degrees of freedom
Geostatistics	0.04383	2
Spatial autoregression	0.21867	2
Interaction	0.01818	4
Error	1.05016	26
Total	1.33084	34

best geostatistical model. Choosing that model with the smallest relative error sum of squares (RSSE), the tabulation of their results appearing in Table 3.1 can be constructed. In no case does the Bessel function achieve the lowest RSSE. In addition, the ANOVA sum of squares decomposition, appearing in Table 3.2, for the spatial autoregression parameter estimate $\hat{\rho}$ suggests that these models do not link to differential levels of SA, either. In other words, most of the variation is within groups.

In summary, theoretically spatial autoregressive models tend to suggest particular geostatistical models, with this linkage being weaker in the other direction. Empirically these linkages remain elusive. Part of the complication here arises from edge effects introduced when moving between these two realms by matrix inversion (i.e., moving from $\mathbf{V}$ to $\mathbf{V}^{-1}$, or vice versa), and part from the discrete /continuous nature of georeferenced data.

3.6 A Major Commonality of Spatial Autoregression and Geostatistics

After SA, the most conspicuous unifying concept common to spatial autoregression and geostatistics is the need to predict attribute values for locations for which data are not present. Both spatial autoregression and geostatistics reside in the realm of multivariate analysis, and interface where they focus on the missing data problem.

Generally speaking, the missing data problem (see Little and Rubin, 1987) exploits redundant information contained in two variables, X and Y. In other words, values missing in variable Y are replaced by their conditional expectations through an iterative regression of Y on X, with the accompanying estimation being nonlinear. The popular *EM* algorithm has been formulated to solve this estimation problem. Letting the subscript "o" denote observed data values, and the subscript "m" denote missing data values, such that $\mathbf{Y}^{\mathrm{T}} = < \mathbf{Y}_o^T \vdots \hat{\mathbf{Y}}_m^T >$, the missing data sub-vector solution may be written as

$$\hat{\mathbf{Y}}_{\mathrm{m}} = \mathbf{X}_{\mathrm{m}}\hat{\boldsymbol{\beta}}. \tag{3.10}$$

When covariations amongst observations are present, Eq. (3.10) becomes

$$\hat{\mathbf{Y}}_{\mathrm{m}} = \mathbf{X}_{\mathrm{m}}\hat{\boldsymbol{\beta}} - \mathbf{A}_{\mathrm{mm}}^{-1}\mathbf{A}_{\mathrm{mo}}(\mathbf{Y}_{\mathrm{o}} - \mathbf{X}_{\mathrm{o}}\hat{\boldsymbol{\beta}}), \tag{3.11}$$

where matrix **A** contains covariation-based terms depicting dependencies amongst observations, and may be written as the following partitioned matrix

$$\begin{pmatrix} \mathbf{A}_{\mathrm{oo}} & \mathbf{A}_{\mathrm{om}} \\ \mathbf{A}_{\mathrm{mo}} & \mathbf{A}_{\mathrm{mm}} \end{pmatrix}.$$

Matrix $\mathbf{A}_{\mathrm{oo}}$ is a function of covariations amongst the observed values, matrices $\mathbf{A}_{\mathrm{om}}$ and $\mathbf{A}_{\mathrm{mo}}$ are functions of covariations between the observed and missing values, and matrix $\mathbf{A}_{\mathrm{mm}}$ is a function of covariations amongst the missing values. In Eq. (3.11), given the *iid* assumption, $\mathbf{A}_{\mathrm{mo}} = \mathbf{A}_{\mathrm{mo}}{}^{\mathrm{T}} = \mathbf{0}$. Accordingly, the missing spatial values problem exploits correlation between variables X and Y as well as correlation between some observation y_i and other observations y_j ($i \neq j$). In the case of spatial autoregression, matrix **A** is constructed from matrix $\rho\mathbf{C}$ or $\rho\mathbf{W}$. Consequently, for the conditional autoregressive (CAR) spatial statistical model, where using traditional multivariate notation the variance-covariance matrix may be written as $\boldsymbol{\Sigma} = (\mathbf{I} - \rho\mathbf{C})^{-1}\sigma^2$, Eq. (3.11) may be rewritten as

$$\hat{\mathbf{Y}}_{\mathrm{m}} = \mathbf{X}_{\mathrm{m}}\hat{\boldsymbol{\beta}} + \rho(\mathbf{I} - \rho\mathbf{C}_{\mathrm{mm}})^{-1}\mathbf{C}_{\mathrm{mo}}(\mathbf{Y}_{\mathrm{o}} - \mathbf{X}_{\mathrm{o}}\hat{\boldsymbol{\beta}}). \tag{3.12}$$

This result can be generalized to other popular spatial autoregressive specifications, such as the simultaneous autoregressive (SAR) model, by considering the inverse-covariance matrix $\mathbf{V}\sigma^{-2}$, which is some function of matrix $\rho\mathbf{C}$, and which allows the variance-covariance matrix to be written as $\boldsymbol{\Sigma} = \mathbf{V}^{-1}\sigma^2$, rendering

$$\hat{\mathbf{Y}}_{\mathrm{m}} = \mathbf{X}_{\mathrm{m}}\hat{\boldsymbol{\beta}} - \mathbf{V}_{\mathrm{mm}}^{-1}\mathbf{V}_{\mathrm{mo}}(\mathbf{Y}_{\mathrm{o}} - \mathbf{X}_{\mathrm{o}}\hat{\boldsymbol{\beta}}). \tag{3.13}$$

Griffith (1992a, 1988) outlines the specification of Eq. (3.13) for other popular spatial autoregressive models. Meanwhile, in the case of geostatistics, matrix **A** is modelled from the semivariogram plot, which depicts the nature, degree, and geographic extent of SA effects. The accompanying kriging equation may be written as

$$\hat{\mathbf{Y}}_{\mathrm{m}} = \mathbf{X}_{\mathrm{m}}\hat{\boldsymbol{\beta}} - \sum\nolimits_{\mathrm{mo}} \sum\nolimits_{\mathrm{oo}}^{-1}(\mathbf{Y}_{\mathrm{o}} - \mathbf{X}_{\mathrm{o}}\hat{\boldsymbol{\beta}}). \tag{3.14}$$

The algebraic relationship between Eqs. (3.13) and (3.14) can be established by inspecting standard partitioned matrix inverse results:

$$\mathbf{\Sigma}_{oo} = (\mathbf{V}_{oo} - \mathbf{V}_{om}\mathbf{V}_{mm}{}^{-1}\mathbf{V}_{mo})^{-1}\sigma^2 \text{ and } \mathbf{\Sigma}_{mo} = -\mathbf{V}_{mm}{}^{-1}\mathbf{V}_{mo}(\mathbf{V}_{oo} - \mathbf{V}_{om}\mathbf{V}_{mm}{}^{-1}\mathbf{V}_{mo})^{-1}\sigma^2$$

and

$$\mathbf{V}_{mm} = (\mathbf{\Sigma}_{mm} - \mathbf{\Sigma}_{mo}\mathbf{\Sigma}_{oo}{}^{-1}\mathbf{\Sigma}_{om})^{-1}\sigma^2 \text{ and } \mathbf{V}_{mo} = -(\mathbf{\Sigma}_{mm} - \mathbf{\Sigma}_{mo}\mathbf{\Sigma}_{oo}{}^{-1}\mathbf{\Sigma}_{om})^{-1}\mathbf{\Sigma}_{mo}\mathbf{\Sigma}_{oo}{}^{-1}\sigma^2.$$

In other words, there is an exact algebraic correspondence between these two results (Griffith 1993b), one highlighting that spatial autoregression directly deals with the inverse variance-covariance matrix while geostatistics directly deals with the variance-covariance matrix itself.

3.7 Implications for Quantitative Human Geography

This chapter presents a general overview of two sub-fields of spatial statistics known as spatial autoregression and geostatistics, with explicit reference to SA. Although some subtle details have been omitted, one aim of the chapter is to emphasize the current status of connections established between geostatistical analysis and spatial autoregression. The most striking feature of progress to date along these lines is the ability to use either of these modelling strategies to do spatial forecasting. For example, Cressie (1991: 160) reports the kriging estimate for a missing coal ash value as 10.27%, using a spherical semivariogram model; Griffith and Layne (1999: 440) report a value of 10.17% using a missing-data SAR model. A second noteworthy feature emphasized in this chapter is that conventional regression actually underlies much of spatial analysis; regression truly is the workhorse of empirical statistical analyses. This latter notion is echoed by a geostatistical capability implementation in the SAS MIXED procedure.

But much work remains to be done! For example, progress still needs to be made in the areas of the auto-logistic and auto-Poisson models. And advantages of each approach—such as geostatistics more easily handles anisotropy—need to be established.

Acknowledgment Contributions made to the development of this chapter by Dr. Larry Layne, University of New Mexico, are greatly appreciated.

Chapter 4
Frequency Distributions for Simulated Spatially Autocorrelated Random Variables

4.1 Introduction

Often quantitative data analysis begins with an inspection of attribute variable histograms. Ratio scale demographic variables, such as population density (which has a natural, meaningful absolute 0 value), are expected to conform, at least approximately, to a normal probability distribution. Frequently this conformity requires that these variables be subjected to a symmetricizing, variance stabilizing transformation, such as the Box-Cox class of power functions or the Manley exponential function. Counts (i.e., aggregated nominal measurement scale) data used to construct ratios, such as the crude fertility rate (i.e., number of births per number of women in the child bearing age cohort), are expected to conform to a Poisson probability distribution. And, counts data that constitute some subset of a total, such as the percentage of people at least 100 years of age or the percentage of a population that is the women in the child bearing age cohort, are expected to conform to a binomial probability distribution. Until the advent of implemented generalized linear models (GLMs), these latter two categories of data also were subjected to variable transformations in order to secure normal probability distribution approximations. Various scholars today argue that GLM technology has made the use of such previously popular variable transformations as the square root for Poisson counts, or the arcsine for percentages, obsolete.

Most spatial statistical work to date addresses impacts of spatial autocorrelation (SA) on parameter estimates, with the general conclusion that positive SA tends to have little or no impact on first moment types of parameter estimates, while inflating their respective standard errors. SA also tends to improve model prediction capabilities, serving remarkably well as a surrogate for missing covariates displaying particular geographic map patterns. This result implies that as SA in a random variable (RV) increases, its tails should become heavier and its center should become flatter. Dutilleul and Legendre (1992) appear to be about the only researchers to systematically investigate this topic, although they do so in a rather pseudo-geographic context.

As is widely acknowledged, positive SA is a source of variance inflation for normal RVs, and a source of overdispersion (i.e., excess variance) for Poisson and binomial RVs. But how does this increased variation impact upon a variable's

D.A. Griffith, J.H.P. Paelinck, *Non-standard Spatial Statistics and Spatial Econometrics*, Advances in Geographic Information Science 1,
DOI 10.1007/978-3-642-16043-1_4,

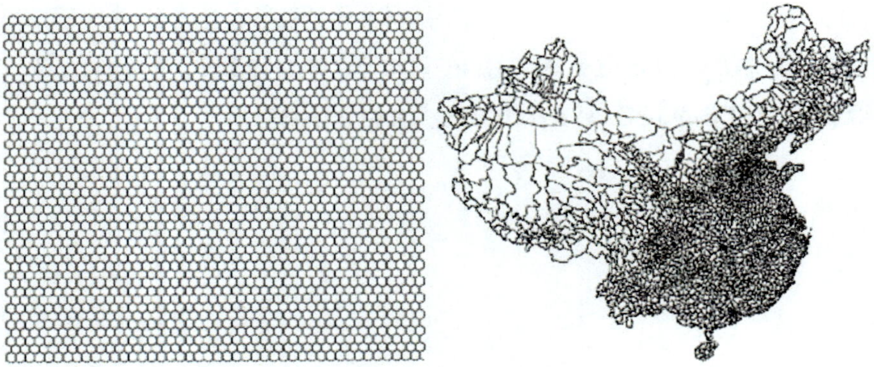

Fig. 4.1 Surface partitionings used for simulation work. *Left* (**a**): a 44-by-54 regular hexagonal tessellation forming a rectangular region. *Right* (**b**): the China county outline map

histogram? This is the question addressed in this chapter. Intuitively speaking, variance increases as increasingly extreme values (i.e., outliers) appear in a histogram. SA-generated heavy tails in a normal distribution are consistent with this data feature. But a binomial RV cannot have extreme values, since its values are constrained by given totals, so that percentages always are contained in the closed interval [0, 100]. A Poisson RV can have extreme large counts; its extreme small counts, however, can only become excessive zeroes. In other words, is some of the quite bothersome noise in or potential dirtiness of data researchers routinely encounter simply a manifestation of SA?

This chapter demonstrates positive SA impacts upon histograms with illustrations based upon simulated data. These data are generated both with autoregressive and with spatial filter (SF; see Sect. 2.5) models (Griffith, 2000a, 2002a, 2004a). Autoregressive models more explicitly focus on SA arising from spatial interaction, whereas SF models more explicitly focus on SA arising from missing variables with specific map patterns—here these map patterns have been selected to represent global, regional, and local spatial effects (Borcard and Legendre, 2002; Borcard et al., 2004). The primary difference is between a variance and a mean response specification that captures SA effects. Furthermore, SF models enable much greater degrees of SA to be explored, primarily because autoregressive models tend to encounter such problems as phase transitions when positive SA becomes excessively strong (Guyon, 1995). The simulated data, which is for an ideal 44-by-54 [$n = 2{,}376$; maximum Moran Coefficient (MC_{max}) of 1.02239] regular hexagonal tessellation (Fig. 4.1a), also is supplemented by simulations for the irregular China county surface partitioning (Fig. 4.1b).

4.2 The Normal Probability Model

Haining et al. (1983) outline a technique, in keeping with normal theory in multivariate statistics, to simulate spatially autocorrelated normal RVs with, for example, the simultaneous autoregressive (SAR; Cliff and Ord, 1973) model. A recent approach sharing a number of the features of their procedure is furnished by Gneiting et al. (2005). Goodchild (1980) offers an alternative procedure that involves permuting independent and identically distributed (iid) values over a map until a prespecified level of SA is attained. Goodchild's method is employed here to remove any conspicuous spurious SA from the simulated data. However, it cannot be used to explore SA impacts upon histograms because histograms are completely insensitive to the locational arrangement of values, simulated or actual, on a map. Furthermore, the resulting observed map would need to have its underlying iid counterpart uncovered in order to explore SA effects.

In keeping with linear statistical models theory, eigenvector-based spatial filtering offers a striking alternative mechanism for simulating spatially autocorrelated normal RVs (see Boots and Tiefelsdorf, 2000, p. 327; Griffith, 2000, p. 146). This technique still begins with a set of iid values.

4.2.1 Simulating Spatially Autocorrelated Normal RVs

The simulated iid normal RV, say n-by-1 vector $\boldsymbol{\varepsilon}$, displays ideal properties (see Fig. 4.2 and Table 4.1). All levels of SA have been embedded into this RV.

Consider a surface that is partitioned into n mutually exclusive and collectively exhaustive areal units. Here these units are regular hexagons forming a 44-by-54 rectangular region (see Fig. 4.1a), or the counties into which China is divided (see Fig. 4.1b). The n-by-n binary geographic connectivity matrix **C** contains the elements $c_{ij} = 1$ if areal units (e.g., hexagons, counties) i and j share a common boundary, and $c_{ij} = 0$ otherwise; $c_{ii} = 0$ by construction (i.e., an areal unit cannot be spatially autocorrelated with itself). This definition of matrix **C** highlights the reason for selecting a hexagonal surface partitioning as the ideal surface, namely the

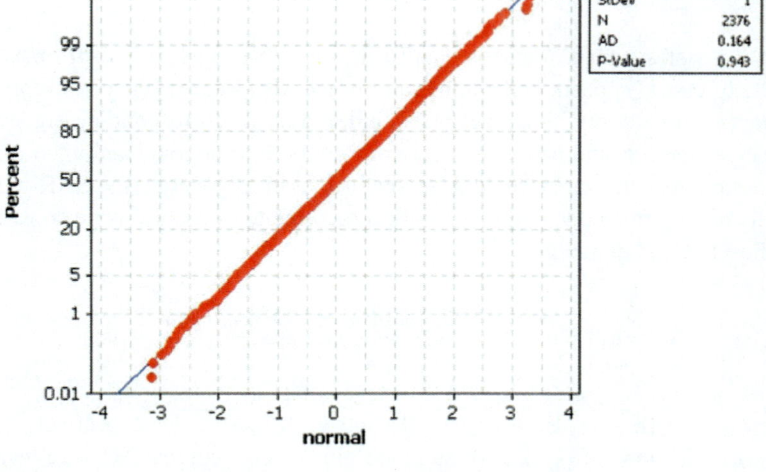

Fig. 4.2 Normal quantile plot for the simulated iid normal RV $\boldsymbol{\varepsilon}$ values

Table 4.1 Descriptive statistics for the SAR model-based simulated data and the hexagonal tessellation geographic configuration

Variable autocorrelation	MC/MC_{max}	GR	$\bar{y}$	s_y	$\lvert z_{skewness}\rvert$ [a]	$\lvert z_{kurtosis}\rvert$ [a]
None (i.e., iid)	–0.01	1.00	–0.000	1.000	0.20	0.30
Weak	0.11	0.89	–0.000	1.025	0.40	0.30
Low-moderate	0.40	0.59	–0.001	1.281	1.19	0
High-moderate	0.60	0.37	–0.006	1.711	1.59	1.29
Strong	0.90	0.07	–0.092	4.707	5.97	4.08

[a]The mean of skewness and kurtosis is 0; the respective standard errors, which can be established using the moment generating function $e^{\mu t+(\sigma^2/2)t^2}$, respectively are $\sqrt{6/n}$ and $\sqrt{24/n}$

lack of areal units sharing only a common point (i.e., a non-zero length boundary)—the difference between rook's and queen's adjacencies, using analogies with chess moves, in the spatial weights matrix literature.

Next, following Haining et al. (1983), matrix **C** was converted to its row-standardized version, matrix **W**, by dividing each c_{ij} value by its row sum $\left(\text{i.e., } \sum_{j=1}^{n} c_{ij}\right)$. Then spatially autocorrelated variables were constructed with the simultaneous autoregressive (SAR)-based equation

$$\mathbf{Y}_j = (\mathbf{I} - \rho_j \mathbf{W})^{-1} \boldsymbol{\varepsilon}, \qquad (4.1)$$

where **I** is the n-by-n identity matrix, and the SA parameter ρ_j was assigned the values 0.30, 0.73, 0.88, and 0.987 (i.e., j = 1, 2, 3, 4) in order to secure the relative MC and Geary Ratio (GR) values reported in Tables 4.1 and 4.3.

Finally, following especially Griffith (2000), the eigenvectors were extracted from matrix

$$(\mathbf{I} - \mathbf{ii}^T/n)\mathbf{C}(\mathbf{I} - \mathbf{ii}^T/n), \qquad (4.2)$$

where T denotes matrix transpose, and **i** denotes an n-by-1 vector of ones. This matrix expression appears in the numerator of the MC. Each eigenvector represents a distinct map pattern with a level of SA indexed by its corresponding eigenvalue. These eigenvectors, and hence map patterns, are both orthogonal and uncorrelated. Then, employing the same random iid vector $\boldsymbol{\varepsilon}$ used to generate the SAR-induced spatially autocorrelated variates, spatially autocorrelated variables were constructed with the SF-based equation

$$\mathbf{Y}_j = \alpha_j \left(\sqrt{n-1}\right) \frac{(a\mathbf{E}_G + b\mathbf{E}_R + c\mathbf{E}_L)/\sqrt{n-1}}{\sqrt{a^2+b^2+c^2}} + \beta\boldsymbol{\varepsilon}, \qquad (4.3)$$

where $\mathbf{E}_G/\sqrt{p_k(n-1)}$, $\mathbf{E}_R/\sqrt{p_k(n-1)}$ and $\mathbf{E}_L/\sqrt{p_k(n-1)}$ respectively denote the z-score versions of p_k (k > 0) summed global (G), regional (R), and local (L) eigenvectors, coefficients a, b, and c are weights that enable a particular level of SA to be induced (Boots and Tiefelsdorf, 2000; Griffith, 2000), and here coefficients $\beta = 1$ and $\alpha_j = \sqrt{\frac{MC_j - MC_\varepsilon}{MC_{eigenvectors} - MC_j}}$, for some target value of MC (i.e., MC_j) for variate Y_j, where $MC_{eigenvectors} = \frac{a^2 MC_G + b^2 MC_R + c^2 MC_L}{a^2 + b^2 + c^2}$ denotes the MC value for a given eigenvector sum. The formula for coefficient α_j assumes that the random error variate and the eigenvectors are uncorrelated.[1] Judiciously selected eigenvectors allow global, regional, and local spatial effects (this interpretation is from Borcard and Legendre, 2002; Borcard et al., 2004) to be simulated with a SF model. The relative MC and GR values obtained with this simulation method are reported in Tables 4.2 and 4.4.

4.2.2 Simulation Results for an Ideal Regular Hexagonal Surface Partitioning

Summary descriptive statistics appear in Table 4.1 for the SAR-induced simulated spatially autocorrelated data. These statistics confirm that the mean essentially is unaffected, while the variance is inflated, by SA. Within the moderate SA range, representing a preponderance of empirical studies to date, variance inflation is problematic, increasing as much as nearly 300%. The histograms appearing in Fig. 4.3 confirm the expectations that low levels of SA have little effect, whereas high levels of SA thicken the tails and squash the center of a normal frequency distribution; this trend is less noticeable with SAR models (see Fig. 4.4). But this latter outcome primarily is because of variance inflation.

Somewhat less noticeable skewness and kurtosis features are better portrayed by inspecting standardized normal curves. Z-score test statistics reported in Table 4.1 reveal that skewness and kurtosis increase as positive SA increases. Skewness becomes more problematic because, similar to a product moment correlation coefficient, the SAR SA parameter is restricted to be < 1, causing a truncation effect in the distribution of values. A surprising outcome is best seen when $MC = 0.90$: as SA becomes strong, not only do the tails become thicker, but values become more concentrated about 0 (the mean), resulting in a relative decrease in the number of intermediate values (the histogram columns are shrinking away from the normal curve outline toward the horizontal axis). This squashing toward the center of the distribution increases kurtosis. Moreover, SA produces more extreme and more near-zero values.

[1] The correlations between the simulated random normal variate and the sum of two eigenvectors representing global map, two representing regional, and two representing local map patterns used to construct Table 2 respectively are 0.031, 0.018 and –0.004—essentially 0.

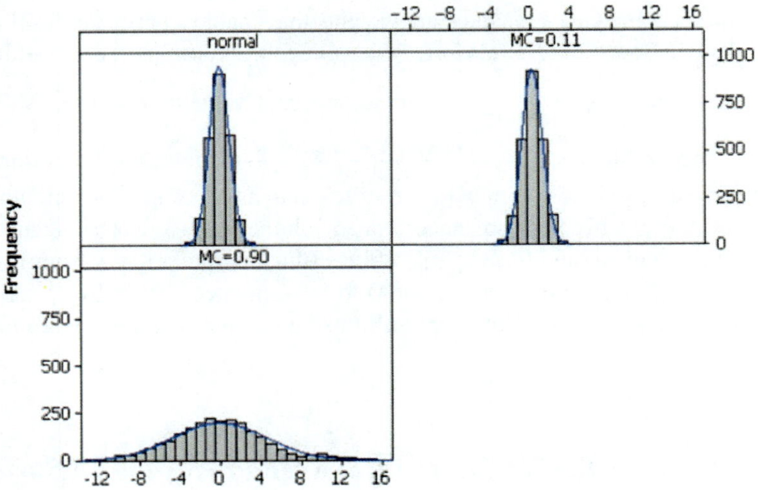

Fig. 4.3 Histograms for iid and two SAR model-induced extreme levels of SA

Because impacts beyond that of variance inflation are difficult to detect visually in the histograms themselves, normal quantile plots also can be inspected (see Fig. 4.5). These plots help highlight the noted changes in the histograms.

Summary descriptive statistics appearing in Table 4.2 for the SF-simulated spatially autocorrelated data (see Fig. 4.6 for maps of the global, regional, and local map patterns employed; as SA decreases in strength, the map patterns become more fragmented) distributed over the hexagonal surface partitioning corroborate findings gleaned from Table 4.1. These statistics confirm that the mean essentially is unaffected, while the variance is inflated, by SA. Again variance inflation is problematic within the moderate SA range, increasing anywhere from 1- to 10-fold, depending upon the mixture of global, regional, and local map patterns. The quantile plots appearing in Fig. 4.7 confirm the expectations that low levels of SA have little effect, whereas high levels of SA thicken the tails and squash the center of a normal frequency distribution—as before, SA produces more extreme and more near-zero values—with this trend being less noticeable with moderate levels of SA. The central tendency concentration is more conspicuous with the global SF-based results than with the preceding SAR-based results. To some degree, local sources of SA seem to dampen more extreme impacts of regional sources, whereas local and regional sources of SA seem to dampen more extreme impacts of global sources. However, a mixture of map patterns—the more common case in practice—appears to produce a more marked impact on variance inflation for moderate levels of SA, but without noticeably affecting kurtosis. Once again, z-score test statistics reported in Table 4.2 reveal that as positive SA becomes marked, kurtosis—but not skewness—increases, with global sources of SA causing the most significant change in kurtosis.

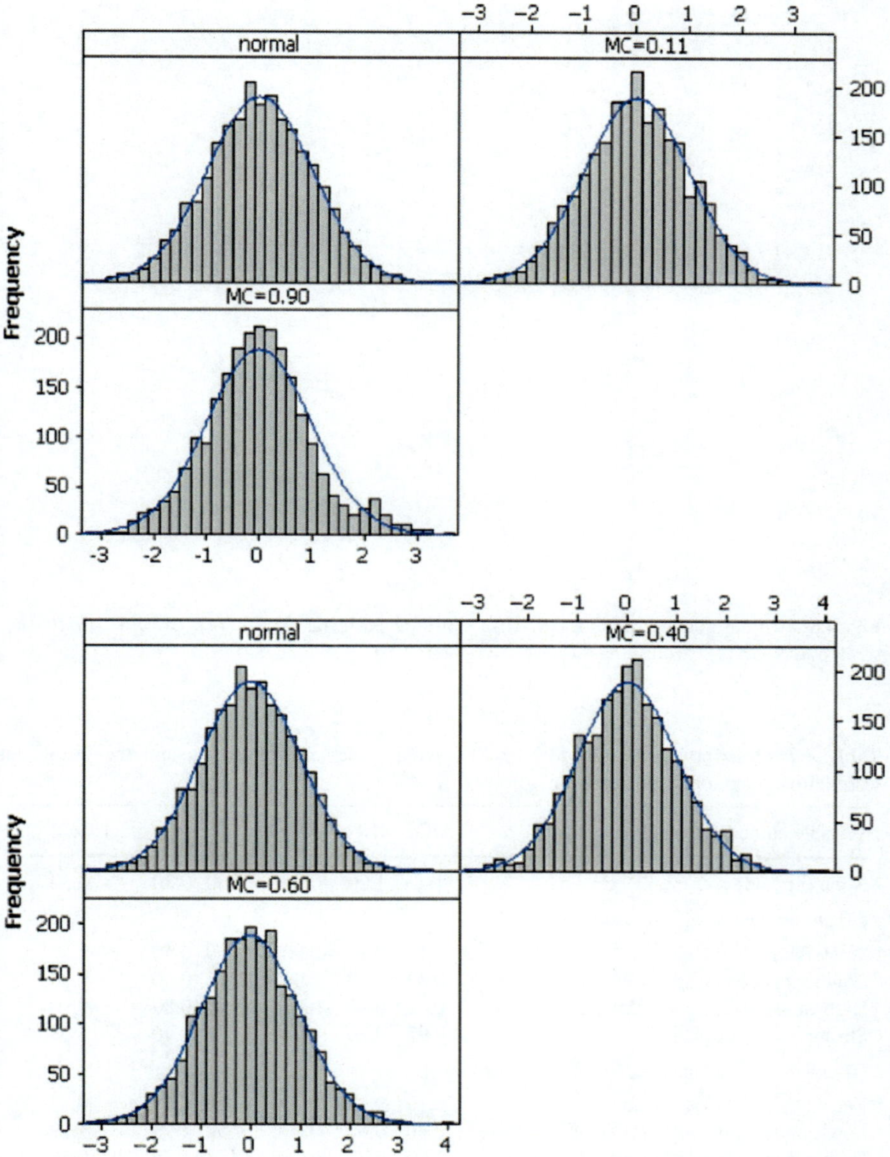

Fig. 4.4 Standard normal deviate histograms for iid and four SAR model-induced levels of SA. *Left* (**a**): induced extreme levels. *Right* (**b**): induced moderate levels

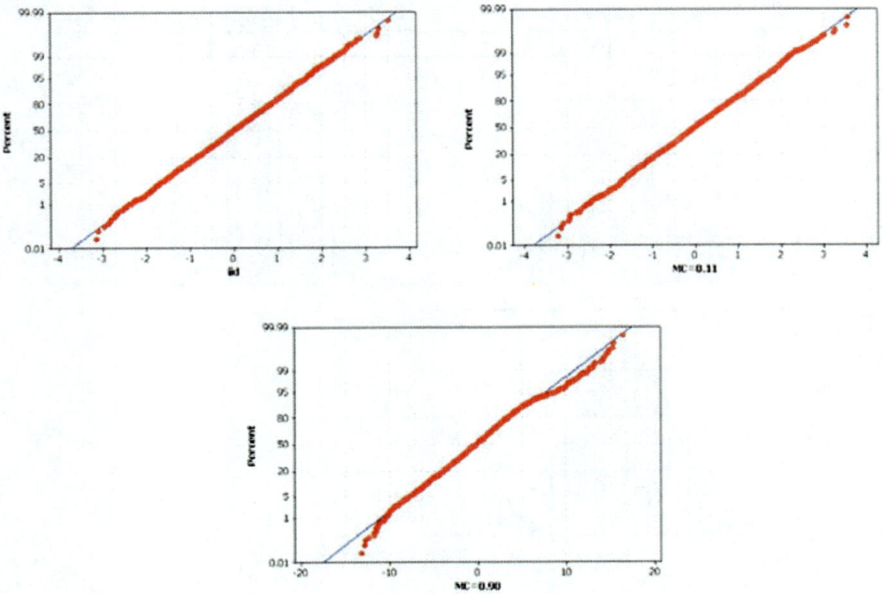

Fig. 4.5 Normal quantile plots for SAR model-induced levels of SA. *Left* (**a**): iid. *Middle* (**b**): weak SA. *Right* (**c**): strong SA

Table 4.2 Descriptive statistics for the SF model-based simulated data and the hexagonal tessellation geographic configuration

Variable autocorrelation	MC	GR	$\bar{y}$	s_y	$\lvert z_{skewness} \rvert$	$\lvert z_{kurtosis} \rvert$
None (i.e., iid)	–0.01	1.00	–0.000	1.000	0.20	0.30
Global map pattern-base results						
Weak (using $0.35\mathbf{E}_G$)	0.11	0.89	–0.000	1.060	0.40	0.50
Low-moderate (using $0.85\mathbf{E}_G$)	0.42	0.59	–0.000	1.306	0.60	0.20
High-moderate (using $1.25\mathbf{E}_G$)	0.62	0.40	–0.000	1.600	0.80	0.90
Strong (using $3.00\mathbf{E}_G$)	0.92	0.10	–0.000	3.181	0.40	3.58
Global + regional map pattern-base results						
Weak (using $0.50\mathbf{E}_R$)	0.10	0.89	–0.000	1.126	1.19	0.70
Low-moderate [using $0.75(\mathbf{E}_G+\mathbf{E}_R)$]	0.40	0.60	–0.000	1.452	1.99	1.39
High-moderate [using $1.33(\mathbf{E}_G+\mathbf{E}_R)$]	0.60	0.41	–0.000	2.123	1.79	0.20
Strong [using $2.50(3\mathbf{E}_G+\mathbf{E}_R)$]	0.91	0.11	–0.000	4.112	0.80	3.08
Global + regional + local map pattern-base results						
Weak (using $0.85\mathbf{E}_L$)	0.10	0.90	–0.000	1.297	0.20	0.10
Low-moderate [using $1.80(1.5\mathbf{E}_R+\mathbf{E}_L)$]	0.40	0.60	–0.000	3.403	0.40	0.99
High-moderate [using $1.10(1.5\mathbf{E}_G+\mathbf{E}_R+\mathbf{E}_L)$]	0.61	0.41	–0.000	2.444	1.59	0.50
Strong [using $1.00(5\mathbf{E}_G+2\mathbf{E}_R+\mathbf{E}_L)$]	0.90	0.12	–0.000	5.624	0.80	3.18

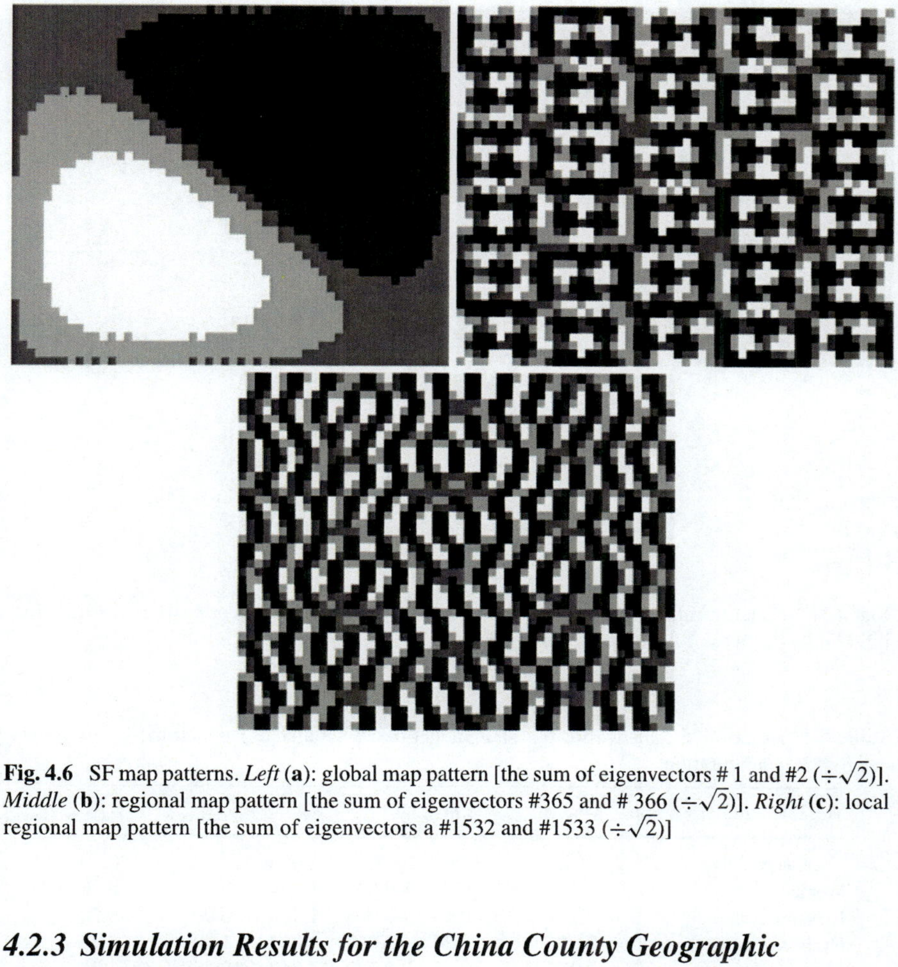

Fig. 4.6 SF map patterns. *Left* (**a**): global map pattern [the sum of eigenvectors # 1 and #2 ($\div\sqrt{2}$)]. *Middle* (**b**): regional map pattern [the sum of eigenvectors #365 and # 366 ($\div\sqrt{2}$)]. *Right* (**c**): local regional map pattern [the sum of eigenvectors a #1532 and #1533 ($\div\sqrt{2}$)]

4.2.3 Simulation Results for the China County Geographic Configuration

The simulated data coupled with the China county geographic configuration (see Fig. 4.1b) includes the 2,376 values used for the regular hexagonal tessellation simulation together with three additional values that were carefully selected so that the descriptive statistics appearing in Table 4.1 and the normal quantile plot appearing in Fig. 4.2 essentially remain unchanged. Once again the SA parameter ρ_j has taken on the values 0.30, 0.78, 0.93, and 0.986 (at this point a phase transition is encountered), rendering the relative MC and GR values reported in Table 4.3. The goal here is to explore impacts in terms of an irregular lattice surface partitioning.

Summary descriptive statistics appear in Table 4.3 for the SAR-induced simulated spatially autocorrelated data. These statistics confirm that the mean essentially

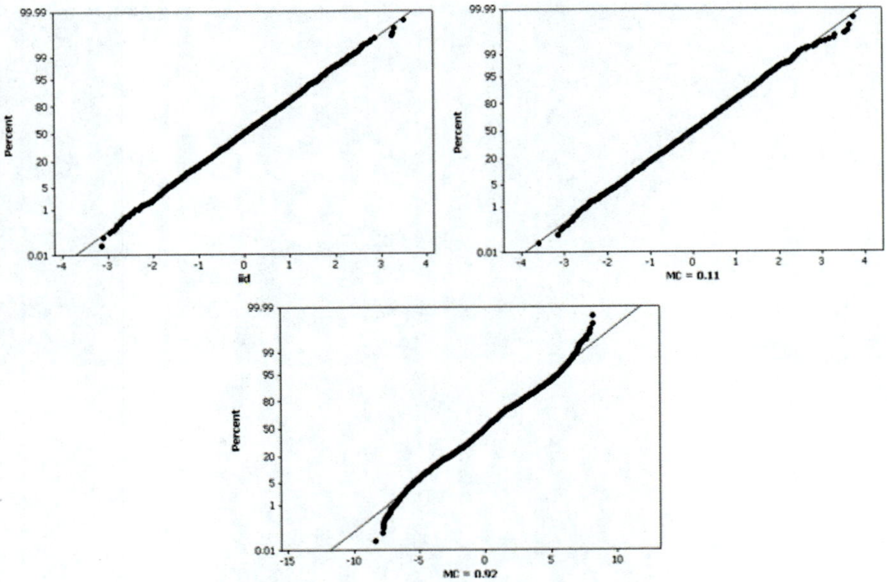

Fig. 4.7 Normal quantile plots for global map pattern SF model-induced levels of SA. *Left* (**a**): iid. *Middle* (**b**): weak SA. *Right* (**c**): strong SA

Table 4.3 Descriptive statistics for the SAR model-based simulated data and the China county geographic configuration

Variable autocorrelation	MC/MC_{max}	GR	$\bar{y}$	s_y	$\lvert z_{skewness} \rvert$	$\lvert z_{kurtosis} \rvert$
None (i.e., iid)	–0.00	1.00	0.000	1.000	0.20	0.30
Weak	0.11	0.86	–0.001	1.031	0.20	0.20
Low-moderate	0.41	0.49	–0.008	1.426	0.00	0.50
High-moderate	0.61	0.24	–0.037	2.253	1.00	3.78
Strong	0.73	0.07	–0.222	4.760	10.75	61.33

is unaffected, while the variance is inflated, by SA. Variance inflation is problematic within the moderate range, increasing as much as nearly 500%. The normal quantile plots appearing in Fig. 4.8 again confirm the expectations that low levels of SA have little effect, whereas high levels of SA thicken the tails and squash the center of a normal frequency distribution. In this case, pronounced levels of SA interact with the irregularness of the surface partitioning to result in the generation of rather dramatic extreme values.

As mentioned previously, less noticeable skewness and kurtosis features are better portrayed by inspecting standardized normal curves. Z-score test statistics reported in Table 4.3 reveal that as positive SA increases, so do skewness and

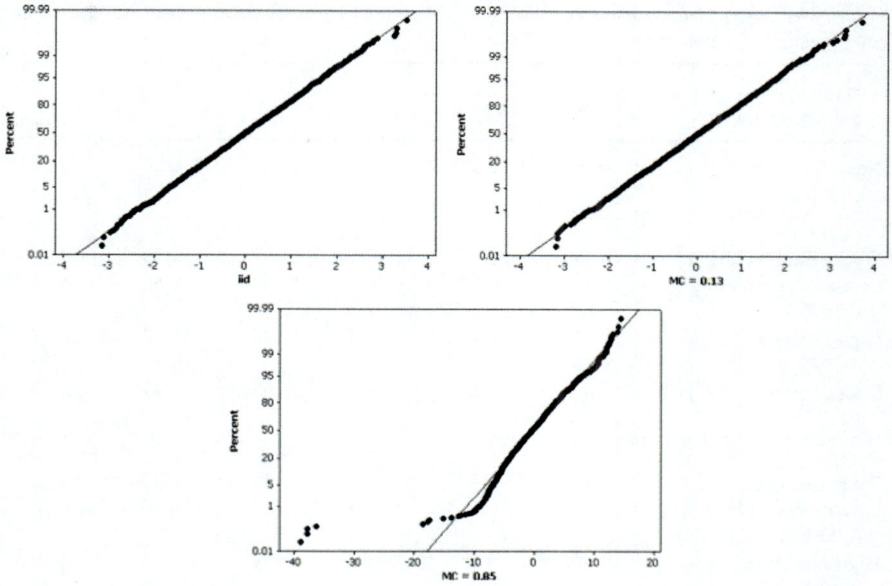

Fig. 4.8 Normal quantile plots for SAR model-induced levels of SA. *Left* (**a**): iid. *Middle* (**b**): weak SA. *Right* (**c**): strong SA

kurtosis. In part, skewness becomes more problematic because of the irregularness of the underlying county geographic configuration. As before, relatively strong levels of SA are accompanied by not only thicker tails, but also values that are more concentrated about 0 (the mean), resulting in a relative decrease in the number of intermediate values (the histogram columns shrink away from the normal curve outline toward the horizontal axis), with this squashing toward the center of the distribution increasing kurtosis.

SF induced SA coupled with the China county geographic configuration, based upon mixtures of global, regional (two levels, R-1 and R-2), and local map pattern eigenvectors render the MC and GR values reported in Table 4.4 (see Fig. 4.9 for maps of the global, two regional, and local map patterns employed here). A lack of impact upon the mean as well as variance inflation continue to characterize these variables. But histogram distortions affiliated with the underlying histogram for the global trend dominate the skewness and kurtosis modifications arising from positive SA. The example normal quantileplots appear in Fig. 4.10 (histograms portray a situation of more extreme values materializing under strong positive SA; a denser concentration about the mean still occurs). Here histogram distortions already become quite apparent at moderate levels of positive SA. Again these tendencies are more apparent visually by inspecting the corresponding standard normal curves.

Table 4.4 Descriptive statistics for the global SF model-based simulated data and the China county geographic configuration

Variable autocorrelation	MC	GR	$\bar{y}$	s_y	$\lvert z_{skewness} \rvert$	$\lvert z_{kurtosis} \rvert$
None (i.e., iid)	–0.00	1.00	0.000	1.000	0.20	0.30
Global map pattern-base results						
Weak (using $0.33\mathbf{E}_G$)	0.11	0.91	0.000	1.052	1.19	0.20
Low-moderate (using $0.75\mathbf{E}_G$)	0.40	0.67	0.000	1.254	8.56	6.17
High-moderate (using $1.10\mathbf{E}_G$)	0.61	0.50	0.000	1.484	15.73	16.13
Strong (using $2.10\mathbf{E}_G$)	0.90	0.26	0.000	2.296	28.47	38.33
Global + regional map pattern-base results						
Weak (using $0.55\mathbf{E}_{R-1}$)	0.10	0.90	0.000	1.136	0.20	0.30
Low-moderate [using $1.00(\mathbf{E}_{R-1}+\mathbf{E}_{R-2})$]	0.41	0.63	0.000	1.758	0.40	0.00
High-moderate [using $1.10(\mathbf{E}_G+\mathbf{E}_{R-1}+\mathbf{E}_{R-2})$]	0.60	0.48	0.000	2.141	4.18	4.48
Strong [using $2.20(\mathbf{E}_G+\mathbf{E}_{R-1})$]	0.90	0.23	0.000	6.229	9.96	17.03
Global + regional + local map pattern-base results						
Weak [using $0.55(\mathbf{E}_{R-1}+\mathbf{E}_L)$]	0.10	0.87	0.000	1.25	1.39	0.70
Low-moderate [using $1.80(\mathbf{E}_{R-1}+\mathbf{E}_{R-2}+\mathbf{E}_L)$]	0.40	0.59	0.000	3.30	0.60	1.10
High-moderate [using $1.375(1.25\mathbf{E}_G+\mathbf{E}_{R-1}+\mathbf{E}_{R-2}+\mathbf{E}_L)$]	0.60	0.46	0.000	3.10	5.77	5.38
Strong [using $1.30(3\mathbf{E}_G+\mathbf{E}_{R-1}+\mathbf{E}_{R-2}+\mathbf{E}_L)$]	0.90	0.25	0.000	4.60	23.10	30.27

4.2.4 Implications

The conceptual discussions allow expectations to be posited with regard to impacts of SA on histograms of normal RVs. In absolute terms, variance inflation generated by positive SA makes a histogram appear flatter. Positive SA also encourages more extreme values (thickening of the tails) to materialize.

The principal implications for normal RVs are: (1) positive SA generates variance inflation, which flattens a frequency distribution; (2) kurtosis tends to be dramatically altered when positive SA becomes very strong; and, (3) tail thickening and

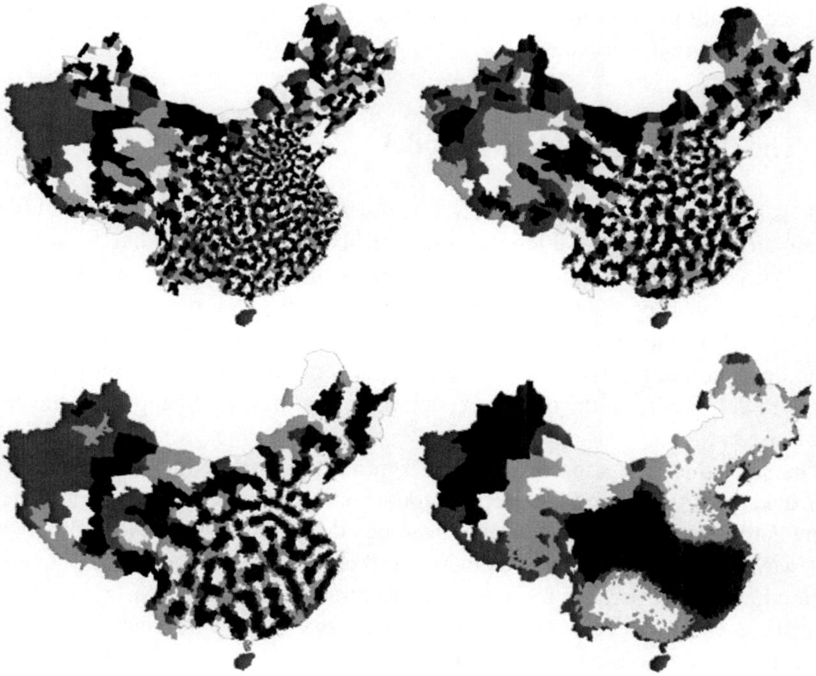

Fig. 4.9 SF map patterns. *Top left* (**a**): local (MC = 0.11). *Top right* (**b**): regional (MC = 0.47). *Bottom left* (**c**): regional (MC = 0.73). *Bottom right* (**d**): global (MC = 1.11)

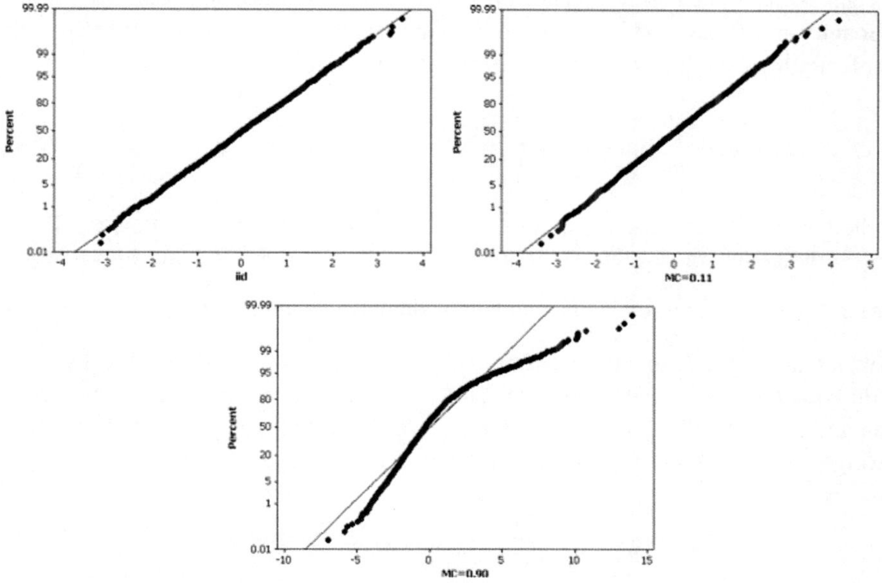

Fig. 4.10 Normal quantile plots for global map pattern SF model-induced levels of SA. *Left* (**a**): iid. *Middle* (**b**): weak SA. *Right* (**c**): strong SA

variance inflation are problematic in the moderate positive SA range that often is encountered in real world data.

4.3 The Poisson Probability Model

Little is known about the impacts of SA on Poisson RVs. Because this type of RV is a member of the exponential family of statistical distributions, just like for a normal RV, positive SA should induce variance inflation in Poisson RVs, too. This expectation is further supported by the close similarity between a normal and a Poisson frequency distribution when the latter's mean, μ, becomes very large. Thus, one should expect that positive SA will create extra-Poisson variation, a notion consistent with discussions in the overdispersion literature. But what happens to the mean of a Poisson RV?

One way that a Poisson RV differs from a normal RV is that its lower tail is truncated at 0. A Poisson RV describes counts of rare events, which naturally yields many zeroes as the event in question becomes increasingly rarer. Accordingly, the best way for Poisson variance to increase, then, is for extremely large counts to materialize, and/or perhaps for an over-concentration of zero or near-zero values to occur (i.e., excessive zeroes) to balance very large values in order to preserve μ. But what happens to the kurtosis of a Poisson RV? And, because it is a discrete RV (whereas a normal RV is continuous over the entire real number line), what happens to its modal value?

A bivariate regression tool for evaluating the Poissonness of a distribution, which is analogous to a normal quantile plot, is the Poissonness plot (Hoaglin, 1980; Hoaglin and Tukey, 1985). The ideal line for this plot, which can be estimated with ordinary bivariate linear regression techniques, is given by

$$\ln(n_k) + \sum_{j=1}^{k} \ln(j) - \ln\left(\sum_{i=1}^{n} y_i\right) = \beta_0 + \beta_1 k\,,\ k = 0, 1, 2, \ldots, \tag{4.4}$$

where k denotes the discrete non-negative values taken on by some Poisson RV Y, n_k is the count for discrete value k in a dataset, and $\beta_0 = -\mu$ and $\beta_1 = \ln(\mu)$, where μ is the mean of Y; of note is that the term $\sum_{j=1}^{k} \ln(j)$ disappears for $k = 0$ (which corresponds to a $0! = 1$ term, whose logarithm is 0). The left-hand side of this equation is referred to as the metaparameter. The Ord plot (Ord, 1967) furnishes an additional assessment tool. For this second regression analysis, which involves weighted least squares (WLS) estimation, the equation is given by

$$\frac{k\, n_k}{n_{k-1}} = \beta_0 + \beta_1\, k, \tag{4.5}$$

where $\beta_0 = \mu$ and $\beta_1 = 0$ for a Poisson distribution, and the weights are $\sqrt{n_k - 1}$. As a benchmark, a judiciously selected ideal set (n = 2,376) of independent and

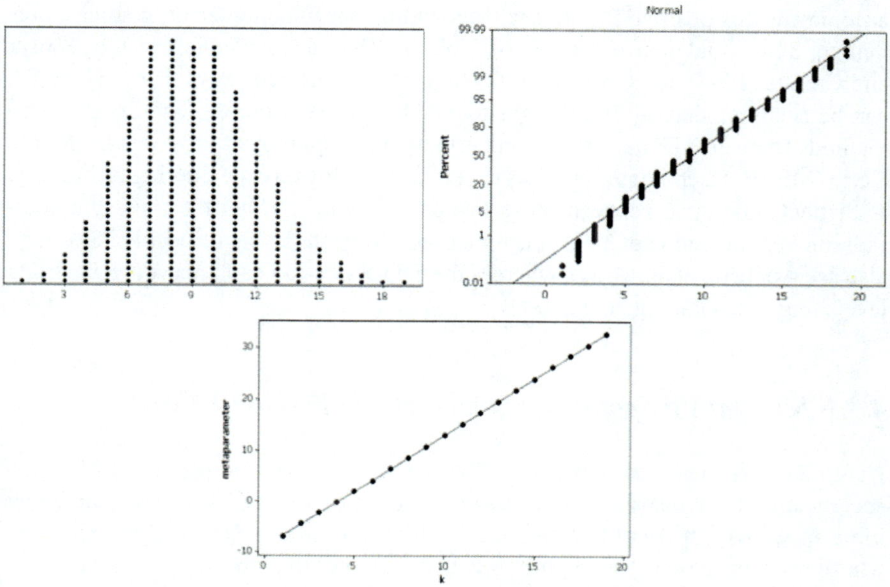

Fig. 4.11 Graphical diagnostic tools for a Poisson RV. *Left* (**a**): a dot-plot histogram. *Middle* (**b**): a normal quantile plot. *Right* (**c**): a Poissonness plot

identically distributed (iid) Poisson RVs with $\mu = 9$ was simulated; the mean and standard deviation for this set are $\bar{y} = 8.99790$ and $s_y = 3.00701$. Graphics portraying it appear in Fig. 4.11. Normal curve theory states that as μ increases beyond some sufficiently large value (e.g., 1,000), a Poisson probability distribution increasingly resembles a normal probability distribution, a feature that already is becoming visible in Fig. 4.11a. But with a mean of only 9, the normal quantile plot (Fig. 4.11b) confirms that the frequency distribution for this simulated Poisson RV deviates substantially from mimicking the form of a bell-shaped curve, particularly in its tails. A Poissonness plot (Fig. 4.11c) confirms that this is a Poisson RV. Its accompanying regression equation yields $\hat{\beta}_0 = -9.01112$ and $e^{\hat{\beta}_1} = 9.00932$. Meanwhile, the Ord plot results in bivariate linear WLS estimates of $\hat{\beta}_0 = 9.00614$ and $\hat{\beta}_1 = 0.00273$. All of these Poisson diagnostics confirm that this is a Poisson RV. In addition, when distributed across the 44-by-54 regular hexagonal surface partitioning employed in this study, this RV yields MC = 0.00143 ($z_{MC} = 0.15$) and GR = 0.99698; at most it contains only a trace amount of positive SA.

One spatial autoregression[2] theoretical statistical difficulty is that the auto-Poisson model can handle only negative SA; this drawback is problematic both because most georeferenced Poisson-distributed data contain positive SA, and because the normal and binomial approximations to an auto-Poisson model can

[2] Auto- models have values of the response variable, Y, on both sides of the equation. The right-hand side, which relates to a probability model, contains a linear combination of values of Y for other than the observation in question.

account for this positive SA. But when avoiding specification error, neither a normal nor a binomial approximation to a Poisson RV is desirable. Fortunately, Kaiser and Cressie (1997) and Griffith (2002) suggest two different ways that positive SA can be accommodated—and hence simulated—in a Poisson RV. The first of these methods truncates the auto-Poisson distribution and employs Markov chain Monte Carlo (MCMC) techniques, whereas the second employs spatial filtering techniques. A distinct difference between these two specifications is that the truncated auto-Poisson version can capture at most weak-to-moderate levels of positive SA (e.g., also see Augustin et al., 2004), whereas the SF version can capture even very strong levels (e.g., see Haining et al., 2009).

4.3.1 Simulating Spatially Autocorrelated Poisson RVs

Kaiser and Cressie (1997) circumvent the negative SA limitation of an auto-Poisson specification by Winsorizing counts to a finite set of integers, which sets an upper limit on the largest count that can occur. This adjustment yields an approximation whose probabilities sum to slightly less than 1, rather than to exactly 1, and allows the following auto-Poisson mean specification to be posited, using matrix notation:

$$\ln(\boldsymbol{\mu}) = \left[\alpha - \ln\left(\frac{1}{n}\sum_{i=1}^{n} e^{\rho\sum_{j=1}^{n} w_{ij}(y_j - e^{\alpha})}\right) \Big/ K\right]\mathbf{i} + \rho\mathbf{W}(\mathbf{Y} - e^{\alpha}\mathbf{i}), \tag{4.6}$$

where α is the regression intercept term ($\mu = e^{\alpha}$ is the mean of the Poisson RV in question), ρ is the spatial autoregression parameter, and this second term corrects for artificial inflation of the intercept term (i.e., an adjustment for trend)—K takes on the value of 1 until the mean begins to explode (see Augustin et al., 2004), at which point it increases to further compensate for this explosion. Equation (4.6) has a functional form very similar to an SAR model; here because matrix **W** mathematically is required to be symmetric, its (i, j) entry is defined as $w_{ij} = c_{ij} \Big/ \sqrt{\left(\sum_{i=1}^{n} c_{ij}\right)\left(\sum_{j=1}^{n} c_{ij}\right)}$.

Explosion of the mean occurs as the autoregressive parameter ρ becomes relatively large. This same outcome can be observed with the SAR model just as ρ approaches the boundary of its feasible parameter space (e.g., see Tables 4.1 and 4.3). Winsorizing the auto-Poisson probability distribution does not control for this explosion of the mean value; rather, it seeks to avoid entering a transition phase of instability, which tends to coincide with this explosion. However, because SA encourages relatively large counts to materialize (with the resulting contrasts with nearby values leading to local negative SA), the truncation point becomes critical. If it is too low, impacts on the mean and variance become more a function of it than of positive SA; if it is too high, phase transitions can be encountered. The two lowest levels of positive SA simulated for Sect. 4.3.2 employed a truncation point of $3y_{max}$, where y_{max} denotes the maximum count from each MCMC initial set of iid randomly generated counts (i.e., because the maximum count by chance when $\mu = 9$ is approximately 29, this truncation point is 26 deviations above the mean); no truncations had to be performed during chain generation. In contrast, for the highest level of positive SA, this truncation point was set to $6y_{max}$ (i.e., 55 deviations above the mean), resulting in roughly 20 million truncations being performed during chain generation.

4.3.1.1 MCMC Map Simulation

A Markov chain is a stochastic process consisting of a finite number of *states (i.e.,* for a Poisson RV, a vector of length n containing integer-valued counts corresponding to n locations) and known transition probabilities of moving from state i to state j at each computational iteration. Here, the matrix of transition probabilities, **M**, is defined by a Winsorized auto-Poisson model probability mass function. An important part of Markov chain theory is based on the *Ergodicity Theorem*, which requires **M** to be irreducible (i.e., any state can be reached from any other state)—the geographic weights matrix used is irreducible—recurrent non-null (the average return time to a given state is finite), and aperiodic (a state cannot be returned to repeatedly after a specific finite number of transition steps)—each areal unit in a hexagonal tessellation has at most 6 neighbors. If a Markov chain is ergodic, then a unique steady state distribution exists, say $\mathbf{M}^*$, which is independent of the initial state. This steady state distribution is given by $\mathbf{M}^* = \underset{k\to\infty}{\text{limit}}\,\mathbf{M}^k$, where k represents transition steps. Monte Carlo simulation is a technique for obtaining realizations of the limiting steady state distribution of a stochastic process through the use of a Poisson random number generator.

MCMC provides a mechanism for taking *dependent* samples from probability distributions in situations where the usual sampling is difficult, if not impossible. A case in point is where the normalizing constant for a joint probability distribution is either too difficult to calculate or analytically intractable. This is exactly the case for the auto-Poisson model. MCMC is used to simulate from some joint probability distribution **p** known only up to a constant factor, C. That is, $\mathbf{p} = C\mathbf{q}$, where **q** is known but C is unknown and an intractable mathematical expression (see Cressie, 1991, p. 428, for a mathematical statement of C for the auto-Poisson model). MCMC sampling begins with conditional (marginal) probability distributions, and with parameter estimates for the auto-Poisson model that can be obtained in practice using pseudo-likelihood estimation. This exercise involves estimating α and ρ as though observations are independent. MCMC outputs a sample of values for each parameter drawn from the joint probability distribution. Gibbs sampling is a MCMC scheme for simulation from **p** where the Markov chain transition matrix (i.e., **M**) is defined by the n *conditional* probability distributions of **p**. It is a stochastic process that returns a different result with each execution, a method for generating a joint empirical distribution of several variables from a set of modeled conditional distributions for each variable when the structure of data is too complex to implement mathematical formulae or directly simulate. It is a recipe for producing a Markov chain that yields simulated data that have the correct unconditional model properties, given the conditional distributions of those variables under study (Robert and Casella, 1999). The principal idea behind it is to convert a multivariate problem into a sequence of univariate problems, which then are iteratively solved to produce a Markov chain. The following Gibbs sampling algorithm description (see Haining et al., 2009; Augustin et al., 2004) for a Winsorized auto-Poisson model begins with pre-specified values of the parameters α and ρ (e.g., pseudo-likelihood parameter estimates in the ensuing China data analysis):

- Step 1: initialize a map ($\tau = 0$, where τ denotes the number of iterations) by taking $i = 1, \ldots, n$ independent random samples $\{y_{i,\tau=0}\}$ from a Poisson probability distribution and determine y_{max};
- Step 2: obtain new values (initially $\tau = 1$) $y_{i,\tau}$ by sequentially moving from one location (i) to another (j) on the initial map and randomly sampling from the Winsorized auto-Poisson probability distribution [i.e., Eq. (4.6) coupled with a truncation value that is a function of y_{max}] using pre-specified parameter values—site selection for this process of obtaining $\{y_{i,\tau=1}\}$ from $\{y_{i,\tau=0}\}$ can follow random permutations of location sequences or simply a systematic sweep across a map;
- Step 3: obtain new values (initially $\tau = 2$) $y_{i,\tau+1}$ by sequentially moving from one location to another on the τth map, again randomly sampling from the Winsorized auto-Poisson distribution, and immediately updating the value at each location; and,
- Step 4: repeat Step 3 for iterations $\tau = 3, 4, 5, \ldots,$ until convergence of the sufficient statistics of the parameters of interest occurs.

Once a Markov chain transition matrix is constructed, a sample of (correlated) drawings from a target distribution can be obtained. This is done by *simulating* the Markov chain a large number of times (say, 100,000) and recording its sufficient statistics after removing a burn-in set (e.g., the first 25,000) of iterations. Convergence needs to be monitored (e.g., time series plots and correlograms need to be inspected), and hence the sufficient statistics need to be recorded. This recording should be done after each iteration. A suitable burn-in period is needed in order to generate $\mathbf{M}^*$, and hence before collecting statistics, and because iteration outcomes may well be correlated, the chain needs to be weeded (e.g., only every hundredth iteration result is retained).

The sufficient statistics for the estimators of the simple auto-Poisson model parameters here are $1 \times \sum_{i=1}^{n} y_i$ and $\sum_{i=1}^{n} y_i \sum_{j=1}^{n} c_{ij}\, y_j$; this first statistic is required for a Poisson model intercept term, whereas this second statistic is required for an auto-Poisson model autoregressive parameter term. Once convergence has been attained (e.g., the accompanying trend line for a time series plot is flat, and the accompanying correlogram displays no significant serial autocorrelation), the last map in the chain is the simulated Winsorized auto-Poisson realization.

4.3.1.2 SF Map Simulation

Meanwhile, SF versions of the Poisson model involve specifying a geographically heterogeneous mean and variance in order to capture positive SA. This implementation requires the usual set of covariates, $X_1, \ldots, X_p$, to be replaced by the eigenvectors of matrix expression (4.2) in order to embed SA in a response counts variable. Compared with the auto- models, spatial dependence effects are shifted to the mean, resulting in the spatial autoregressive parameter [i.e., ρ in Eq. (4.6)] being forced to 0. Accordingly, a realization can be obtained by sampling from a Poisson distribution with mean

$$\mathrm{LN}(\mu) = \left[\alpha - \ln\left(\frac{1}{n}\sum_{i=1}^{n} e^{\sum_{k=1}^{K}\mathbf{E}_k\beta_k}\right)\right]\mathbf{1} + \sum_{k=1}^{K}\mathbf{E}_k\beta_k, \tag{4.7}$$

where $\mathbf{E}_k$ denotes eigenvector k of matrix expression (4.2), β_k is its relative weight (somewhat similar to ρ in the autoregressive specification), and this second term corrects for artificial inflation of the intercept term α (i.e., an adjustment for trend) due to the presence of covariates. An additional adjustment for α in the third term is unnecessary here because the mean of each $\mathbf{E}_k\beta_k$ is zero, by construction.

4.3.2 Simulation Results for an Ideal Regular Hexagonal Surface Partitioning

Figure 4.12, which characterizes all three chains, furnishes strong evidence that the generated MCMC chains converged, rendering useful maps with positive SA embedded in them. The time series plots exhibit random stability. For example, for $\rho = 0.06$—the maximum positive SA that was successfully embedded into simulated data here—the trend line has not converged within the burn-in set of iterations, but does converge long before the end of the chain; here this situation is acceptable since only the last map of the chain is used here. Meanwhile, the correlograms reveal that virtually no serial autocorrelation is present in the three chains.

The Poissonness plots for the autoregressive model results appear in Fig. 4.13. These plots begin to exhibit slight but detectable tail disturbances beginning with low-weak positive SA. Moderate positive SA results in a complete deterioration of linearity.

Summary descriptive statistics appear in Table 4.5 for the Winsorized auto-Poisson simulated data containing positive SA. These statistics confirm that the (controlled for trend) mean essentially is unaffected, while the variance is inflated (i.e., overdispersion), quite noticeably by moderate positive SA. Corresponding histograms confirm the expectations that low levels of positive SA have little effect, whereas moderate levels tend to stretch the right-hand tail and shift the concentration of values toward 0. This same pattern is displayed by: the maximum values, the mode, and kurtosis. Plot diagnostic statistics begin detecting deviation from

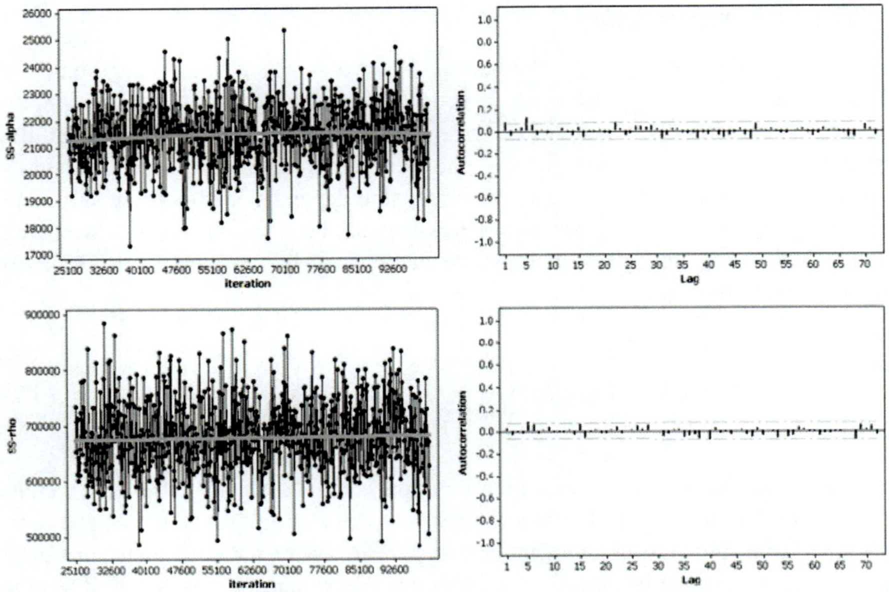

Fig. 4.12 MCMC time series plot and correlogram diagnostic graphics based on the ideal hexagonal surface partitioning when $\rho = 0.06$ (*bottom*). *Left* (**a**): for the intercept term. *Right* (**b**): for the autoregressive term

a Poisson RV at weak levels of positive SA; the Ord plot statistics emphatically detect deviation at the low-moderate level, whereas both sets of these statistics unambiguously detect deviation at the moderate level, of positive SA.

The Poissonness plots for the SF model-embedded positive SA results are illustrated in Fig. 4.14. Not only do these results confirm those found for the Winsorized auto-Poisson model, but the SF model, because it is able to capture much stronger levels of positive SA, extends the autoregressive findings. Furthermore, the corresponding summary descriptive statistics, which appear in Table 4.6, corroborate those trends detected in Table 4.5. Overall, as positive SA increases in a Poisson RV, variance increases, both near-zero and extreme values become more likely, kurtosis increases, and the Ord plot bivariate regression parameter estimates provide a very good diagnostic of its presence, one that furnishes superior diagnostics to those associated with the Poissonness plot.

4.3.3 Simulation Results for the China County Geographic Configuration

As before, MCMC simulation of Winsorized auto-Poisson model-based maps employing the China county irregular surface partitioning at most could embed only moderate positive SA. A bifurcation point appears to be present because of

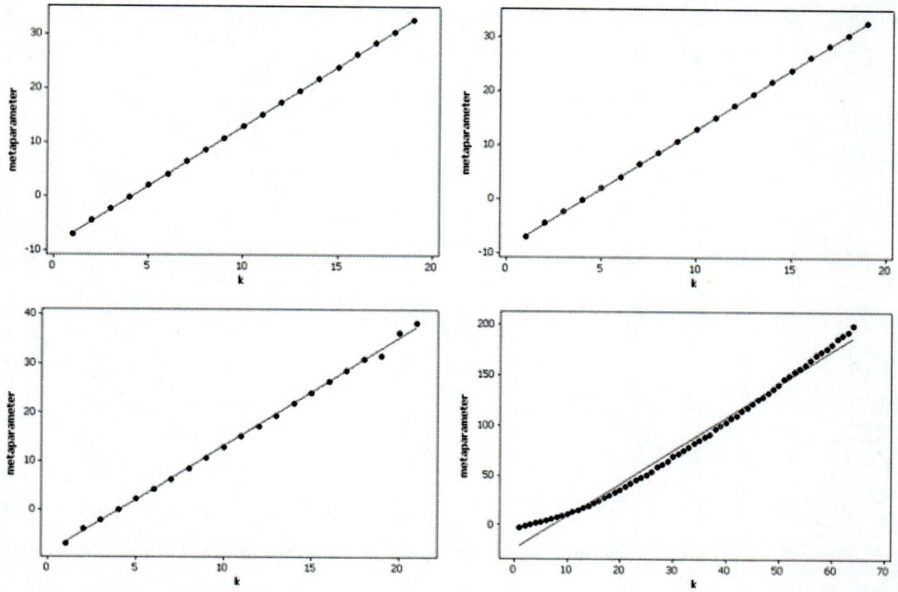

Fig. 4.13 Poissonness plots for the Winsorized auto-Poisson-model induced levels of positive SA based upon the ideal hexagonal surface partitioning. *Left* (**a**): iid with no SA. *Middle left* (**b**): weak positive SA. *Middle right* (**c**): low-moderate positive SA. *Right* (**d**): moderate positive SA

the irregularness of the surface partitioning; MCMC simulation produces maps containing either weak or moderate positive SA, without a transition between them. Nevertheless, graphical diagnostics indicate that the resulting maps are properly generated. In addition, summary descriptive statistics reported in Table 4.7 are consistent with those appearing in Tables 4.5 and 4.6: overdispersion is induced, outliers are generated, relatively small values become more likely, and kurtosis is affected

Table 4.5 Descriptive statistics for the Winsorized auto-Poisson model-based simulated data and the hexagonal tessellation geographic configuration

								Poissonness plot		Ord plot	
Variable autocorrelation	MC	GR	$\bar{y}$	s_y	y_{max}	mode	$\lvert z_{kurtosis} \rvert$ [a]	$-\hat{\beta}_0$	$e^{\hat{\beta}_1}$	$\hat{\beta}_0$	$\hat{\beta}_1$
None (i.e., iid)	0.00	1.00	9.00	3.007	19	8	0.04	9.01	9.01	9.01	0.00
Weak	0.11	0.89	8.99	3.050	21	8	0.19	9.29	9.27	9.88	–0.04
Low-moderate	0.19	0.82	8.90	3.210	21	8	0.23	9.08	9.19	2.48	0.85
moderate	0.47	0.55	9.05	16.477	123	4	67.77	–3.32	1.05	–6.83	2.26

[a]The mean of kurtosis is $1/\mu = 1/9 = 0.111$; the standard error, which can be established using the moment generating function $e^{\mu(e^t - 1)}$, is $\sqrt{151.23594 / n}$ for $\mu = 9$

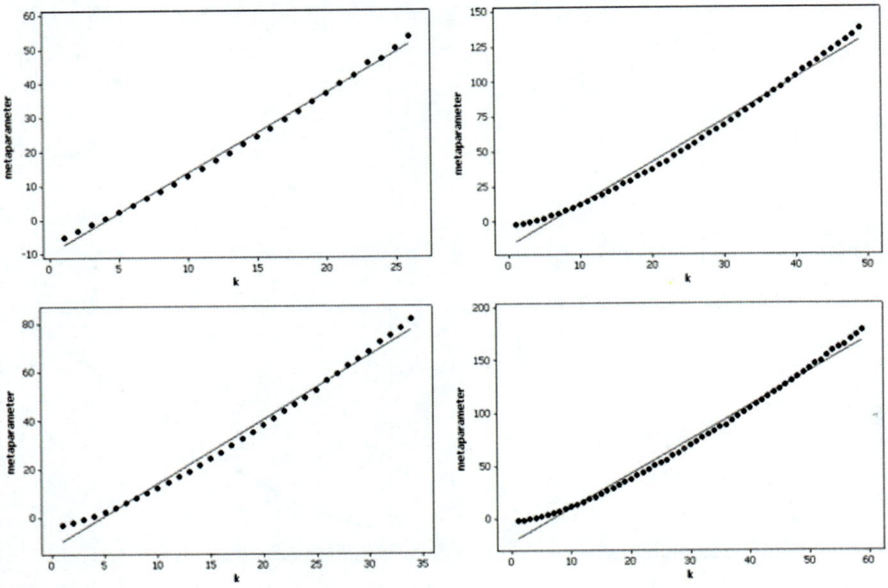

Fig. 4.14 Poissonness plots for SF-model induced positive SA using an ideal hexagonal surface partitioning and a mixture of global, regional and local map patterns. *Left* (**a**): weak positive SA. *Middle left* (**b**): low-moderate positive SA. *Middle right* (**c**): high-moderate positive SA. *Right* (**d**): strong positive SA

by even moderate amounts of positive SA. Meanwhile, corresponding histograms once more confirm the expectations that low levels of positive SA have little effect, whereas moderate levels tend to stretch the right-hand tail and shift the concentration of counts toward 0.

Poissonness plots for the SF model-embedded positive SA results appearing in Fig. 4.15 reveal that the irregularness of the China county surface partitioning introduces additional skewness into count distributions; the upper tail becomes increasingly separated from the middle and lower tail as positive SA increases. In other words, positive SA and the irregularness of a geographic configuration appear to interact. Meanwhile, Tables 4.7 and 4.8 exhibit the same histogram trends detectable in Tables 4.5 and 4.6: low levels of positive SA have little effect, whereas moderate and strong levels tend to stretch the right-hand tail and shift the concentration of values toward 0 (i.e., the mode tends to decrease), while maximum values and kurtosis increase with increasing positive SA. Plot diagnostic statistics begin detecting deviation from a Poisson RV at weak levels of positive SA, again with the Ord plot statistics being more sensitive to the presence of positive SA. A distinction between Tables 4.7 and 4.8 is that SF-induced positive SA can cover the entire range of SA, while a Winsorized auto-Poisson model encounters difficulties and phase transition problems at moderate levels. SF simulations also do not encounter a bifurcation point, and because they lack truncation, they allow much larger counts to materialize.

Table 4.6 Descriptive statistics for the SF auto-Poisson model-based simulated data and the hexagonal tessellation geographic configuration

								Poissonness plot		Ord plot	
Variable autocorrelation	MC	GR	$\bar{y}$	s_y	y_{max}	mode	$\lvert z_{kurtosis}\rvert$	$-\hat{\beta}_0$	$e^{\hat{\beta}_1}$	$\hat{\beta}_0$	$\hat{\beta}_1$
None (i.e., iid)	0.00	1.00	9.00	3.007	19	8	0.04	9.01	9.01	9.01	0.00
Global map pattern-base results											
Weak (using 0.125$\mathbf{E}_G$)	0.11	0.90	9.08	3.149	21	9	0.12	9.23	9.31	6.28	0.35
Low-moderate (using 0.285$\mathbf{E}_G$)	0.40	0.61	8.96	3.749	26	9	1.98	9.18	9.68	3.53	0.58
High-moderate (using 0.4$\mathbf{E}_G$)	0.60	0.41	9.03	4.629	30	6	2.29	10.90	11.89	1.67	0.79
Strong (using 0.9$\mathbf{E}_G$)	0.91	0.11	9.06	8.990	48	2, 4	10.14	16.80	19.24	–2.05	1.14
Global + regional map pattern-base results											
Weak (using 0.17$\mathbf{E}_R$)	0.10	0.90	8.97	3.336	29	8	2.06	9.29	9.50	6.47	0.29
Low-moderate [using 0.225($\mathbf{E}_G$+$\mathbf{E}_R$)]	0.41	0.60	9.08	4.317	30	7	2.53	9.96	10.76	1.82	0.77
High-moderate [using 0.5($\mathbf{E}_G$+$\mathbf{E}_R$)]	0.61	0.41	8.98	7.626	65	4	28.42	13.83	15.59	–3.45	1.35
Strong [using 0.3(4$\mathbf{E}_G$+$\mathbf{E}_R$)]	0.90	0.12	9.02	12.803	99	1	34.72	18.84	21.58	–1.54	1.10
Global + regional + local map pattern-base results											
Weak (using 0.3$\mathbf{E}_L$)	0.11	0.88	8.99	4.082	29	7	1.98	9.80	10.49	3.40	0.59
Low-moderate [using 0.1(9$\mathbf{E}_R$+$\mathbf{E}_L$)]	0.39	0.62	9.02	9.586	67	3	25.68	17.79	20.37	–1.94	1.15
High-moderate [using 0.25(2$\mathbf{E}_G$+$\mathbf{E}_R$+$\mathbf{E}_L$)]	0.61	0.40	8.97	6.485	50	4, 5	13.67	12.39	13.89	–0.05	0.93
Strong [using 0.15(9$\mathbf{E}_G$+2$\mathbf{E}_R$+$\mathbf{E}_L$)]	0.89	0.13	9.02	14.705	150	0	60.72	21.73	25.36	–4.65	1.50

Table 4.7 Descriptive statistics for the Winsorized auto-Poisson model-based simulated data and the China county geographic configuration

Variable autocorrelation	MC	GR	$\bar{y}$	s_y	y_{max}	mode	$\lvert z_{kurtosis} \rvert$ [a]	Poissonness plot		Ord plot	
								$-\hat{\beta}_0$	$e^{\hat{\beta}_1}$	$\hat{\beta}_0$	$\hat{\beta}_1$
None (i.e., iid)	0.02	0.98	8.99	3.032	20	8	0.64	9.05	9.04	8.97	0.02
Very weak	0.13	0.87	9.10	3.249	23	8, 9	0.40	9.05	9.27	6.76	0.23
Low-moderate	0.53	0.56	9.79	14.990	120	5	79.99	40.47	48.23	–10.67	2.32

[a]The mean of kurtosis is $1/\mu = 1/9 = 0.111$; the standard error, which can be established using the moment generating function $e^{\mu(e^t - 1)}$, is $\sqrt{151.23594 / n}$ for $\mu = 9$

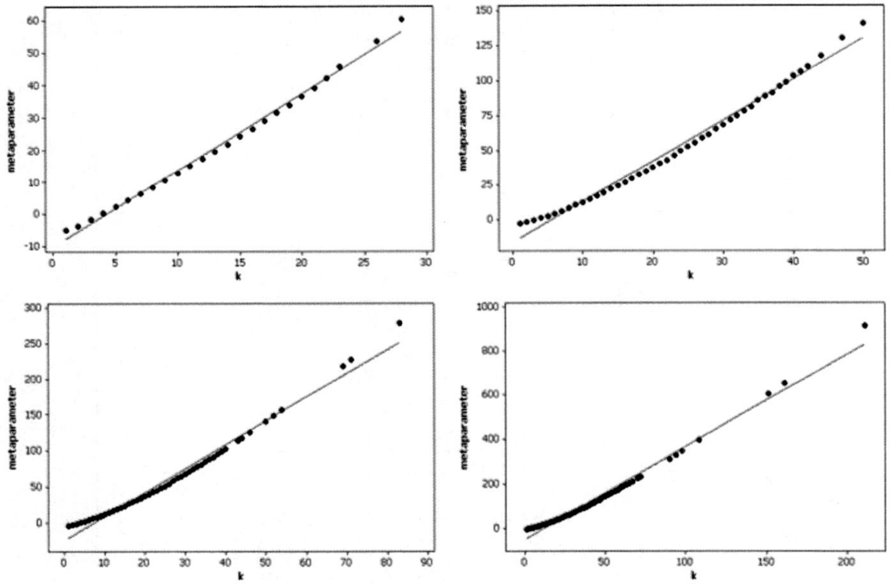

Fig. 4.15 Poissonness plots for SF-model induced levels of positive SA using the China county surface partitioning for a mixture of global, regional, and local map patterns. *Left* (**a**): weak positive SA. *Middle left* (**b**): low-moderate positive SA. *Middle right* (**c**): high-moderate positive SA. *Right* (**d**): strong positive SA

4.3.4 Implications

In conclusion, numerical results reported in this section suggest the following implications about a georeferenced Poisson RV:

(1) by controlling for trend in data when estimating a mean, positive SA has no impact upon the resulting estimated mean value;
(2) positive SA increases the chances of much larger counts materializing;

Table 4.8 Descriptive statistics for the SF auto-Poisson model-based simulated data and the China county geographic configuration

Variable autocorrelation	MC	GR	$\bar{y}$	s_y	y_{max}	mode	$\lvert z_{kurtosis} \rvert$	Poissonness plot		Ord plot	
								$-\hat{\beta}_0$	$e^{\hat{\beta}_1}$	$\hat{\beta}_0$	$\hat{\beta}_1$
None (i.e., iid)	0.02	0.98	8.99	3.032	20	8	0.64	9.05	9.04	8.97	0.02
Global map pattern-base results											
Weak (using 0.07$\mathbf{E}_G$)	0.11	0.92	9.03	3.103	24	8	0.75	9.60	9.65	4.07	0.61
Low-moderate (using 0.145$\mathbf{E}_G$)	0.41	0.67	8.94	3.639	36	8, 9	11.74	11.80	12.41	4.31	0.51
High-moderate (using 0.19$\mathbf{E}_G$)	0.62	0.53	9.03	4.676	57	7	55.84	18.10	19.80	–2.48	1.32
Strong (using 0.35$\mathbf{E}_G$)	0.90	0.46	8.92	10.175	190	5	424.57	48.67	63.28	–6.81	1.73
Global + regional map pattern-base results											
Weak (using 0.12$\mathbf{E}_{R\text{-}1}$)	0.11	0.90	8.90	3.474	23	7	0.55	9.19	9.57	5.60	0.35
Low-moderate [using 0.25($\mathbf{E}_{R\text{-}1}$+$\mathbf{E}_{R\text{-}2}$)]	0.41	0.64	8.91	5.687	52	6	15.62	16.12	17.94	0.01	0.94
High-moderate [using 0.20($\mathbf{E}_G$+$\mathbf{E}_{R\text{-}1}$+$\mathbf{E}_{R\text{-}2}$)]	0.61	0.52	9.04	6.536	83	7	56.16	22.61	25.13	–6.29	1.63
Strong [using 0.09(4$\mathbf{E}_G$+$\mathbf{E}_{R\text{-}1}$)]	0.91	0.43	9.06	10.550	201	6	340.97	47.65	59.64	–7.88	1.89
Global + regional + local map pattern-base results											
Weak [using 0.125($\mathbf{E}_{R\text{-}1}$ +$\mathbf{E}_L$)]	0.11	0.88	8.97	3.780	28	5	2.34	10.09	10.71	3.67	0.56
Low-moderate [using 0.15(2$\mathbf{E}_{R\text{-}1}$+2$\mathbf{E}_{R\text{-}2}$+$\mathbf{E}_L$)]	0.42	0.61	9.01	6.816	50	5	16.38	16.68	18.84	–4.99	1.47
High-moderate [using 0.15(1.5$\mathbf{E}_G$+$\mathbf{E}_{R\text{-}1}$+$\mathbf{E}_{R\text{-}2}$ +$\mathbf{E}_L$)]	0.60	0.57	9.00	6.357	83	4	78.72	24.49	27.75	–1.32	1.09
Strong [using 0.05(7$\mathbf{E}_G$+ $\mathbf{E}_{R\text{-}1}$+$\mathbf{E}_{R\text{-}2}$ +$\mathbf{E}_L$)]	0.90	0.45	9.07	10.544	211	6	404.94	49.50	64.12	–6.12	1.68

(3) especially strong positive SA increases the chances of counts toward 0 materializing;
(4) as positive SA increases, a histogram moves toward the exponential distribution in form;
(5) strong positive SA increases kurtosis;
(6) as positive SA increases, the linearity of a Poissoness plot deteriorates, especially in the tails of an empirical distribution;
(7) the Ord-plot appears very sensitive to the presence of positive SA, and appears to out-perform the Poissonness plot as a diagnostic tool in this context;
(8) a particular mixture of eigenvectors in a SF plays an important role in terms of the impacts of positive SA that materialize (see Table 4.6); and,
(9) the Winsorized auto-Poisson model is unable to capture more than weak-to-moderate positive SA.

In other words, even modest amounts of positive SA do make a difference!

The general importance of these findings concerns data analysis problems, such as excessive zeroes and outliers, that spatial scientists frequently encounter with real world data. These implications should cause a spatial researcher to think more earnestly about the georeferenced nature of his/her data when faced with such problems. In addition, particularly results for the SF-model-based simulations presented here demonstrate that georeferenced Poisson RVs are capable of containing markedly high levels of positive SA.

4.4 The Binomial Probability Model, N > 1

As with Poisson RVs, little is known about the impacts of SA on binomial RVs.[3] Because these RVs also are a member of the exponential family of statistical distributions, just like the normal and Poisson RVs, positive SA should induce variance inflation in them, too. This expectation is further supported by the close similarity between a normal and a binomial frequency distribution when the binomial probability of an event occurring is $p = 0.5$, and the number of events N becomes very large. Thus, one should expect that positive SA will create extra-binomial variation, a notion consistent with discussions in the overdispersion literature.[4] But what happens to the mean of a binomial RV?

One way that a binomial RV differs from both a normal and a Poisson RV is that its values are restricted to the range [0, N], where N is the maximum number of items that can occur at a location. In other words, it is a count with both a lower and

[3] More work has been done on the Bernoulli, vis-à-vis the autologistic model, than on the general binomial RV.

[4] This is not the case for binary 0–1 Bernoulli RVs, which by their very nature cannot exhibit extra variation. The concept of extra variation in a logistic regression has to be teased out of data by, for example, grouping values in order to have an $N > 1$.

an upper bound. The best way for binomial variance to increase is for the relative frequencies of 0 and N to increase when $p = 0.5$, or for the frequency of 0 to increase when $p > 0.5$, or of N when $p < 0.5$. The restricted range should help preserve the mean, μ.

The Ord plot (Ord, 1967) also can be used here for diagnostic purposes. In this context the slope parameter, β_1, becomes negative (< 0). Through the Poisson approximation of a binomial distribution when p is very small (or by symmetry, very large) and $Np < 5$, the preceding Poisson analysis reveals impacts of SA on binomial histograms when p becomes very small; hence, only the case of $p = 0.5$ is treated here. So that more direct comparisons can be made with the preceding findings, N is set to 18 (i.e., $\mu = 18/2 = 9$). The simulated iid values have the following descriptive statistics:

	Mean	*Standard deviation*	*Skewness*	*Kurtosis*
theoretical	9	2.1232	0	–0.11111
observed: n = 2,376	8.9933	2.1220	–0.01	–0.10
observed: n = 2,379	8.9975	2.1196	0.00	–0.11

The MCs and GRs for the simulated data are as follows:

n = 2,376: MC = 0.00431, GR = 0.99488
n = 2,379: MC = –0.00168, GR = 1.00731

In other words, these simulated binomial RVs display the necessary characteristics of iid.

Illustrative graphic portrayals of these values appear in Fig. 4.16. Of note is that weighted least squares regression estimation yields b = –0.9995 for the theoretical Ord plot, and b = –0.9981 and –1.0178 for the two simulation data Ord plots, confirming that the values are for binomial RVs. These slope parameter estimates can be converted to their corresponding binomial probability estimates with the equation $p = \frac{\hat{\beta}_1}{\hat{\beta}_1 - 1}$, respectively yielding 0.49988 for the theoretical binomial data, and 0.50441 and 0.49953 for the simulated data; the true value is 0.5.

Bernoulli RVs (i.e., N = 1) are not be treated in this section, since their histograms tend to be too simple to display conspicuous impacts of SA.

4.4.1 Simulating Spatially Autocorrelated Binomial RVs

The simulation of either multivariate binary or multivariate binomial georeferenced data has not been given as much attention in the literature as has the simulation of spatially autocorrelated normal or Poisson RVs. Dolan et al. (2000), for example, simulate a spatially autocorrelated log-normal RV and then do a back-transformation, an approach not endorsed here. Heagerty and Lele (1998), for

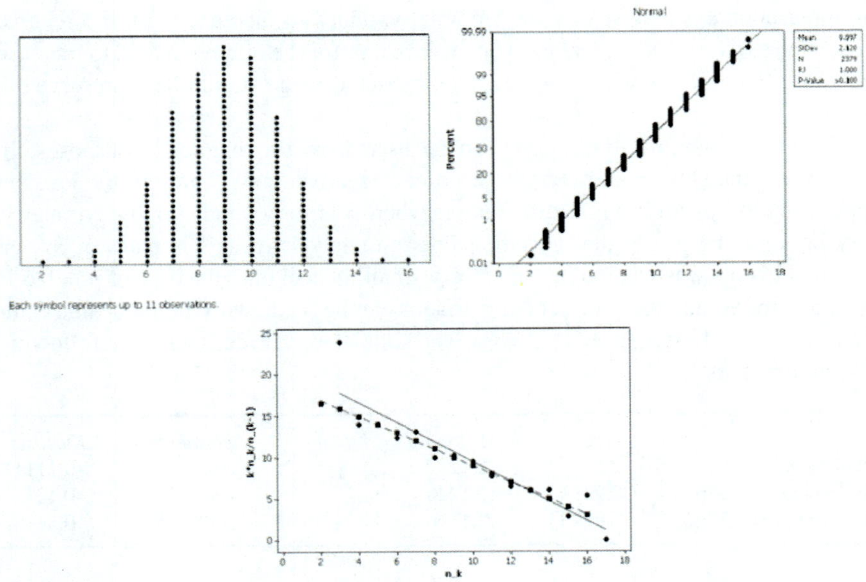

Fig. 4.16 Graphical diagnostic tools for the iid binomial RVs; $n = 2{,}379$. *Left* (**a**): a dot-plot histogram. *Middle* (**b**): a normal quantile plot. *Right* (**c**): an Ord plot with its trend line (*solid line with solid circles*) together with and Ord plot of the theoretical counterpart (*broken line with solid squares*)

instance, promote the use of a generalized linear mixed model coupled with a geostatistical perspective for binary georeferenced data. And, Augustin et al. (1998), for example, promote the use of the autologistic model. As in the Sects. 4.2 and 4.3, auto-binomial model RVs are simulated here with MCMC and SF techniques. The autoregressive equation employed with MCMC is given by

$$\begin{aligned} P(Y_i = y|\alpha_i, \mathbf{C}_i\mathbf{Y}) \\ = \exp\left(\alpha_i + \rho\sum_{j=1}^{n} c_{ij}(y_j - \bar{y})\right) \Big/ \left[1 + \exp\left(\alpha_i + \rho\sum_{j=1}^{n} c_{ij}(y_j - \bar{y})\right)\right], \end{aligned} \tag{4.8}$$

where y is contained in the interval [0, N], and including subtraction of the mean $\bar{y}$ in parallel with Kaiser and Cressie's (1997) specification for the Winsorized auto-Poisson model specification. Meanwhile, the SF equation employed is given by

$$P\left(Y_i = y|\mathbf{E}_{i,K}\right) = \exp(\alpha + \mathbf{E}_{i,K}\boldsymbol{\beta})/[1 + \exp(\alpha + \mathbf{E}_{i,K}\boldsymbol{\beta})], \tag{4.9}$$

where $\mathbf{E}_K$ is the n-by-K matrix of SF eigenvectors. The procedural steps for using these equations to simulate geographic distributions are exactly the same as those outlined in the preceding section for Poisson RVs, except that the Poisson probability model is replace with the binomial probability model.

4.4.2 *Simulation Results for an Ideal Regular Hexagonal Surface Partitioning*

MCMC map simulation can exploit the particular relationship between the intercept and autoregressive parameter when $p = 0.5$, namely asymptotically $\alpha = -3\rho$, which was done here. As with Poisson RVs, phase transitions tend to be encountered beyond moderate SA. Consequently, only weak and low-moderate SA have been simulated for analysis purposes.

Figure 4.17 furnishes strong evidence that the generated MCMC chains converged, rendering useful maps with positive SA embedded in them. The time series plots exhibit random stability. Meanwhile, the correlograms reveal that virtually no serial autocorrelation is present in the three chains. As with the Poisson RVs simulated in the preceding section, only the last map of a chain is used here.

Summary descriptive statistics appear in Table 4.9 for the auto-binomial simulated data containing positive SA. These statistics confirm that the (controlled for trend) mean essentially is unaffected, while the variance is inflated (i.e., overdispersion). Dot plot versions of histograms appearing in Fig. 4.18 confirm the expectations that low levels of positive SA have little effect, whereas low-moderate levels already tend to redistribute counts to the two tails. In addition, the Kolmogorov-Smirnov (K-S) statistic quantifies a movement away from the corresponding theoretical binomial distribution as positive SA increases. Unfortunately,

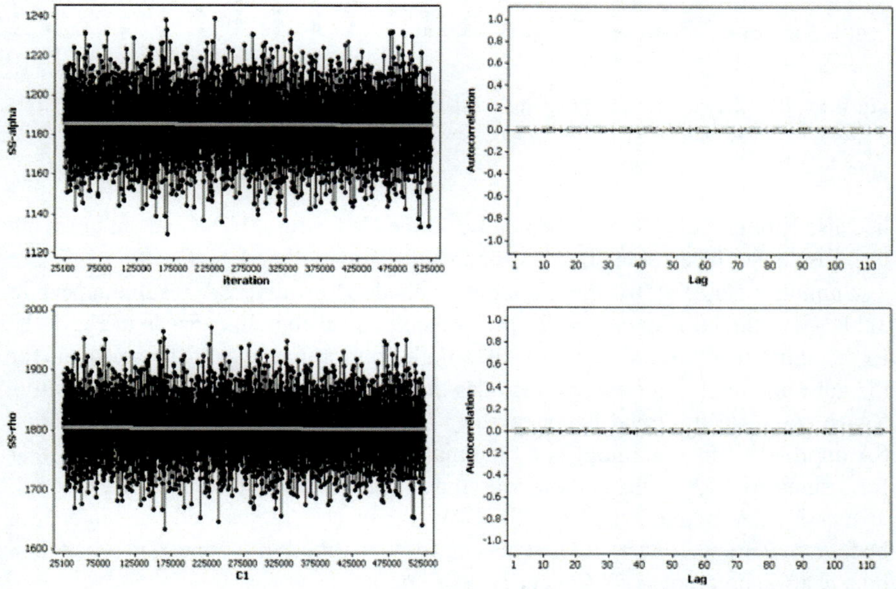

Fig. 4.17 MCMC time series plot and correlogram diagnostic graphics based on the ideal hexagonal surface partitioning when $\rho = 0.60$ (*bottom*). *Left* (**a**): for the intercept term. *Right* (**b**): for the autoregressive term

Table 4.9 Descriptive statistics for the auto-binomial model-based MCMC simulated data and the hexagonal tessellation geographic configuration

Variable autocorrelation	MC	GR	$\bar{y}$	s_y	y_{min}	y_{max}	Skewness	Kurtosis	K-S[a]
None (i.e., iid)	0.00	0.99	8.99	2.122	2	16	–0.01	–0.10	0.0038
Weak	0.10	0.89	8.81	2.212	2	15	–0.02	–0.19	0.0400
Low-moderate	0.39	0.55	9.07	2.759	0	18	–0.24	–0.07	0.0700

[a]K-S denotes the Kolmogorov-Smirnov statistic, used here to index deviation from the theoretical binomial distribution for which N = 18 and p = 0.5

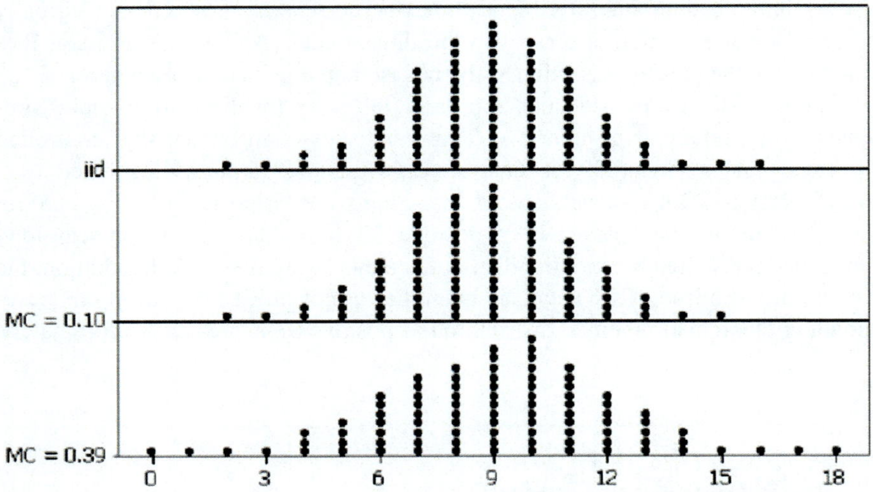

Fig. 4.18 Dot plot versions of histograms for the MCMC auto-binomial simulated data. *Top* (**a**): iid. *Middle* (**b**): weak positive SA. *Bottom* (**c**): low moderate positive SA

because strong positive SA cannot be embedded with MCMC techniques, its impacts cannot be assessed in terms of an auto-binomial model.

Summary statistics for the SF model-embedded positive SA results appear in Table 4.10; the corresponding dot plot versions of histograms appear in Fig. 4.19 for a global map pattern. These results both confirm and extend those found for the auto-binomial. Figure 4.19a includes the dot plot for extremely strong positive SA to complete the trend being revealed by these illustrative results: as positive SA approaches its maximum, the binomial histogram increasingly resembles that for a sinusoidal RV—this is the reason for change in the kurtosis statistic. Overall, as positive SA increases in a binomial RV, variance increases, and the center of a histogram flattens, converging first on a uniform distribution in appearance, and then on a near-dichotomous 0/N frequency distribution.

Dot plot versions of histograms for SFs constructed with global and regional map patterns appear in Fig. 4.19b. As with the global map pattern results, the mean remains unaffected, variance is inflated and kurtosis is impacted upon by positive

Table 4.10 Descriptive statistics for the auto-binomial model-based SF simulated data and the hexagonal tessellation geographic configuration

Variable autocorrelation	MC	GR	$\bar{y}$	s_y	y_{min}	y_{max}	Skewness	Kurtosis	K-S[a]
None (i.e., iid)	0.00	0.99	8.99	2.122	2	16	−0.01	–0.10	0.0038
Global map pattern-base results									
Weak (using $0.0035\mathbf{E}_G$)	0.10	0.90	9.01	2.239	2	16	−0.02	–0.19	0.0187
Low-moderate (using $0.008\mathbf{E}_G$)	0.39	0.61	9.03	2.673	1	18	0.00	–0.24	0.0625
High-moderate (using $0.0125\mathbf{E}_G$)	0.65	0.37	9.02	3.278	0	18	0.02	–0.47	0.1164
Strong (using $0.03\mathbf{E}_G$)	0.90	0.13	8.97	5.165	0	18	0.03	–1.07	0.2351
Global + regional map pattern-base results									
Weak (using $0.005\mathbf{E}_R$)	0.09	0.91	9.09	2.309	2	16	−0.04	–0.24	0.0413
Low-moderate [using $0.008(\mathbf{E}_G+\mathbf{E}_R)$]	0.42	0.59	8.94	3.061	1	18	0.06	–0.43	0.1063
High-moderate [using $0.015(\mathbf{E}_G+\mathbf{E}_R)$]	0.61	0.40	8.97	4.321	0	18	0.01	–0.92	0.2036
Strong [using $0.0095(4\mathbf{E}_G+\mathbf{E}_R)$]	0.90	0.12	9.01	5.763	0	18	0.01	–1.27	0.2802
Global + regional + local map pattern-base results									
Weak (using $0.008\mathbf{E}_L$)	0.10	0.90	8.98	2.710	1	17	0.01	–0.26	0.0633
Low-moderate [using $0.003(9\mathbf{E}_R+\mathbf{E}_L)$]	0.44	0.56	8.99	4.937	0	18	−0.02	–1.11	0.2457
High-moderate [using $0.007(2\mathbf{E}_G+\mathbf{E}_R+\mathbf{E}_L)$]	0.61	0.40	9.04	3.880	0	18	−0.01	–0.71	0.1643
Strong [using $0.005(9\mathbf{E}_G+2\mathbf{E}_R+\mathbf{E}_L)$]	0.91	0.11	8.95	6.147	0	18	0.01	–1.38	0.3008

[a]K-S denotes the Kolmogorov-Smirnov statistic, used here to index deviation from the theoretical binomial distribution for which $N = 18$ and $p = 0.5$

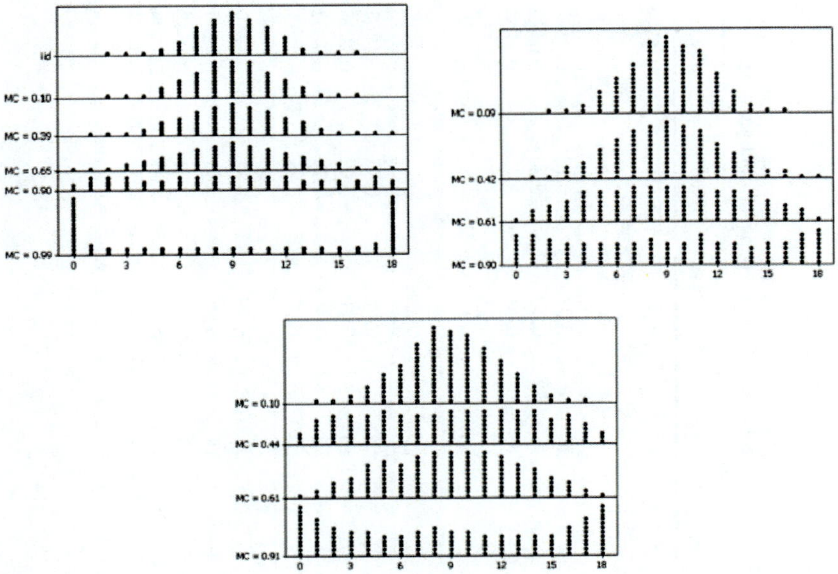

Fig. 4.19 Dot plot versions of histograms for the SF binomial simulated data using the regular hexagonal surface partitioning. *Left* (**a**): SF results from a global map pattern. *Middle* (**b**): SF results from a global combined with a regional map pattern. *Right* (**c**): SF results from a global combined with a regional and a local map pattern

SA. Again, these results both confirm and extend those found for the auto-binomial. Furthermore, the tendency toward a sinusoidal RV shaped histogram already is becoming apparent here for MC = 0.90. The same histogram patterns appear for SFs constructed with global, regional and local map patterns. As with the Poisson case, Fig. 4.19c (as well as its corresponding part of Table 4.10) indicates that the mixture of map patterns constituting a SF, rather than only the level of positive SA, plays an important role, too.

4.4.3 Simulation Results for the China County Geographic Configuration

As with the hexagonal surface partitioning, MCMC simulation of auto-binomial model-based maps employing the China county irregular surface partitioning at most could embed only moderate positive SA. The graphical diagnostics appearing in Fig. 4.20 indicate that the resulting maps are properly generated. In addition, summary descriptive statistics reported in Table 4.11 are consistent with those appearing in Table 4.10: overdispersion is induced, and the distribution appears more uniform in shape. Meanwhile, dot plot versions of histograms appearing in Fig. 4.21 once more confirm the expectations that low levels of positive SA have little effect,

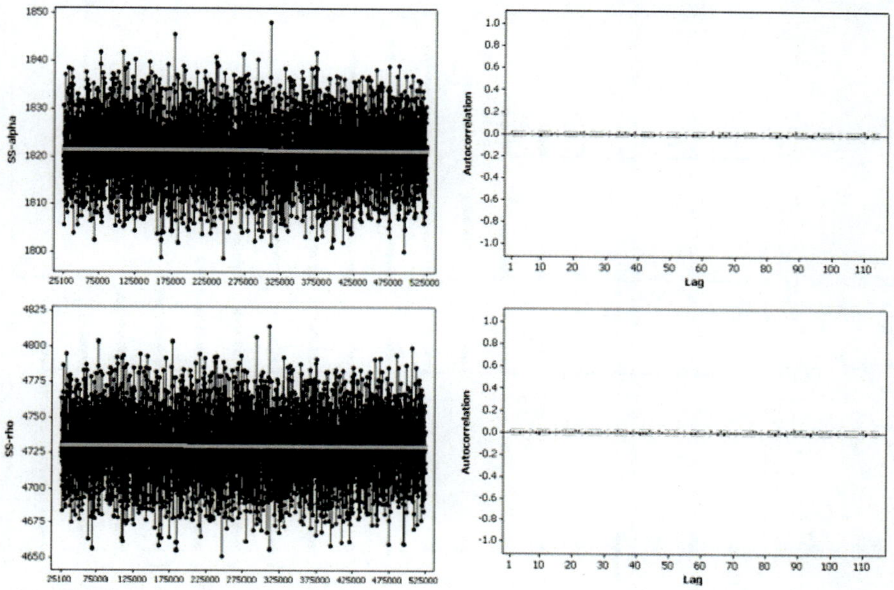

Fig. 4.20 MCMC time series plot and correlogram diagnostic graphics based on the China county surface partitioning when ρ = 1.00 (*bottom*). *Left* (**a**): for the intercept term. *Right* (**b**): for the autoregressive term

Table 4.11 Descriptive statistics for the auto-binomial model-based MCMC simulated data and the China irregular county geographic configuration

Variable autocorrelation	MC	GR	$\bar{y}$	s_y	y_{min}	y_{max}	Skewness	Kurtosis	K-S[a]
None (i.e., iid)	0.00	0.99	8.99	2.122	2	16	–0.01	–0.10	0.0038
Weak	0.12	0.86	9.18	2.511	1	16	–0.11	–0.27	0.0697
Low-moderate	0.40	0.55	11.96	4.167	0	18	–0.73	–0.73	0.5007
moderate	0.51	0.25	13.79	5.128	0	18	–1.37	0.61	0.6963

[a]K-S denotes the Kolmogorov-Smirnov statistic, used here to index deviation from the theoretical binomial distribution for which N = 18 and p = 0.5

whereas moderate levels tend to squash the center of a distribution and thicken its tails—in this case the irregularity of the surface partitioning distorts this tail thickening by skewing it to one side of its distribution. Of note is that the irregular surface partitioning introduces some trend in the mean, indicating that the relationship between α and ρ most likely needs to be more carefully articulated for irregular surface partitionings. Furthermore, the Kolmogorov-Smirnov statistics reported in Table 4.11 are indexing this deviation from p = 0.5 as much, if not more, than the change in the shape of the histogram. Skewness distortion with increasing positive SA appears in both the MCMC auto-binomial and the SF model-based simulation data.

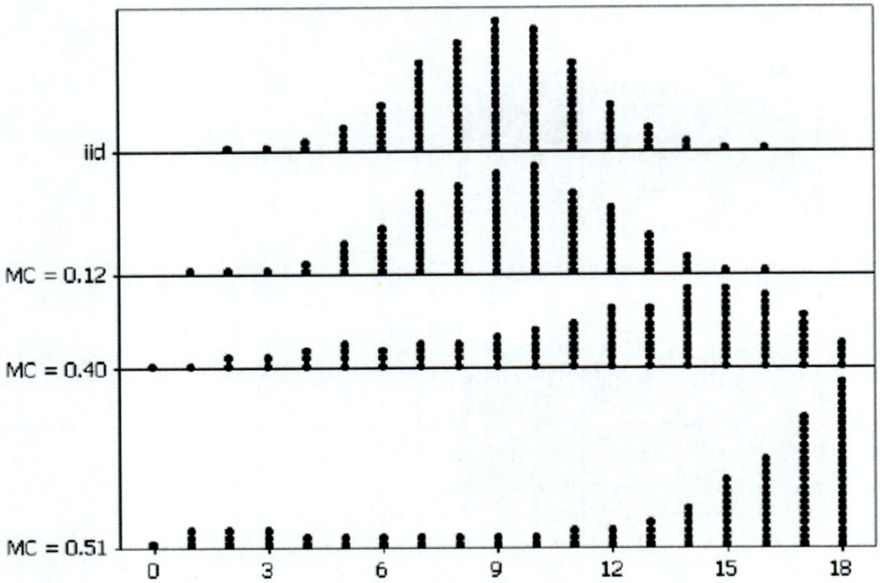

Fig. 4.21 Dot plot versions of histograms for the MCMC auto-binomial simulated data. *Top* (**a**): iid. 2nd from *top* (**b**): weak positive SA. 2nd from *bottom* (**c**): low moderate positive SA. *Bottom* (**d**): moderate positive SA

As is also seen with the Poisson RV analysis, one conspicuous difference between Figs. 4.18 and 4.19, and Figs. 4.21 and 4.22, is the interaction effect between positive SA impacts and the irregularness of the geographic configuration. One outcome of this interaction is that the flattening of a binomial histogram is followed by less of a sinusoidal RV shape as positive SA approaches its maximum value.

4.4.4 Implications

In conclusion, numerical results reported in this section suggest the following implications about a georeferenced binomial RV:

(1) by controlling for trend in data when estimating a mean (apparently this only needs to be done with MCMC simulation, not with SF simulation), positive SA has no impact upon the resulting estimated mean value;
(2) positive SA increases the chances of a histogram resembling that for a uniform distribution, and in the extreme, for a sinusoidal distribution;
(3) especially strong positive SA increases the chances of most counts being only 0 or N;
(4) as positive SA increases, the Kolmogorov-Smirnov test statistic tends to increase;

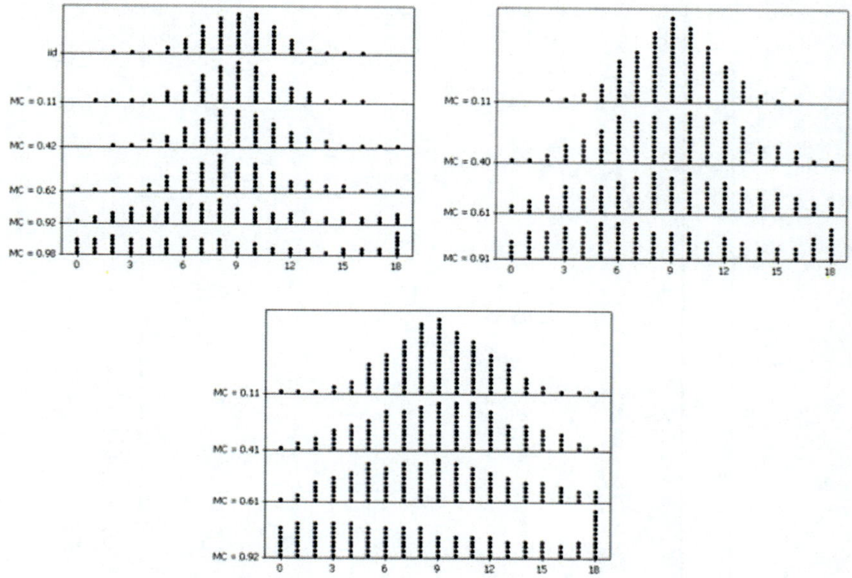

Fig. 4.22 Dot plot versions of histograms for the SF binomial simulated data using the China irregular surface partitioning. *Left* (**a**): SF results from a global map pattern. *Middle* (**b**): SF results from a global combined with a regional map pattern. *Right* (**c**): SF results from a global combined with a regional and a local map pattern

(5) a particular mixture of eigenvectors in a SF plays an important role in terms of the impacts of positive SA that materialize (see Tables 4.10 and 4.12);
(6) an interaction effect appears to occur between SA and the irregular nature of a surface partitioning; and,
(7) the conventional auto-binomial model is able to capture only weak-to-moderate positive SA.

In other words, just as with a Poisson RV, even modest amounts of positive SA do make a difference!

4.5 Discussion

This chapter indicates what a spatial scientist should expect from commonly encountered levels of SA when inspecting histograms constructed with georeferenced data. Regardless of whether a RV is normal, binomial, or Poisson in nature, its variance will tend to be inflated, with inflation increasing as positive SA increases. This is the single most common impact, which results in histograms being flatter than they would otherwise be if the data observations were iid. It leads to heterogeneity for normal RVs, excessive 0 s and extreme values (i.e., overdispersion) for

Table 4.12 Descriptive statistics for the auto-binomial model-based SF simulated data and the China irregular county geographic configuration

Variable autocorrelation	MC	GR	$\bar{y}$	s_y	y_{min}	y_{max}	Skewness	Kurtosis	K-S[a]
None (i.e., iid)	0.00	0.99	8.99	2.122	2	16	−0.01	−0.10	0.0038
global map pattern-base results									
Weak (using $0.112\mathbf{E}_G$)	0.11	0.91	8.98	2.256	1	16	−0.00	−0.14	0.0212
Low-moderate (using $0.275\mathbf{E}_G$)	0.42	0.63	9.06	2.632	2	18	0.23	0.03	0.0536
High-moderate (using $0.42\mathbf{E}_G$)	0.62	0.47	8.92	3.030	0	18	0.48	0.09	0.1002
Strong (using $1.10\mathbf{E}_G$)	0.92	0.19	8.40	4.644	0	18	0.47	−0.59	0.2674
global + regional map pattern-base results									
Weak (using $0.18\mathbf{E}_{R-1}$)	0.11	0.91	9.00	2.379	2	16	−0.02	−0.07	0.0345
Low-moderate [using $0.35(\mathbf{E}_{R-1}+\mathbf{E}_{R-2})$]	0.40	0.64	9.05	3.505	0	18	0.01	−0.55	0.1405
High-moderate [using $0.46(\mathbf{E}_G+\mathbf{E}_{R-1}+\mathbf{E}_{R-2})$]	0.61	0.45	8.93	4.416	0	18	0.07	−0.80	0.1991
Strong [using $0.35(4\mathbf{E}_G+\mathbf{E}_{R-1})$]	0.91	0.18	8.19	5.272	0	18	0.40	−0.91	0.3212
global + regional + local map pattern-base results									
Weak [using $0.2(\mathbf{E}_{R-1}+\mathbf{E}_L)$]	0.11	0.87	9.02	2.701	0	18	0.02	−0.17	0.0648
Low-moderate [using $0.2(2\mathbf{E}_{R-1}+2\mathbf{E}_{R-2}+\mathbf{E}_L)$]	0.41	0.63	9.05	3.846	0	18	−0.03	−0.64	0.1560
High-moderate [using $0.3(2\mathbf{E}_G+\mathbf{E}_{R-1}+\mathbf{E}_{R-2}+\mathbf{E}_L)$]	0.61	0.44	8.84	4.276	0	18	0.19	−0.74	0.2073
Strong [using $0.25(7\mathbf{E}_G+\mathbf{E}_{R-1}+\mathbf{E}_{R-2}+\mathbf{E}_L)$]	0.92	0.17	8.00	5.692	0	18	0.39	−1.07	0.3617

[a] K-S denotes the Kolmogorov-Smirnov statistic, used here to index deviation from the theoretical binomial distribution for which $N = 18$ and $p = 0.5$

Poisson RVs, and overdispersion for binomial RVs. Positive SA corrupts quantile plots when assessing normality, Poissonness plots, and other goodness-of-fit test, even when only its most commonly encountered moderate levels are present.

SF model specifications furnish an efficient and effective way of capturing SA effects, and render simulation results that are consistent with those obtained with the more conventional auto- model specifications. Because these models are constructed with stepwise regression techniques when an empirical analysis is being undertaken, they signal that Gaussian approximations actually are not obsolete. The role of these approximations is to supply a first glimpse of SA, as well as a first screening of a large number of candidate eigenvectors when constructing a SF.

Finally, the lessons to be learned from this chapter may be summarized as follows: *caution SA at work!* Cursory initial graphical inspections of empirical data can be misleading when SA is present. Spatial scientists need to heed this warning.

Acknowledgment This research was completed while the author was a visiting scientist at the Max Planck Institute for Demographic Research, Rostock, Germany, 2005.

Chapter 5
Understanding Correlations Among Spatial Processes

5.1 Introduction

The Pearson product-moment, Spearman's rank, point biserial and phi correlation coefficients are calculated to quantify the nature and degree of linear correspondence between observation pairs of attributes. Bivand (1980) and Griffith (1980) were among the very first spatial analysts to address the impacts of spatial autocorrelation (SA) on conventional Pearson correlation coefficients. In the decades since their studies, an increasing understanding has been attained about correlation coefficients computed with georeferenced data. This understanding includes how: SA alters conventional degrees of freedom and sample size, the nature and degree of SA affects correlation coefficients, and SA can simultaneously inflate and deflate correlation coefficients. The primary objective of this chapter is to review each of these topics, adding some extensions when possible.

5.2 Two Illustrative Examples

Two georeferenced data sets have been selected for illustrative purposes here. One is the famous geocoded (by district) Scottish lip cancer data reported by Clayton and Kaldor (1987, pp. 676–677), which comprises the number of observed lip cancer cases (O_i) geographically aggregated (i.e., post-stratified) into 56 district, expected values (E_i) computed on the basis of age and sex compositions of district populations, and the percentage of each district's outdoor labor force employed in agriculture, fishing, or forestry. Six districts consist of multiple islands, and sometimes parts of the mainland; centroids for these districts were approximated by selecting a central location for georeferencing purposes. The other data set comprises a number of surface (0–2") soil samples collected in and around the vacant industrially contaminated Murray (Utah, USA) smelter site (see Griffith, 2002). Three samples failed to have a geocode recorded, 173 samples were collected from the smelter site itself, and a total of 101 samples were collected from two adjacent residential neighborhoods, yielding 253 locations from which soil samples were taken. Each locationally tagged sampling quantity is a pooled composite result of

D.A. Griffith, J.H.P. Paelinck, *Non-standard Spatial Statistics and Spatial Econometrics*, Advances in Geographic Information Science 1,
DOI 10.1007/978-3-642-16043-1_5, © Springer-Verlag Berlin Heidelberg 2011

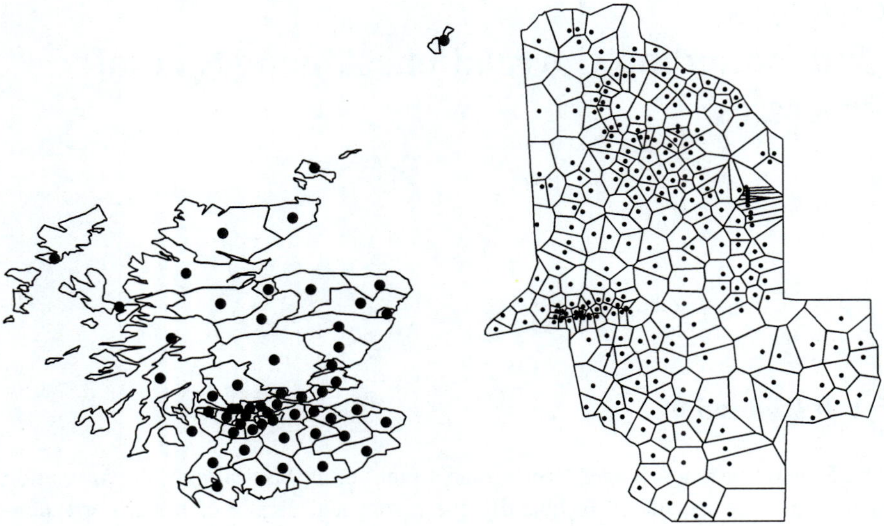

Fig. 5.1 Georeferencing of the illustrative data by surface tessellations (areal unit outlines) and Cartesian coordinates (solid circles, •). *Left* (**a**): Scotland districts.*Right* (**b**): Murray smelter site Thiessen polygons

assays for a number (usually four) of nearly adjacent soil samples for which lead (Pb) and arsenic (As) were measured. A tessellation for these points was established by generating Thiessen polygons. Maps of the surface partitionings and locational coordinates for these two landscapes appear in Fig. 5.1.

For the Scottish lip cancer data, one variable is the standardized mortality ratio (SMR), which, after application of a suitable Box-Cox power transformation, results in the variable $Y_{SMR} = LN\left(\frac{O+0.5}{E+0.5}\right)$ conforming closely to a normal distribution [Shapiro-Wilk statistic (S-W) = 0.971, p = 0.20]. The estimated optimal translation parameter, 0.5, is necessary here because some counts are 0. The detected SA changes little when this transformation is applied: the Moran Coefficient (MC) is 0.49652 and the Geary Ratio (GR) is 0.43830. Meanwhile, the Box-Tidwell linearization transformation identified for the outdoor labor percentage covariate is $X_{OLP} = LN(\% + 1.2)$. The S-W for this transformed variable is 0.860 ($p < 0.01$); but its linear relationship with Y_{SMR} is near optimal. The estimated translation parameter, 1.2, is necessary here because some percentages are 0. The detected SA is indexed as follows: MC = 0.39787, and GR = 0.56661. Significant ($\hat{\sigma}_{MC} \approx 0.086$), moderate positive SA is detected in each of these two variables.

The following optimal heterogeneous Box-Cox power transformations were employed with the Murray smelter site geographic landscape:

$$Y_{As} = LN(As - 38.5 + 98.1\left(\frac{\eta_{As}}{n+1}\right)^{0.31}\left(1 - \frac{\eta_{As}}{n+1}\right)^{2.90},$$

and

$$X_{Pb} = LN\left(Pb - 25.9 + 2287.5\left(\frac{\eta_{Pb}}{n+1}\right)^{1.45}\left(1 - \frac{\eta_{Pb}}{n+1}\right)^{-0.93}\right),$$

where η_{As} and η_{Pb} respectively denote the ranking of As and Pb values, in descending order. These two transformations are consistent with the common finding that pollution data are log-normally distributed. Each of the two translation parameters relates to its variable's minimum contaminant value, as well as heterogeneity associated with size of values (captured by letting the translation parameter be a function of rank order). The primary impact of the translation parameters is better alignment of distributional tails in a normal quantile plot. The respective S-W statistics are 0.972 ($p < 0.01$) and 0.999 ($p \approx 1.00$). The detected SA is indexed as follows: for Y_{As}, MC = 0.29114 and GR = 0.60855; and, for Y_{Pb}, MC = 0.25534 and GR = 0.77245. Significant ($\hat{\sigma}_{MC} \approx 0.038$), weak positive SA is detected in each of these two variables.

5.3 Geostatistical Semivariogram Model Implications

Clifford, Richardson and Hémon (1989), and Richardson (1990), use semivariogram modeling to link the correlation coefficient, r, to its correct sampling distribution. They develop the notion of *effective degrees of freedom* (edfs), which are the equivalent number of degrees of freedom for n^* independent and identically distributed (iid) observations. In other words, SA represents redundant, or duplicated, information contained in georeferenced data due to the relative locations of observations. One way of adjusting for this redundancy is to prorate the number of observations, n, to n^*. Calculations are based upon the formula

$$n^* = 1 + \sigma_r^{-2} \tag{5.1}$$

where σ_r^2 denotes the variance of the sampling distribution of the correlation coefficient, r. If zero SA is present, then $n^* = n$; if perfect positive SA is present, then $n^* = 1$.

Dutilleul (1993) further clarified this notion of edfs by incorporating impacts of estimating means and variances for a correlation coefficient, an adjustment that is set aside here for simplicity (these types of adjustments also are outlined in Griffith and Zhang, 1999), and rewriting σ_r^{-2} in matrix notation as

$$\frac{TR(\mathbf{V}_X^{-1})TR(\mathbf{V}_Y^{-1})}{TR(\mathbf{V}_X^{-1}\mathbf{V}_Y^{-1})}, \tag{5.2}$$

where $\mathbf{V}_X^{-1}$ and $\mathbf{V}_Y^{-1}$ respectively are the n-by-n SA covariance structure matrices for variables X and Y, and TR denotes the matrix trace operator. If the n observations contain zero SA, then $\mathbf{V}_X^{-1} = \mathbf{V}_Y^{-1} = \mathbf{I}$, the n-by-n identity matrix, and expression

(5.2) becomes $\dfrac{\mathrm{TR}(\mathbf{I})\mathrm{TR}(\mathbf{I})}{\mathrm{TR}(\mathbf{I}\times\mathbf{I})} = \dfrac{n^2}{n} = n$; if the n observations contain perfect positive SA, then conceptually $\mathbf{V}_X^{-1} = \mathbf{V}_Y^{-1} = \mathbf{1}\mathbf{1}^T$ and hence expression (5.2) becomes $\dfrac{\mathrm{TR}(\mathbf{1}\mathbf{1}^T)\mathrm{TR}(\mathbf{1}\mathbf{1}^T)}{\mathrm{TR}(\mathbf{1}\mathbf{1}^T\mathbf{1}\mathbf{1}^T)} = \dfrac{n^2}{n^2} = 1$.

For a semivariogram model of the form

$$C_0 + C_1 f(r_d, \bar{d}_k)$$

where C_0 denotes the nugget effect (e.g., variation due to measurement error and/or model misspecification), C_1 is the SA adjusted variance when $C_0 = 0$, r_d is a range parameter (i.e., an index of the distance beyond which SA is trivial), f denotes one of the various valid semivariogram model distance decay functions, and $\bar{d}_k$ denotes the average inter-location distance for some distance range k, expression (5.2) becomes

$$\frac{n^2}{\mathrm{TR}(\mathbf{V}_X^{-1}\mathbf{V}_Y^{-1})}, \tag{5.3}$$

where cell entry (i, j) in the covariance structure matrix $\mathbf{V}_j^{-1}$ (j = X, Y) may be defined by

$$1 - \left[C_0/C_1 + f\left(r, d_{ij}\right)\right]^1.$$

Accordingly, matrices $\mathbf{V}_X^{-1}$ and $\mathbf{V}_Y^{-1}$ in expression (5.3) have 1s in their diagonal entries. Again, if zero SA prevails, this expression reduces to n. An assortment of semivariogram models has been estimated with the two illustrative data sets. Iteratively customized distance intervals were constructed, and maximum distance included was determined by visual inspection of the semivariogram plots. Georeferencing coordinates were standardized to a near-unit square, for ease of comparison between the two geographic landscapes, resulting in a maximum distance of 1.12625 for the Scotland landscape, and of 1.05010 for the Murray smelter landscape. One guiding principle for estimation purposes was that any distance interval k had a sample size of at least 30. Estimation results appear in Tables 5.1 and 5.2. Practical ranges were computed following Griffith and Layne (1999, p. 468). Because the MC and GR indices indicate the presence of positive SA, the wave-hole semivariogram model was not estimated. And, because it has neither a range nor a practical range, the power semivariogram model was not estimated.

Estimation results for Scotland appear in Table 5.1. Because excessively large nugget estimates were obtained for it, the cubic semivariogram model was set aside. Both the stable and the Cauchy semivariogram model specifications yielded unreliable results, perhaps because geographically aggregated data are being assigned to tessellation centroids here. Noticeable variation is apparent in the ranges/practical

[1] See Cressie (1991) for the specific equations for many of the semivariogram models.

Table 5.1 Selected semivariogram modeling results for Scottish lip cancer data using a maximum standardized distance of 0.50 for the standardized mortality rate (SMR), and 0.25 for the percentage of outside workers

Model	C_0	$C_1 \mid C_0 = 0$	r_d	100×RESS	Range/practical range	$n^* \mid C_0 = 0$
Standardized mortality rate (SMR)						
Spherical	0.1275	1.1091	0.4255	17.49	0.4255	33.33
Exponential	0.1165	1.4349	0.2807	17.42	0.8421	33.44
Stable	Unreliable results					
Penta-spherical	0.1326	1.1285	0.5269	17.77	0.5269	33.83
Rational quadratic	0.2708	1.2025	0.1509	25.14	0.6578	30.13
Bessel function	0.2143	1.1803	0.1201	21.26	0.4804	33.36
Gaussian	0.2856	1.0623	0.1791	30.21	0.3102	36.53
Circular	0.1358	1.1127	0.3796	17.48	0.3796	34.79
Cauchy	Unreliable results					
Percentage of outside workers						
Spherical	0.0272	0.9392	0.0759	48.45	0.0759	33.33
Exponential	−0.0421	0.9468	0.0268	48.08	0.0804	33.44
Stable ($\hat{\lambda} = 1.4263$)	−0.0127	0.9407	0.0310	47.27	0.0670	NA
Penta-spherical	0.0029	0.9388	0.0887	47.91	0.0887	33.83
Rational quadratic	−0.0194	0.9616	0.0219	48.25	0.0955	30.13
Bessel function	−0.0189	0.9435	0.0181	47.31	0.0724	33.36
Gaussian	−0.0082	0.9329	0.0311	47.58	0.0539	36.53
Circular	0.0223	0.9360	0.0647	49.13	0.0647	34.79
Cauchy ($\hat{\lambda} = 2.8802$)	−0.0136	0.9400	0.0478	47.06	0.0647	NA

NOTE: NA denotes "not applicable"

Table 5.2 Selected semivariogram modeling results for soil samples from the Murray smelter site using a maximum standardized distance of 0.20

Model	C_0	$C_1 \mid C_0 = 0$	r_d	100×RESS	Range/ practical range	$n^* \mid C_0 = 0$
Arsenic (As)						
Spherical	0.5865	4.4286	0.0763	45.98	0.0763	119.64
Exponential	0.1654	4.5108	0.0296	43.47	0.0888	126.25
Stable ($\hat{\lambda} = 0.8160$)	0.0102	4.5903	0.0284	43.18	0.1092	123.70
Penta-spherical	0.5030	4.4266	0.0891	45.55	0.0891	123.05
Rational quadratic	0.3460	4.5606	0.0224	44.53	0.0976	119.95
Bessel function	0.4235	4.4579	0.0184	44.82	0.0736	124.17
Gaussian	0.9554	4.4016	0.0320	49.45	0.0554	116.30
Circular	0.7201	4.4324	0.0696	46.74	0.0696	114.80
Cauchy ($\hat{\lambda} = 0.3623$)	0.0472	5.1645	0.0090	43.36	0.5620	36.36
Lead (Pb)						
Spherical	0.3490	3.7921	0.0802	33.94	0.0802	119.64
Exponential	0.0975	3.9162	0.0346	31.06	0.1038	126.25
Stable ($\hat{\lambda} = 0.8623$)	–0.0133	3.9839	0.0343	30.86	0.1226	123.70
Penta-spherical	0.3358	3.7996	0.0969	33.63	0.0969	123.05
Rational quadratic	0.3040	3.9606	0.0263	32.03	0.1146	119.95
Bessel function	0.3309	3.8501	0.0211	32.39	0.0844	124.17
Gaussian	0.7071	3.7824	0.0356	36.97	0.0617	116.30
Circular	0.4072	3.7911	0.0719	34.31	0.0719	114.80
Cauchy ($\hat{\lambda} = 0.2484$)	0.0466	4.5525	0.0116	30.90	0.6794	36.36

ranges for SMRs, while these values are very similar for the worker percentages; the same patterns appear in the relative error sums of squares (REESs) and the C_1 values. Nevertheless, all of the calculated effective sample sizes, denoted by n^*, are roughly the same. In other words, for the most part, the model specification makes little difference when numerically evaluating expression (5.3). The exponent for the stable semivariogram model specification that is estimated with the percentage of outside workers suggests that this model specification is better than the Gaussian specification. The MC and GR values suggest that the exponential or Bessel function model specifications should be preferable (see Griffith and Layne, 1999, pp. 142–152), an implication largely corroborated by the RESSs.

See Cressie (1991) for the mathematical equations for these model specifications.

Estimation results for the Murray smelter site appear in Table 5.2. Once more, because excessively large nugget estimates were obtained for it, the cubic model was set aside. Of note is that, except for the Cauchy model specification, all of the ranges/practical ranges are very similar, as are the REESs and the C_1 values. Estimated exponents for the stable model specification suggest that the exponential model specification is better than the Gaussian specification. The MC and GR values suggest that the exponential or Bessel function model specifications should be preferable, an implication largely corroborated by the RESSs. Except for the Cauchy model specification, all of the effective sample sizes (n^*) are roughly the same. In other words, again, the model specification tends to make little difference when numerically evaluating expression (5.3).

A better understanding of correlations among spatial processes, in terms of sampling distribution degrees of freedom, exists because of this work. But in many empirical cases the reduction from n to n^* makes little difference in resulting t-distribution values. For example, for the Scotland data, $t_{0.95,30-2} = 1.70113$ differs little from $t_{0.95,56-2} = 1.67356$; an even smaller difference exists between $t_{0.95,\,120\text{-}2} = 1.65781$ and $t_{0.95,253\text{-}2} = 1.65095$ for the Murray smelter site. Meanwhile, what already is known may be supplemented by considering the relationship between n^* obtained for a correlation coefficient and individual variable values, say n^*_X and n^*_Y. Griffith and Zhang (1999), for instance, report that the univariate values of n^* are given by $\frac{\mathrm{TR}(\mathbf{V}_j^{-1})}{\mathbf{1}^T\mathbf{V}_j^{-1}\mathbf{1}}$ (j = X, Y), based on the sampling distribution of a sample mean. Suppose the value produced by expression (5.3) is denoted by $n^*_{XY}-1$. A simulation experiment based upon the Bessel function semivariogram model, the Scotland coordinates, and independent pairs of range parameters randomly selected from the uniform distribution, U(0, 0.25), and replicated 10,000 times, rendered the following equation:

$$\begin{aligned}\hat{n}^*_{XY} = 1 - 1.3594\frac{n^{1.7842} - n}{n^{1.7842} - 1} + \left[(n^*_X)^{0.8075} + (n^*_Y)^{0.8075}\right]^{1.2384} \\ - 1.3594\frac{n^{1.7842} - 1}{n^{1.7842} - 1}(n^*_X n^*_Y)^{0.8921}, \ n = 56 \text{ and } R^2 = 0.9999.\end{aligned} \tag{5.4a}$$

A graph of Eq. (5.4a) appears in Fig. 5.2a. A second simulation experiment based upon the Bessel function semivariogram model, the Murray smelter site coordinates,

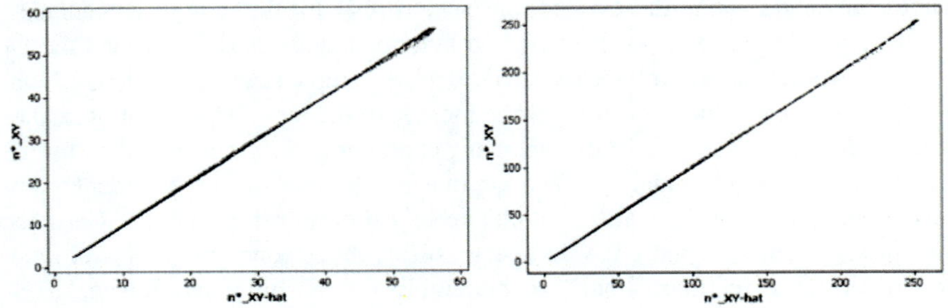

Fig. 5.2 Simulation-based relationships between a composite of individual univariate effective sample sizes and their corresponding bivariate effective sample size. *Left* (**a**): Scotland results based upon a Bessel function and Eq. (5.4a). *Right* (**b**): Murray smelter site results based upon a Bessel function and Eq. (5.4b)

and independent pairs of range parameters randomly selected from the uniform distribution, U(0, 0.25), and replicated 10,000 times, rendered the following equation:

$$\begin{aligned}\hat{n}^*_{XY} = 1 - 1.4731\frac{n^{1.6212}-n}{n^{1.6212}-1} + [(n^*_X)^{0.7655} + (n^*_Y)^{0.7655}]^{1.3063} \\ - 1.4731\frac{n^{1.6212}-1}{n^{1.6212}-1}\left(n^*_X n^*_Y\right)^{0.8106}, \ n = 253 \text{ and } R^2 \approx 1.0000.\end{aligned} \qquad (5.4b)$$

A graph of Eq. (5.4b) appears in Fig. 5.2b. These two graphs reveal a close correspondence between univariate and bivariate results, for the Bessel function semivariogram model. A third simulation, in which the model also was randomly selected, produced very similar output, but, as expected, output displaying more variation. One important implication here is that impacts of SA can be mitigated, to some degree, by incorporating redundant georeferenced attribute information, a natural form of which arises in space-time series data. Lahiri (1996) notes that this is one way of regaining estimator consistency when employing infill asymptotics (i.e., the sample size increases by keeping the study area size constant and increasing the sampling intensity). The interplay between spatial auto- and attribute correlation also is addressed by, among others, Wartenberg (1985) and Lee (2001).

5.4 Spatial Autoregressive Model Implications

Haining (1991) follows the prewhitening, impulse-response function approach of time series analysis to adjust for SA effects on correlation coefficients (i.e., the spatial version of a Cochrane-Orcutt time-series transformation). In doing so, he compares this prewhitening approach with that of Clifford, Richardson and Hémon (1989), and extends findings to the Spearman's rank correlation coefficient. One of his findings is that the Clifford-Richardson- Hémon adjustment discussed in the preceding section improves the large sample test for Spearman's rank correlation.

Prewhitening two georeferenced variables often takes a spatial analysis into the realm of spatial autoregressive models. As such, the inverse covariance matrix $\mathbf{V}\sigma^{-2}$, rather than the covariance matrix itself, is modeled. Accordingly, for purely spatial processes (i.e., not including covariates),

$$\mathbf{Y} = \mu_Y \mathbf{1} + \mathbf{V}_X^{-1} \xi_Y, \text{ and} \tag{5.5a}$$

$$\mathbf{X} = \mu_X \mathbf{1} + \mathbf{V}_Y^{-1} \xi_X \tag{5.5b}$$

where μ_Y and μ_X respectively denote the means of variables Y and X, the n-by-1 vectors $\mathbf{X}$ and $\mathbf{Y}$ respectively are the observed values of variables X and Y, $\mathbf{1}$ is an n-by-1 vector of ones, and ξ_X and ξ_Y are n-by-1 vectors of iid random variables frequently assumed to be distributed $N(0, \sigma_{\xi_j}^2)$, j = X or Y. Two model specifications are popular for matrices $\mathbf{V}_X^{-1}$ and $\mathbf{V}_Y^{-1}$. The conditional autoregressive (CAR) model includes the definition $\mathbf{V} = (\mathbf{I} - \rho\, \mathbf{C})$, where ρ is an autocorrelation parameter indicating the nature and degree of SA $\left(\frac{1}{\lambda_{min}} < \rho < \frac{1}{\lambda_{max}}\right.$, where λ_{min} and λ_{max} are the extreme eigenvalues of matrix $\mathbf{C}$), and the simplest version of the n-by-n geographic weights matrix $\mathbf{C}$ is a binary form whose entries are $c_{ij} = 1$ if areal units i and j are neighbors, and $c_{ij} = 0$ otherwise. The simultaneous autoregressive (SAR) model includes the definition $\mathbf{V} = [(\mathbf{I} - \rho\, \mathbf{W})^T(\mathbf{I} - \rho\, \mathbf{W})]$, where matrix $\mathbf{W}$ frequently is the row standardized (i.e., each row sums to 1) version of matrix $\mathbf{C}$ $\left(\frac{1}{\lambda_{min}} < \rho < 1\right.$, where λ_{min} and 1 are the extreme eigenvalues of matrix $\mathbf{W}$). Based upon the SAR model, Eqs. (5.5a) and (5.5b) produce the following pair of equations:

$$\mathbf{Y} = \rho_Y \mathbf{WY} + (1 - \rho_Y)\mu_Y \mathbf{1} + (\mathbf{I} - \rho_Y \mathbf{W})[(\mathbf{I} - \rho_Y \mathbf{W})^{-1}]\xi_Y, \tag{5.6a}$$

and

$$\mathbf{X} = \rho_X \mathbf{WY} + (1 - \rho_X)\mu_X \mathbf{1} + (\mathbf{I} - \rho_X \mathbf{W})\, [(\mathbf{I} - \rho_X \mathbf{W})^{-1}]\xi_X, \tag{5.6b}$$

where ρ_X and ρ_Y respectively denote the autoregressive parameters for variables X and Y. The matrix multiplications $[(\mathbf{I} - \rho_Y\, \mathbf{W})\, [(\mathbf{I} - \rho_Y\, \mathbf{W})^{-1}]$ and $[(\mathbf{I} - \rho_X\, \mathbf{W})\, [(\mathbf{I} - \rho_X\, \mathbf{W})^{-1}]$ remove SA from the error terms, which is the process of prewhitening.

Griffith and Layne (1999) establish numerical links between the CAR and exponential, and the SAR and Bessel function, semivariogram models. Although an increase in the range of semivariogram models accompanies an increase in the degree of SA, one functional advantage Eqs. (5.6a) and (5.6b) have over the preceding semivariogram models is that n*, for instance, can be written explicitly in terms of the nature and degree of SA as indexed by ρ. Haining (1991) also notes that the semivariogram approach may well require a detrending of georeferenced data prior to model estimation. This necessity is easily accommodated with Eqs. (5.6a) and (5.6b) by including covariates in their specifications.

Table 5.3 Summary statistics that are informative about spatial autocorrelation and bivariate attribute correlation

	Scottish lip cancer data		Murray smelter site data	
Statistic	SMR	% Outdoor workers	As	Pb
r	0.54087		0.74775	
$r_{\xi_X \xi_Y}$	0.26979		0.70058	
s_{ξ_j} (j = X, Y)	0.61169	0.77922	1.87642	1.65854
SAR $\hat{\rho}$	0.72021	0.58082	0.53180	0.49363
Residual MC	–0.03685	0.02085	–0.02978	–0.03362
Residual GR	1.01248	0.89913	1.04530	1.08512
Residual R-J	0.9693 (p = 0.01)	0.9850 (p > 0.10)	0.9980 (p > 0.10)	0.9979 (p > 0.10)

NOTE: R-J denotes the Ryan-Joiner normality test statistic

Table 5.3 summarizes descriptive statistics affiliated with spatial autoregressive models for the two empirical data sets being used here for illustrative purposes. The autoregressive parameter estimates indicate the presence of moderate, positive SA in all four variables. Although the MC and GR values for the residuals are only approximations, they do suggest the absence of all but trace SA in the SAR residuals. And, each of the four sets of residuals appears to conform reasonably well to a normal distribution. For the Scottish lip cancer data, the correlation is markedly inflated by the presence of SA; for the Murray smelter data, the correlation is only moderately inflated by the presence of SA.

Simulation and resampling experiments, involving 10,000 replications, were conducted based upon the vectors $\hat{\xi}_X$ and $\hat{\xi}_Y$. The simulation experiment involved sampling from a bivariate normal distribution with the following attribute covariance matrices:

$$\underline{\text{Scotland}}\begin{pmatrix} 0.61169 & 0.12859 \\ 0.12859 & 0.77922 \end{pmatrix};\underline{\text{Murray}}\begin{pmatrix} 1.87642 & 2.19585 \\ 2.19585 & 1.65854 \end{pmatrix}.$$

The bootstrapping experiment involved simple random sampling, with replacement, of pairs of estimated errors ($\hat{\xi}_X$, $\hat{\xi}_Y$). And, the permutation experiment (see, for example, Costanzo, 1983) involved randomly permuting pairs of estimated errors ($\hat{\xi}_X$, $\hat{\xi}_Y$). Summary results from these experiments are reported in Table 5.4. During each replication, after experimental errors were obtained for each location, the estimated versions of Eqs. (5.6a) and (5.6b) were used to compute the X and Y values for which correlation coefficients were calculated. Of note is that the sampling distribution means are corroborated by the different experiments, and are closely related to the values obtained by prewhitening. None of the standard errors are very close to the theoretical value of $\sqrt{\frac{1-r^2_{\xi_X \xi_Y}}{n-2}}$, which is 0.13104 for the Scottish data, and 0.04504 for the Murray data. All three of the Scottish lip cancer data computer-generated sampling distributions tend to have marked deviations in their upper tails,

Table 5.4 Simulation and resampling experiment sampling distribution results using Eqs. (5.6a) and (5.6b); 10,000 replications

Error source	Scottish lip cancer data (r = 0.54087)			Murray smelter site data (r = 0.74775)		
	$\hat{\mu}_r$	$\hat{\sigma}_r$	R-J	$\hat{\mu}_r$	$\hat{\sigma}_r$	R-J
Bivariate normal	0.2622	0.1777	0.9976 ($p < 0.01$)	0.70379	0.03794	0.9970 ($p < 0.01$)
Permutation	0.2625	0.1264	0.9988 ($p < 0.01$)	0.70500	0.02065	0.9989 ($p < 0.01$)
Bootstrap	0.2723	0.1870	0.9973 ($p < 0.01$)	0.70437	0.04885	0.9960 ($p < 0.01$)

resulting in the Ryan-Joiner (R-J) normality test statistics implying non-normality. The Murray data tend to have marked deviations in both tails.

Figure 5.3 portrays the impact of SA on the sampling distribution of r, which contains graphs of the simulated bivariate normal data for the Murray smelter pollution case. As these graphs demonstrate, positive SA deflates the central part and inflates the tails of the sampling distribution. In other words, it increases the variance of the sampling distribution. In this example—which involves moderate, positive SA—the variance is inflated by a factor of 1.41578. These graphs also show that SA basically does not alter the mean of the sampling distribution of r.

5.4.1 *Variance and Covariance Inflation Attributable to Spatial Autocorrelation*

Variance inflation may be written for attribute variables X and Y, in terms of spatial autoregressive models, as $\frac{TR(\mathbf{V}_X^{-1})}{n}$ and $\frac{TR(\mathbf{V}_Y^{-1})}{n}$ [see expression (5.2)]. When $\mathbf{V} = (\mathbf{I} - \rho\,\mathbf{C})$ or $\mathbf{V} = [(\mathbf{I} - \rho\,\mathbf{W})^T(\mathbf{I} - \rho\,\mathbf{W})]$, these traces no longer equal n, as is the case with the semivariogram modeling [see expression (5.3)]. The covariance term

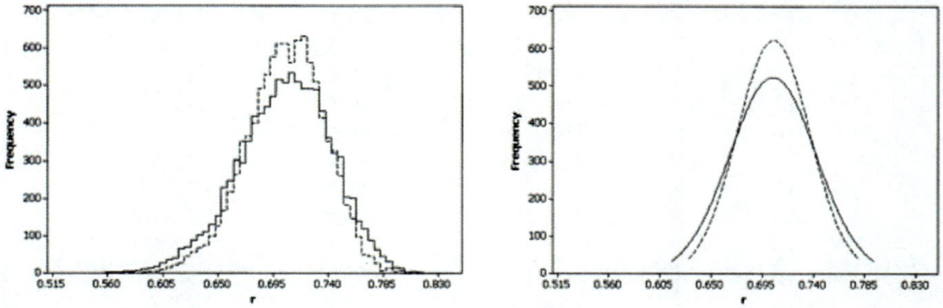

Fig. 5.3 Computer-generated sampling distributions of the correlation coefficient, r. Broken lines (---) denote r computed with spatially unautocorrelated data values; solid lines (—) denote r computed with spatially autocorrelated data values. *Left* (**a**): histograms. *Right* (**b**): normal curve approximations to the histograms

of the correlation coefficient numerator becomes $\frac{\mathrm{TR}[(\mathbf{V}_X^{-1/2})^T\mathbf{V}_Y^{-1/2}]}{n}$; this and the two univariate variance inflation factors (VIFs) are the same forms found in the preceding discussion about semivariogram models. The ratio of these terms that is part of the calculation of the correlation coefficient, namely

$$\frac{\mathrm{TR}\left[(\mathbf{V}_X^{-1/2})^T\mathbf{V}_Y^{-1/2}\right]}{\sqrt{\mathrm{TR}(\mathbf{V}_X^{-1})}\sqrt{\mathrm{TR}(\mathbf{V}_Y^{-1})}},$$

highlights how the variance and covariance inflation factors compensate for each other; a correlation coefficient cannot exceed 1 in absolute value, constraining the joint effects of variance and covariance inflations.

A simulation experiment, involving 10,000 replications, was conducted in which n pairs of random numbers were drawn from a bivariate normal distribution with attribute correlations ranging from –1 to 1, and using the SA covariance structure matrices for the Murray smelter site data (n = 253). Results of this experiment appear in Table 5.5 and Fig. 5.4. The resulting average spatially autocorrelated correlations are indistinguishable from their unautocorrelated counterparts. Of note is that a slight amount of variation appears in the extreme cases of ±1 for the spatially autocorrelated case. Except for correlations very close to the near-degenerate case of zero, the VIF is approximately constant (i.e., 1.22570 from the simulation

Table 5.5 Simulation experiment sampling distribution results for cross-product terms; 10,000 replications

Correlation	$\hat{\mu}_{r_{XY}}$	$\hat{\mu}_{r_{\xi_X\xi_Y}}$	VIF
1.0	0.99957	1.00000	1.22604
0.9	0.89918	0.89966	1.22577
0.8	0.79840	0.79923	1.22347
0.7	0.69922	0.69926	1.22664
0.6	0.59907	0.59926	1.22718
0.5	0.49884	0.49894	1.22641
0.4	0.39945	0.39981	1.22482
0.3	0.29932	0.29922	1.22615
0.2	0.19954	0.19932	1.22806
0.1	0.09971	0.09967	1.22879
0.0	0.00061	0.00018	4.84194
–0.1	–0.09931	–0.09954	1.22497
–0.2	–0.19923	–0.19929	1.22617
–0.3	–0.30004	–0.30025	1.22510
–0.4	–0.39865	–0.39921	1.22467
–0.5	–0.49814	–0.49870	1.22503
–0.6	–0.59836	–0.59880	1.22626
–0.7	–0.69923	–0.69977	1.22568
–0.8	–0.79916	–0.79960	1.22523
–0.9	–0.89886	–0.89958	1.22307
–1.0	–0.99957	–1.00000	1.22455

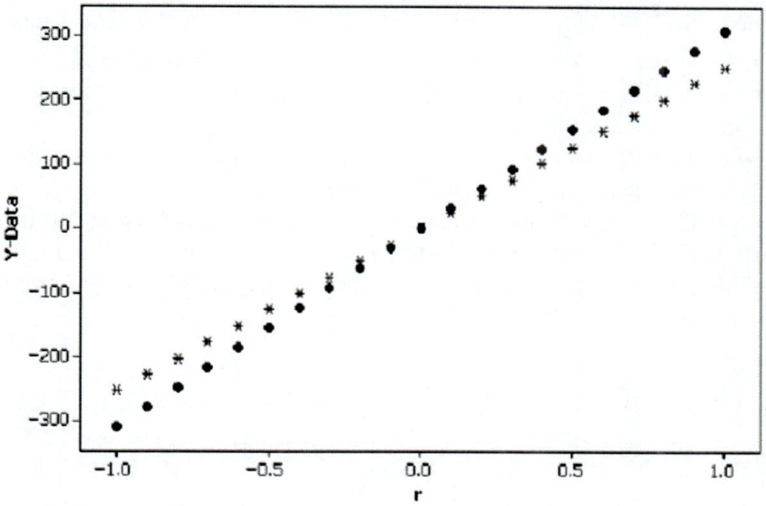

Fig. 5.4 Average cross-product terms for simulated data based upon the Murray smelter site. *Asterisks* (∗) denote spatially autocorrelated data; *solid circles* (•) denote spatially unautocorrelated data

experiment, excluding the case of 0 correlation, and 1.23701 from the covariance inflation factor formula). Figure 5.4 reveals that the cross-product terms remain a linear function of attribute correlation, regardless of whether or not SA is present. The numerical results demonstrate that the associated univariate VIFs completely compensate for the covariance inflation factor.

5.4.2 Effective Sample Size as a Function of ρ_X and ρ_Y

One advantage of the spatial autoregressive over the semivariogram modeling approach is that specific natures and degrees of SA can be specified with parameter ρ. The same effect can be obtained with semivariogram models by altering the range parameter, but the degree of SA change is not obvious from this manipulation. Another advantage is that the full gamut of SA, from strong negative to strong positive, can be studied with autoregressive models. The wave-hole semivariogram model is one of the very few that captures negative SA. One disadvantage is that when the inverse covariance matrix is modeled, boundary effects introduce more complications in spatial autoregression. This in part is why, for the Murray smelter site, $\hat{n}^*_{XY} = 177.60$ obtained with the SAR model specification differs from the range of semivariogram model results (i.e., 115 to 126) reported in Table 5.2.

A simulation experiment, involving 10,000 replications, was conducted in which each entry of an independent pair of SA values (ρ_X, ρ_Y) was randomly drawn from the uniform distribution U(–1, 1), and using the **W** matrix for the Murray smelter site data (n = 253). Although the sampling range does not span the full range of negative SA $\left(\frac{1}{-0.57823} = -1.72294 < -1\right)$, if furnishes a practical range for assessment purposes. These simulated data were used to evaluate expression (5.2) in order to establish an effective sample size equation for spatial autoregression that is comparable with Eq. (5.4b).

Impacts of SA appear to vary by its nature. For example, n^* exceeds n for negative SA, whereas it lies in the range [1, n] for positive SA. Let I_{++} denote the case where $\rho_X > 0$ and $\rho_Y > 0$; approximately 25% of the simulated cases fall into this category. Let I_{--} denote the case where $\rho_X < 0$ and $\rho_Y < 0$; again, approximately 25% of the simulated cases fall into this category. A reasonably good description of $\hat{n}^*_{XY}$ (e.g., pseudo-$R^2 = 0.9458$) is furnished by

$$\begin{aligned}\hat{n}^*_{XY} = 1 + a + \{253[1 - \Bigg(&\{253[1 - \frac{1}{1-e^{-\alpha}}\frac{253-1}{253}(1-e^{-\alpha\times\rho_X})\}^c \\ &+\{253[1 - \frac{1}{1-e^{-\alpha}}\frac{253-1}{253}(1-e^{-\alpha\times\rho_Y})\}^c\Bigg) - \\ d\{253[1-\frac{1}{1-e^{-\alpha}}\frac{253-1}{253}(1-e^{-\alpha\times\rho_X})\}\,&\{253[1-\frac{1}{1-e^{-\alpha}}\frac{253-1}{253}(1-e^{-\alpha\times\rho_Y})\},\end{aligned} \tag{5.7}$$

where the coefficients depend upon the nature of the autocorrelation (see Table 5.6). Although Eq. (5.7) needs further refinement, especially when SA differs in nature

Table 5.6 Estimation results for Eq. (5.7)

Coefficient	$\rho_X > 0,\ \rho_Y > 0$	$\rho_X < 0,\ \rho_Y < 0$	$\rho_X > 0,\ \rho_Y < 0$	$\rho_X < 0,\ \rho_Y > 0$
a	–33.4318	50.4059	35.2668	28.0208
c	0.7412	1.3069	1.1032	1.0726
d	–0.0058	–0.0035	–0.0038	–0.0038
α	0.5519	–0.5784	1.2554; –4.4807	1.2590; –4.3611
Pseudo-R^2	0.9971	0.9997	0.9886	0.9890

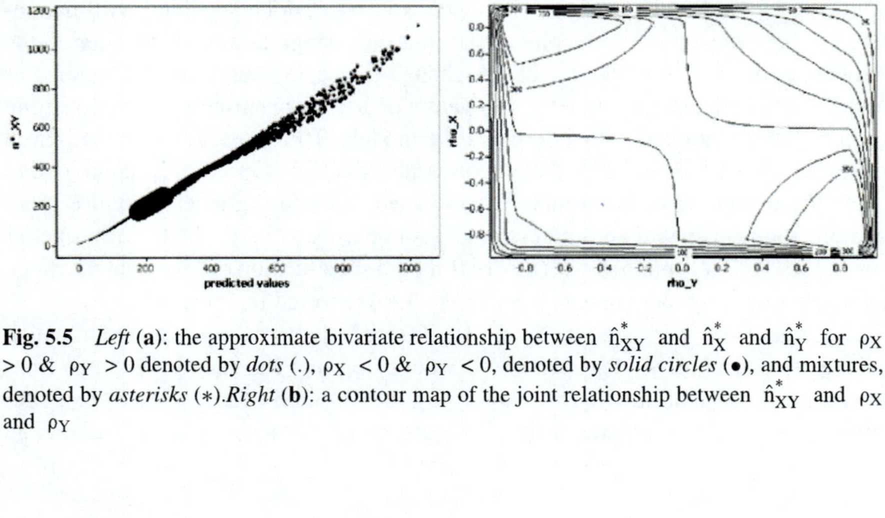

Fig. 5.5 *Left* (**a**): the approximate bivariate relationship between $\hat{n}^*_{XY}$ and $\hat{n}^*_X$ and $\hat{n}^*_Y$ for $\rho_X > 0$ & $\rho_Y > 0$ denoted by *dots* (.), $\rho_X < 0$ & $\rho_Y < 0$, denoted by *solid circles* (•), and mixtures, denoted by *asterisks* (∗).*Right* (**b**): a contour map of the joint relationship between $\hat{n}^*_{XY}$ and ρ_X and ρ_Y

for a pair of variables, it provides a reasonably good description of the positive only and negative only situations, and a slightly poorer description of mixed situations (see Fig. 5.5a). Furthermore, the mixture situations are the ones that tend to result in dramatically larger n^* values (see Fig. 5.5b).

The counterintuitive feature of SA more clearly revealed through spatial autoregression model specifications is that effective sample size, n^*, exceeds actual sample size, n, for negative SA. In other words, more information is contained in the geographic sample than would be contained in n iid random sample values. Of note, however, is that negative SA rarely is encountered in practice.

5.5 Spatial Filtering Model Implications

Spatial filtering techniques (Getis, 1990, 1995; Griffith 2000a, 2003; Getis and Griffith, 2002) allow spatial analysts to employ traditional regression techniques while insuring that regression residuals behave according to the traditional model assumption of no SA in residuals. One spatial filtering method exploits an eigenfunction decomposition associated with the MC. A spatial filter (SF) is constructed from the eigenfunctions of the following modified geographic weights matrix that represents the configuration of areal units in the MC, and is used to capture the covariation among attribute values of one or more georeferenced random variables:

$$\left(\mathbf{I} - \mathbf{1}\mathbf{1}^{\mathrm{T}}/\mathrm{n}\right)\mathbf{C}\left(\mathbf{I} - \mathbf{1}\mathbf{1}^{\mathrm{T}}/\mathrm{n}\right), \tag{5.8}$$

where $(\mathbf{I} - \mathbf{1}\mathbf{1}^{\mathrm{T}}/\mathrm{n})$ is the projection matrix commonly found in conventional multivariate and regression analysis that centers the n-by-1 vector of attribute values. The eigenvectors of this modified form of matrix $\mathbf{C}$ are both orthogonal and uncorrelated (Griffith, 2000c).

Spatial filtering uses the geographic configuration information contained in expression (5.8) to partition georeferenced data into synthetic spatial variates, containing the SA, and synthetic aspatial variates that are free of SA. For two georeferenced attribute variables X and Y, this decomposition may be written, using matrix notation, as

$$\mathbf{Y} = \mu_{\mathrm{Y}}\mathbf{1} + \mathbf{E}_{\mathrm{c}}\boldsymbol{\beta}_{\mathrm{cY}} + \mathbf{E}_{\mathrm{uY}}\boldsymbol{\beta}_{\mathrm{uY}} + \boldsymbol{\varepsilon}_{\mathrm{Y}}, \tag{5.9a}$$

and

$$\mathbf{X} = \mu_{\mathrm{X}}\mathbf{1} + \mathbf{E}_{\mathrm{c}}\boldsymbol{\beta}_{\mathrm{cX}} + \mathbf{E}_{\mathrm{uX}}\boldsymbol{\beta}_{\mathrm{uX}} + \boldsymbol{\varepsilon}_{\mathrm{X}}, \tag{5.9b}$$

where $\mathbf{E}$ is an n-by-H matrix for X and an n-by-K matrix for Y (with H and K not necessarily equal) of selected eigenvectors, subscripts c and u respectively denote common and unique sets of eigenvectors, $\boldsymbol{\beta}$ is a vector of regression coefficients, and $\boldsymbol{\varepsilon}_{\mathrm{Y}}$ and $\boldsymbol{\varepsilon}_{\mathrm{X}}$ respectively are the iid $N(0, \sigma^2_{\varepsilon_j})$, j = X or Y, aspatial variates for variables X and Y. The linear combinations of eigenvectors are the SFs. These linear combinations can be constructed with stepwise regression selection procedures.

5.5.1 Correlation Coefficient Decomposition

Equations (5.9a) and (5.9b) allow a correlation coefficient to be decomposed into components associated with the eigenfunctions of expression (5.1). To do so, the product moment correlation coefficient formula can be rewritten so that its covariance numerator term becomes

$$(\mathbf{E}_c\boldsymbol{\beta}_{cX} + \mathbf{E}_{uX}\boldsymbol{\beta}_{uX} + \boldsymbol{\varepsilon}_X)^T(\mathbf{E}_c\boldsymbol{\beta}_{cY} + \mathbf{E}_{uY}\boldsymbol{\beta}_{uY} + \boldsymbol{\varepsilon}_Y),$$

and its two denominator standard deviation terms become

$$\sqrt{(\mathbf{E}_c\boldsymbol{\beta}_{cX} + \mathbf{E}_{uX}\boldsymbol{\beta}_{uX} + \boldsymbol{\varepsilon}_X)^T(\mathbf{E}_c\boldsymbol{\beta}_{cX} + \mathbf{E}_{uX}\boldsymbol{\beta}_{uX} + \boldsymbol{\varepsilon}_X)}$$
$$\text{and} \quad \sqrt{(\mathbf{E}_c\boldsymbol{\beta}_{cY} + \mathbf{E}_{uY}\boldsymbol{\beta}_{uY} + \boldsymbol{\varepsilon}_Y)^T(\mathbf{E}_c\boldsymbol{\beta}_{cY} + \mathbf{E}_{uY}\boldsymbol{\beta}_{uY} + \boldsymbol{\varepsilon}_Y)}.$$

The respective expected values associated with these terms render $\left[\frac{\sum_{c=1}^{K^*}\rho_{cX}\rho_{cY}}{\sqrt{1-R_X^2}\sqrt{1-R_Y^2}} + \rho_{\varepsilon X\varepsilon Y}\right]\sigma_{\varepsilon X}\sigma_{\varepsilon Y}$, $\sqrt{\frac{\sigma_{\varepsilon X}^2}{(1 - R_X^2)}}$, and $\sqrt{\frac{\sigma_{\varepsilon Y}^2}{(1 - R_Y^2)}}$, where ρ_{cX} and ρ_{cY} respectively are the correlations between common eigenvector c and variables X and Y, R_X^2 and R_Y^2 respectively are the SF multiple correlation coefficients for variables X and Y, $\sigma_{\varepsilon X}^2$ and $\sigma_{\varepsilon Y}^2$ respectively are the aspatial variate variances for variables X and Y, K^* is the number of common eigenvectors, and ρ_{exeY} is the correlation between the synthetic aspatial (i.e., SA free) variates. Accordingly, $r = \sum_{c=1}^{K^*}\rho_{cX}\rho_{cY} + \rho_{\varepsilon X\varepsilon Y}\sqrt{1 - R_X^2}\sqrt{1 - R_Y^2}$. This result indicates that the range of possible observed values spans [–1, 1], regardless of the SA free value of r. This outcome is illustrated in Fig. 5.6a for the case of no common eigenvectors. Figure 5.6b illustrates that the presence of common eigenvectors shifts the center of the scatterplot to $\sum_{c=1}^{K^*}\rho_{cX}\rho_{cY}$, and shrinks each range of observed correlation possibilities while spanning a subinterval of [–1, 1]. The general shapes of the graphs are the same, but with a shift in the interval [–1, 1] when $\sum_{c=1}^{K^*}\rho_{cX}\rho_{cY} \neq 0$ that causes some possibilities to be eliminated (i.e., the plot becomes truncated).

This decomposition reveals that SA can both inflate and deflate a correlation coefficient. If neither of the SFs contains unique eigenvectors, then $\sum_{c=1}^{K^*}\rho_{cX}\rho_{cY}$ can introduce considerable inflation. If no eigenvectors are common to the SFs, then SA deflates the correlation through the product term $\sqrt{1 - R_X^2}\sqrt{1 - R_Y^2}$. In practice, a mixture of these two effects occurs.

For illustrative purposes again consider the Scottish lip cancer and Murray smelter site soil pollution data sets, where SF decomposition results appear in

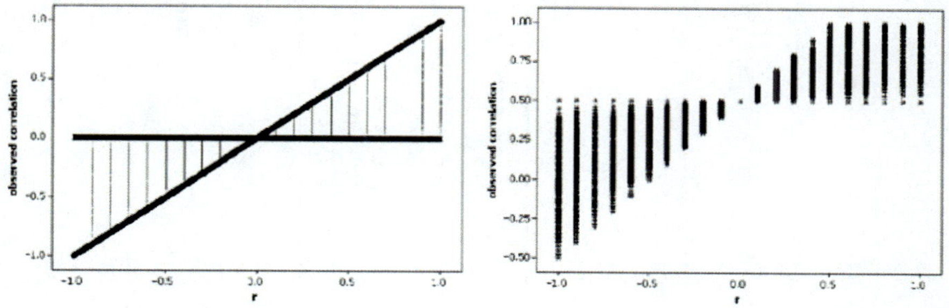

Fig. 5.6 Relationships between the spatially adjusted correlation coefficient, r, and the spatially unadjusted correlation coefficient. *Left* (**a**): the case of no common eigenvectors in the spatial filters. *Right* (**b**): the case of common eigenvectors in the spatial filter that account for 50% of the variance

Table 5.7. The correlation between the lip cancer SMR and the percentage of outdoor labor, for the Scottish lip cancer data set, is 0.54087. But latent SA has dramatically inflated this coefficient from a value of 0.16077. Each SF contains 5 eigenvectors, of which two are unique. The three common eigenvectors introduce considerable inflation, accounting for nearly 75% of the value of the observed correlation coefficient. The following are the components of this correlation coefficient:

$$0.16077\sqrt{0.42485 \times 0.48840} + 0.90768\sqrt{0.43487 \times 0.44569} + 0\sqrt{0.14028 \times 0.06591}$$
$$+0.28643\sqrt{0.14028 \times 0.48840} + (-0.04146)\sqrt{0.42485 \times 0.06591}.$$

The correlation between As and Pb contamination concentrations for the Murray superfund site data set is 0.74775. Here latent SA has modestly inflated this coefficient from a value of 0.64256. One SF contains 19, while the other contains 20, eigenvectors. Only 11 of these eigenvectors are common to both SFs, accounting for roughly 40% of the value of the observed correlation coefficient. The following are the components of this correlation coefficient:

$$0.64256\sqrt{0.58025 \times 0.57275} + 0.94627\sqrt{0.31781 \times 0.33324} + 0\sqrt{0.10194 \times 0.09401} +$$
$$0.16646\sqrt{0.10194 \times 0.57275} + 0.12480\sqrt{0.58025 \times 0.09401}.$$

In each empirical example, the roles of the common and unique SF components are revealed.

Results of a set of simulation experiments, each involving 10,000 replications, are summarized in Table 5.8. The estimated SFs were combined with the following random error terms: (1) bivariate normally distributed with mean, variance and correlation of the observed error terms; (2) permutation of the observed error term pairs; and, (3) resampling, with replacement, of the observed error term pairs (i.e., bootstrapping). The results confirm that: (1) the unique parts of SFs deflate the correlation coefficient ($\hat{\mu}_{r_{u\&\varepsilon}}$); (2) the common parts of SFs inflate the correlation

Table 5.7 Eigenvector spatial filter regression results using a 10% level of significance selection criterion

	Standard mortality ratio (Y) and percentage outdoor labor (X), for Scottish lip cancer (n = 56); $K^* = 3$, $K_X = K_Y = 2$		As and Pb concentration is soil samples, Murray superfund site (n = 253); $K^* = 11$, $K_X = 8$, $K_Y = 9$	
component	Transformed Y	Transformed X	Transformed Y	Transformed X
common eigenvectors	$R^2 = 0.43487$	$R^2 = 0.44569$	$R^2 = 0.31781$	$R^2 = 0.33324$
unique eigenvectors	$R^2 = 0.14028$	$R^2 = 0.06591$	$R^2 = 0.10194$	$R^2 = 0.09401$
all selected eigenvectors	$R^2 = 0.57515$	$R^2 = 0.51160$	$R^2 = 0.41975$	$R^2 = 0.42724$
residual MC	$z_{MC} \approx -0.01$	$z_{MC} \approx -0.37$	$z_{MC} \approx -0.29$	$z_{MC} \approx -1.45$
Shapiro-Wilk (S-W) statistic	0.950 (p = 0.023)	0.983 (p = 0.590)	0.996 (p = 0.833)	0.993 (p = 0.253)
MC for linear combination of eigenvectors	0.662	0.545	0.796	0.757

coefficient ($\hat{\mu}_{r_{c\&\varepsilon}}$); and, (3) the estimated aspatial correlation coefficient is reasonably well behaved ($\hat{\mu}_{r_\varepsilon}$ and $\hat{\sigma}_{r_\varepsilon}$). These results also reveal that: (1) the bootstrap standard errors for the aspatial correlation coefficient are closer to the standard theoretical value given by $\sqrt{\frac{1-r^2}{n-2}}$; (2) both tails of the sampling distributions tend to be heavy, which is the source of deviation from normality; (3) the mean of the sampling distribution of the spatially filtered correlation coefficient ($\hat{\mu}_{r_\varepsilon}$) appears to be unbiased; and, (4) the mean of the sampling distribution of the unfiltered correlation coefficient ($\hat{\mu}_r$) appears to be biased downward.

5.5.2 Variance Inflation

One well-known impact of positive SA is that it inflates the variance of a georeferenced variable. With regard to the aspatial variates $\boldsymbol{\varepsilon}_Y$ and $\boldsymbol{\varepsilon}_X$, this inflation is given by the standard multiple linear regression VIF result of $\frac{1}{1-R^2}$, yielding $\sigma_X^2 = \frac{\sigma_{\varepsilon X}^2}{(1-R_X^2)}$ and $\sigma_Y^2 = \frac{\sigma_{\varepsilon Y}^2}{(1-R_Y^2)}$.

The covariation between variables X and Y also can be rewritten in this VIF form (see Griffith and Zhang, 1999):

$$\sigma_{XY} = \frac{\rho_{\varepsilon X \varepsilon Y}\sigma_{\varepsilon X}\sigma_{\varepsilon Y}}{\sqrt{(1-R_X^2)}\sqrt{(1-R_Y^2)}}, \tag{5.10}$$

Therefore, the aspatial correlation itself is impacted by SA as follows:

$$r = \hat{\rho}_{\varepsilon X \varepsilon Y}\sqrt{(1-R_X^2)(1-R_Y^2)} \tag{5.11}$$

The correlation coefficient r is further modified by adding the common eigenvectors effect to expression (5.11). In other words, when two georeferenced random variables X and Y contain no SA, then $R_X^2 = R_Y^2 = 0$, and $r = \hat{\rho}_{\varepsilon X \varepsilon Y}$. In contrast, as the limiting, degenerate case of perfect positive SA (i.e., a constant) is approached, for either variable X or Y, expression (5.11) goes to 0.

5.6 Discussion

Results reviewed in this chapter emphasize that overlooking latent SA in georeferenced data can lead to a misinterpretation of conventional correlation coefficients, while acknowledging and accounting for SA can help furnish a better understanding of correlations among spatial processes. Equations (5.4a), (5.4b), and (5.7) illuminate how SA alters conventional degrees of freedom and sample size in a bivariate context. The mystifying feature of this result that needs to be better understood is how negative SA can inflate n*, as seen here with the autoregressive model

Table 5.8 Simulation and resampling experiment sampling distribution results for spatial filtering; 10,000 replications

Error source	$\hat{\mu}_{r_\varepsilon}$	$\hat{\sigma}_{r_\varepsilon}$	R-J	$\hat{\mu}_{r_{u\&\varepsilon}}$	$\hat{\sigma}_{r_{u\&\varepsilon}}$	$\hat{\mu}_{r_{c\&\varepsilon}}$	$\hat{\sigma}_{r_{c\&\varepsilon}}$	$\hat{\mu}_r$	$\hat{\sigma}_r$	R-J
Scottish lip cancer data ($r = 0.5409$)										
Bivariate normal	0.1612	0.1309	0.9996 (p=0.07)	0.1326	0.1310	0.5108	0.0901	0.4596	0.0936	0.9975 (p=<0.01)
Permuta-tion	0.1608	0	NA	0.1308	0.0748	0.5244	0.0579	0.4701	0.0706	0.9964 (p<0.01)
Boot-strap	0.1624	0.1357	0.9999 (p>0.10)	0.1305	0.1322	0.5315	0.0931	0.4755	0.0943	0.9977 (p<0.01)
Murray superfund site data ($r = 0.7478$)										
Bivariate normal	0.6417	0.0376	0.9981 (p<0.01)	0.5506	0.0433	0.7399	0.0277	0.6693	0.0326	0.9984 (p<0.01)
Permuta-tion	0.6426	0	NA	0.5493	0.0244	0.7519	0.0155	0.6783	0.0236	0.9992 (p=0.02)
Boot-strap	0.6411	0.0469	0.9973 (p<0.01)	0.5476	0.0494	0.7516	0.0323	0.6777	0.0353	0.9979 (p<0.01)

NOTE: NA denotes "not applicable"; subscripts "u& ε " and "c& ε " respectively denote addition of the random error term to the unique and to the common eigenvector part of the spatial filter specifications. Interestingly, this finding is not corroborated with spatial filtering results. Another puzzle that needs to be solved is a reconciliation between n^* values obtained with semivariogram and spatial autoregressive model specifications.

Although varying the range, via r_d, of a semivariogram model allows variation in observed correlation coefficients to be monitored, the manner in which the nature and degree of SA affects correlation coefficients is better illuminated by studying spatial autoregressive model specification results. By doing so, variance and covariance inflation that is attributable to the presence of SA can be linked explicitly to the autoregressive parameters ρ_X and ρ_Y, with a one-to-one correspondence between each of these parameters and SA.

Spatial filtering allows particular map patterns, reflecting specific natures and degrees of SA, to be identified and connected to inflation and deflation of correlation coefficients. This specific dissection furnishes an even better understanding of the change in sampling distribution variance for r (see Fig. 5.3) induced by SA.

All in all, modern spatial statistics supplies considerable understanding of the nuances and idiosyncrasies introduced into correlation coefficients by SA.

Chapter 6
Spatially Structured Random Effects: A Comparison of Three Popular Specifications

6.1 Introduction

Random effects models are increasing in popularity (see, for example, Demidenko, 2004), partially because they have become estimable. One common specification is for the intercept term to be cast as a random effects, resulting in it representing variability about the conventional single-value, constant mean. The role of a random effects in this context may be twofold: (1) supporting inferences beyond the specific fixed values of covariates employed in an analysis; and, (2) accounting for correlation in a non-random sample of data being analyzed. Including a random effects term moves a frequentist analysis a bit closer to a Bayesian analysis, given that, for instance, the intercept term becomes a random variable rather than being a constant, and has a prior probability distribution (usually normal) attached to it. Nevertheless, a *bone fide* Bayesian analysis would have a random variable for each of the n intercept term components comprising such a random effects, maintaining some degree of differentiation here between the frequentist and Bayesian approaches.

Georeferenced data—data that are tagged to the Earth's surface—contain spatial autocorrelation (SA; i.e., nearby values are more related to one another than are distant values), resulting in the additional feature of spatial structuring of a random effects. The linear mixed model (LMM) procedure in SAS employs semivariogram models (see Cressie, 1991), and WinBUGS (see Cowles, 2004) employs a conditional autoregressive (CAR) model, in order to incorporate this type of structuring. A spatial filter (SF; Griffith, 2000, 2002, 2004) offers an appealing alternative to either of these formulations, and also can be used with the generalized linear mixed model (GLMM) procedure of SAS or with WinBUGS. The purpose of this chapter is to present an assessment of the utility of these three forms of spatial structuring by summarizing selected empirically-based comparisons between them.

6.2 Modeling Spatial Structure

Various methods for modeling spatial structure have emerged. The first to be developed was autoregression, which was popularized by Ord (1975), after Cliff and Ord (1973). A more comprehensive discussion of the full family of auto- models can be

D.A. Griffith, J.H.P. Paelinck, *Non-standard Spatial Statistics and Spatial Econometrics*, Advances in Geographic Information Science 1,
DOI 10.1007/978-3-642-16043-1_6, © Springer-Verlag Berlin Heidelberg 2011

found in Besag (1974), who eventually developed the CAR specification, in both its proper and improper (ICAR[1]) forms, for hierarchical modeling purposes; this is the version available in WinBUGS. The initial autoregressive model specification contains Y on both sides of a regression equation, indicating that observed values themselves are either directly (i.e., the autoregressive response or spatial lag model) or indirectly (i.e., the simultaneous autoregressive or autoregressive errors model) correlated. The hierarchical version embeds SA into a parameter, which is feasible because the parameter is a variable, rather than having observations directly correlated; it is a feature of the upper- rather than the lower-level of a two-tier hierarchical model. More precisely, its common implementation in WinBUGS is with a random effects. Here spatial structure most often is portrayed by the way a surface is partitioned into areal units, resulting in it being topological in nature.

In addition, spatial structure can be described with a semivariogram model. This approach differs from the autoregressive one by focusing on the n-by-n inter-observation SA covariance matrix, whereas autoregression focuses on the inverse of this matrix. Modeling begins by calculating squared differences between data values, and distances between their affiliated location points. Next, these squared data value differences are aggregated according to bins established with their separating distances, and then averaged. A semivariogram plot is constructed by graphing the pairs of averaged grouped squared differences, divided by 2, on the vertical axis, and their corresponding average grouped distance separation on the horizontal axis. Several dozen valid semivariogram models are available to describe trend in this scatterplot. A number of them have been implemented in SAS, including ones that allow for anisotropy (i.e., both direction and separation distance play an important role in dependency structure). Here spatial structure most often is portrayed by interpoint distances between locations, resulting in it being metric in nature.

Finally, spatial structure can be represented by the eigenfunctions of the aforementioned n-by-n covariance matrix. Because this matrix is symmetric, and one approach works with it directly while the other works with its inverse, conceptually the eigenvectors are the same in both cases. In practice, the eigenvectors differ somewhat because the autoregressive approach tends to be topologically based (i.e., determination of neighborhood structure is by areal unit shared common boundaries) while the semivariogram approach is distance based (i.e., determination of neighborhood structure is by distance between areal unit centroids). Nevertheless, Griffith and Peres-Neto (2006) find that both approaches render equivalent eigenfunction depictions. Spatial filtering seeks to partition a response variable into two synthetic variates: a spatial structure component, and a nonspatial variate that is free of spatial dependence. Griffith (e.g., 2000) proposes a transformation procedure that depends on the eigenfunctions of matrix $(\mathbf{I} - \mathbf{11}^T/n)\mathbf{C}(\mathbf{I} - \mathbf{11}^T/n)$—where $\mathbf{I}$ denotes

[1]ICAR denotes an intrinsic version—a generalization to support certain types of non-stationarity—of the conditional autoregressive (CAR) model in which the variance-covariance matrix is positive semi-definite rather than positive-definite, and has a single parameter to control both the strength of and total amount of spatial dependence.

the identity matrix, **1** is an n-by-1 vector of ones, **C** denotes a geographic connectivity matrix (e.g., $c_{ij} = 1$ if areal units i and j are neighbors, and $c_{ij} = 0$ otherwise), and T denotes matrix transpose—a term appearing in the numerator of the Moran Coefficient (MC) SA index, and is based on the following theorem (Griffith, 2003):

> The first eigenvector, say $\mathbf{E}_1$, is the set of numerical values that has the largest MC achievable by any set of real values for the spatial arrangement defined by a geographic connectivity matrix **C**. The second eigenvector is the set of values that has the largest achievable MC by any set of real values that is uncorrelated with $\mathbf{E}_1$. The third eigenvector is the third such set of values. And so on. This sequential construction of eigenvectors continues through $\mathbf{E}_n$, the set of values that has the largest negative MC achievable by any set of real values that is uncorrelated with the preceding (n–1) eigenvectors.

As such, Griffith (2000a) argues that these eigenvectors furnish distinct map pattern descriptions of latent SA in georeferenced variables. Each eigenvector's corresponding MC-indexed level of SA is given by $\frac{n}{\mathbf{1}^T\mathbf{C}\mathbf{1}}$ times its eigenvalue (Tiefelsdorf and Boots, 1995). The SF is constructed by using judiciously selected eigenvectors as regressors (e.g., selected with a stepwise regression routine), which results in SA being accounted for by a linear combination of mutually orthogonal and uncorrelated eigenvectors.

6.3 Linear Mixed Models

Consider the geographic distribution of elevation across the island of Puerto Rico (see Fig. 6.1). Averages ($\overline{elev}$) for the island's 73 municipalities (outlined on the map in Fig. 6.1) were calculated for modeling purposes. A Box-Cox type of power transformation results in

$$(\mathrm{LN}(\overline{elev} + 17.5) \sim \mathrm{N}, \text{ or normally distributed,}$$

where LN denotes the natural logarithm, and the probability of the accompanying Shapiro-Wilk diagnostic statistic is P(S-W) = 0.4758 (a substantial increase from

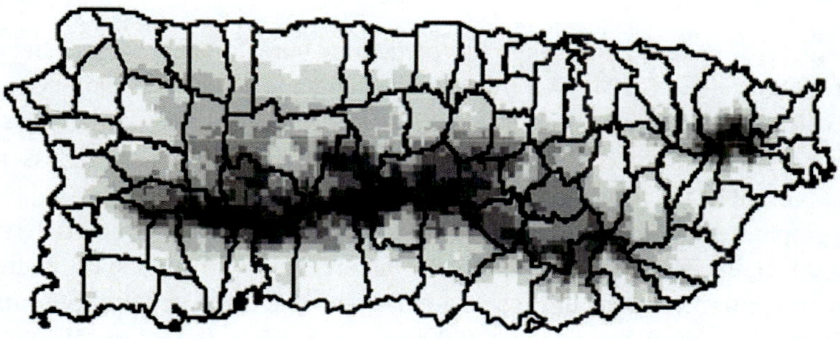

Fig. 6.1 The geographic distribution of elevation across the island of Puerto Rico, from a USGS DEM containing 87,358,136 points. Darkness of gray scale is directly proportional to elevation

< 0.0001 for the raw data). Moderate positive SA is exhibited by these aggregated data: MC = 0.517; and, Geary Ratio (GR) = 0.621.

The following linear SF, linked to a normal probability model, selected from the 18 candidate (from a set of 72 possible) eigenvectors representing at least weak positive SA (i.e., $MC/MC_{max} > 0.25$, where MC_{max} denotes the maximum possible MC value), was constructed for the transformed version of the response variable, $LN(\overline{elev} + 17.5)$:

$$1.71735\mathbf{E}_2 - 1.14043\mathbf{E}_3 + 2.31266\mathbf{E}_4 + 1.19040\mathbf{E}_6 - 1.48295\mathbf{E}_7 + 1.95080\mathbf{E}_{10} + 2.28604\mathbf{E}_{12} - 1.22260\mathbf{E}_{13} - 0.92428\mathbf{E}_{14} + 0.72093\mathbf{E}_{15} + 1.12531\mathbf{E}_{16} - 1.49864\mathbf{E}_{18},$$

for which MC = 0.655. This SF accounts for roughly 75% of the variance in $LN(\overline{elev} + 17.5)$, yielding residuals for which $z_{MC} = 2.36$, and P(S-W) = 0.1891; it also accounts for nearly 75% of the variance in the back-transformed version of $\overline{elev}$. Although marginally significant positive SA remains in the residuals, no evidence could be uncovered indicating the presence of hidden negative SA (see Griffith, 2006a), based upon the 35 candidate eigenvectors representing at least weak negative SA.

A linear model description of $LN(\overline{elev} + 17.5)$ identifies, after centering of each coordinate axis, the north-south (V), the squared north-south (V^2), the squared east-west (U^2), and the crossproduct north-south times east-west (UV) covariates as being statistically significant (accounting for the 3-dimensional elongated mound shape generated by the island's east-west oriented mountain chain). A LMM description of $LN(\overline{elev} + 17.5)$ as a quadratic function of the geocoding coordinates, which has spatial structuring of a random effects intercept term induced with a semi-variogram model, yields the results reported in Table 6.1. Retaining the two squared and the cross-product geocoding terms—the linear north-south term becomes non-significant when a random effects term is introduced—results in the following SF for the linear regression analysis:

$$1.11020\mathbf{E}_1 - 0.75483\mathbf{E}_3 + 2.18301\mathbf{E}_4 - 1.03904\mathbf{E}_5 - 1.58150\mathbf{E}_7 - 0.93133\mathbf{E}_{11} + 1.15186\mathbf{E}_{12} + 1.15631\mathbf{E}_{15} + 0.95847\mathbf{E}_{16} - 0.88066\mathbf{E}_{17} - 0.89963\mathbf{E}_{18},$$

which shares 7 eigenvectors with the preceding pure SF expression, and has a MC of 0.688 (GR = 0.294). All five semivariogram specification estimations (see Table 6.1) based upon inclusion of this SF yield a nugget of 0, a spatial correlation of 0, and a residual variance component of 0.0553; in other words, the SF essentially accounts for all of the SA in these data.

Geographic distributions of the unstructured and spatially structured random effects, which can be written as the sum of the preceding SF and the estimated random intercept term, appear in Fig. 6.2a and 6.2b. The estimated random intercept has a mean of nearly 0, a variance of 0.00005, a P(S-W) of 0.6711, a MC of –0.028 (GR = 0.937), and is almost perfectly uncorrelated with the 14 covariates contained in the mean response equation. The spatially structured random effects has a mean

Table 6.1 SAS PROC MIXED summary results for a LMM quadratic gradient description of LN ($\overline{\text{elev}}$ + 17.5) across Puerto Rico, by municipality

Semivariogram model	None	Spherical	Expo-nential	Gaussian	Power	Bessel[a]
Variance (nugget)	–	0.0331	0.2151	0.2210	0.2514	0.2450
Spatial correlation	–	< 0.0001	0.7643	0.5730	0.2702	0.6089
Residual	0.2171	0.1691	< 0.0001	0.0253	< 0.0001	0.0057
b_0	6.0799***	6.0799***	6.1569***	6.2009***	6.1569***	6.2022***
b_{u^2}	–0.3491***	–0.3491***	–0.4626***	–0.4863***	–0.4626***	–0.4861***
b_{uv}	–0.2626***	–0.2626***	–0.2539**	–0.2546**	–0.2539**	–0.2566**
b_v	–0.2695***	–0.2695***	–0.1142	–0.1684	–0.1142	–0.1678
b_{v^2}	–0.5271***	–0.5271***	–0.5289***	–0.5607***	–0.5289***	–0.5685***

***, **, * respectively denote statistical significance at a 1, a 5 and a 10% level
[a]modified, of 2nd kind
U denotes the east-west geocoding coordinate
V denotes the north-south geocoding coordinate

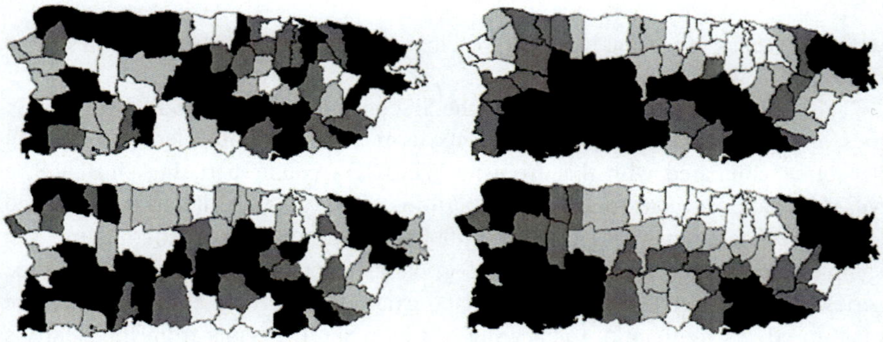

Fig. 6.2 Geographic distributions of unstructured (*left*) and spatially structured (*right*) random effects. Darkness of gray scale is directly proportional to the magnitude of random effects values. *Top left* (**a**): a quantile map of the normal approximation random effects from SAS.*Top right* (**b**): a quantile map of the normal approximation spatially structured random effects from SAS based upon a SF. *Bottom left* (**c**): A quantile map of the normal approximation mean random effects from WinBUGS.*Bottom right* (**d**): a quantile map of the normal approximation mean spatially structured random effects from WinBUGS based upon an ICAR model

of almost exactly 0, a variance of 0.22477, and essentially the same level of SA as displayed by the SF itself.

The hierarchical modeling available via GeoBUGS, in WinBUGS, allows a random effects term to be spatially structured with a CAR model. Because explicit estimation of the autoregressive parameter is excessively numerically intensive, following common practice (see Thomas et al., 2004), an ICAR model was estimated. This specification sets the autoregressive parameter to its maximum value, and then estimates both spatially structured and unstructured random effects terms; their relative proportion of variance indicates the importance of each term. A Markov chain Monte Carlo (MCMC) series (i.e., chain) was generated with a burn-in period of 50,000 iterations, followed by 550,000 iterations of which every 1,100th iteration was retained, yielding 500 sample values. The detected SA accounts for 99.0% of the relative variance in the combined random effects term, signifying the presence of marked positive spatial dependence. All chains display good behavior, as is illustrated by the time series plot and correlogram for a selected example parameter chain (the average total random effects by iteration) appearing in Fig. 6.3a, b. Normal priors with sizeable variance were placed on all regression coefficient parameters; gamma distribution priors were posited for the variance terms. The resulting mean of the total random effects is –0.00011, with a variance of 0.38629, a P(S-W) of 0.1263, and a MC of 0.651 (GR = 0.237). Although the geographic distributions differ to some degree (compare Fig. 6.2b, d), the level of SA in the spatial structuring with an ICAR and a SF model are roughly the same. The regression equation predicting the SAS term ($\hat{\xi}_{SAS}$) from the ICAR term ($\hat{\xi}_{GeoBUGS}$) is

$$\hat{\hat{\xi}}_{SAS} = 0.00006 + 0.57926\,\hat{\xi}_{GeoBUGS},\ R^2 = 0.5767,$$

which indicates a reasonable, but not close, correspondence between these two random effects.

The model specification employing the SF constructed with SAS PROC NLMIX to spatially structure the random effects term also was implemented, and then its output compared with that from its WinBUGS counterpart. In WinBUGS, a MCMC chain was constructed with a burn-in period of 25,000 iterations, followed by 500,000 iterations of which every 1,000th iteration was retained, yielding 500 sample values. Again, all chains display good behavior, as is illustrated by the time series plot and correlogram for a selected example parameter chain (the average random effects by iteration) appearing in Fig. 6.3c, d. Results from this analysis compare very favorably with those from the corresponding SAS analysis: parameter estimates are almost identical, and standard errors are not markedly different

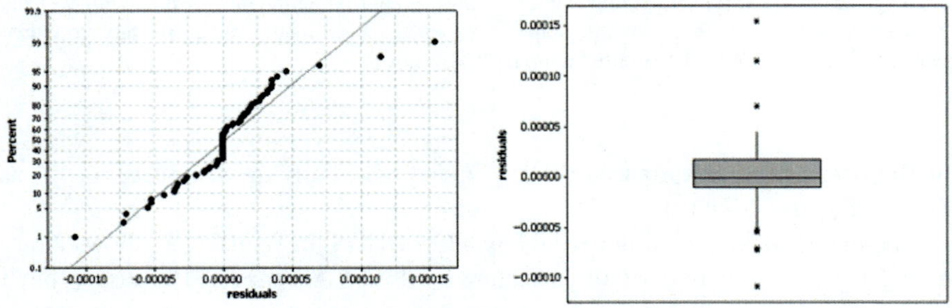

Fig. 6.3 Graphical diagnostics of residuals for the GLMM estimated with SAS. *Left* (**a**): normal quantile plot. *Right* (**b**): boxplot

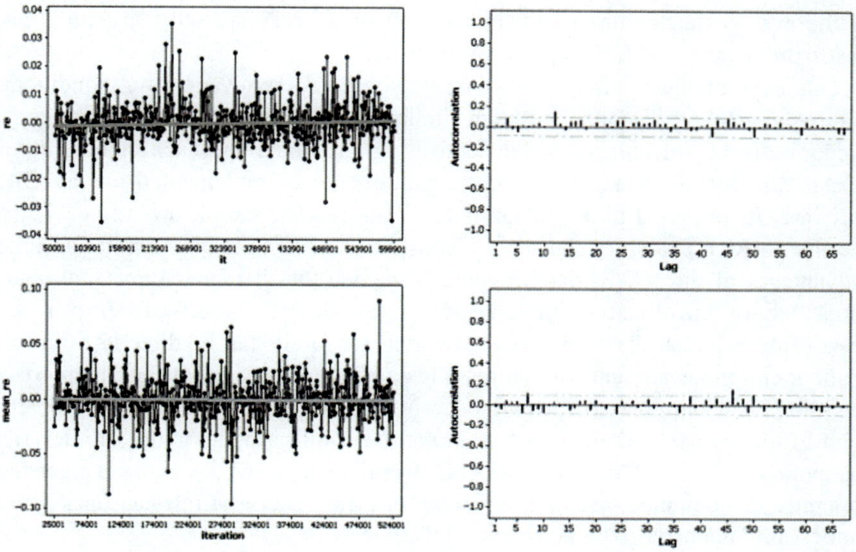

Fig. 6.4 A average random effects term example MCMC chain from a WinBUGS run in which the random effects intercept was spatially structured. *Top left* (**a**): a time series plot from an ICAR model with a trend line (*gray*) superimposed. *Top right* (**b**): a correlogram from an ICAR model with a 95% confidence interval (*gray*) superimposed. *Bottom left* (**c**): a time series plot from a SF model with a trend line (*gray*) superimposed. *Bottom right* (**d**): a correlogram from a SF model with a 95% confidence interval (*gray*) superimposed.

(WinBUGS standard errors are –0.6–11.8% larger)—these differences may well diminish by lengthening the MCMC chain (see Fig. 6.4). Once more, normal priors with sizeable variance were placed on all regression coefficient parameters; gamma distribution priors were posited for the variance terms. Summary statistics for the SAS- and WinBUGS-estimated spatially structured random effects include

Statistic	SAS	WinBUGS
Mean	0.00000	−0.00021
Variance	0.22477	0.23790
P(S-W)	0.6701	0.6031

The regression equation predicting the SAS term ($\hat{\xi}_{SAS}$) from the WinBUGS term ($\hat{\xi}_{WinBUGS}$) is

$$\hat{\hat{\xi}}_{SAS} = 0.00020 + 0.94182\hat{\xi}_{WinBUGS}, R^2 = 0.9388,$$

which indicates a very close correspondence between these two random effects, but with more variability being allocated in WinBUGS. The geographic distributions of the average unstructured and the sum of the average spatially structured and unstructured random effects appear in Fig. 6.2c, d.

One experimental finding from this section is that semivariogram, conditional autoregressive and SF models furnish similar-to-equivalent spatial structuring of a random effects. Advantages of the semivariogram model approach include that it relates directly to geostatistical theory, and already is implemented in, say, SAS software. Its principal disadvantage is that even modest sample sizes (e.g., 73 for the Puerto Rico example) can require considerable amounts of computation time. Advantages of the ICAR model approach include that it relates directly to auto-model theory, and already is implemented in, say, GeoBUGS software. Its principal disadvantage is that the autoregressive parameter rarely can be directly estimated without enormous amounts of computer resources. Advantages of the SF approach include that it requires no special software; for example, it can be used directly with PROC NLMIXED for which a normal probability model is used to describe data, or equivalently with PROC MIXED. Its principal drawback is that n eigenvectors must be computed, and then screened for identification of relevant ones—both numerically intensive, but one-time exercises.

One substantive finding from this section is that inclusion of spatial structuring results in the regression coefficient of the north-south (i.e., V) coordinate becoming nonsignificant. Meanwhile, comparative results reported in Table 6.2 reveal that the SF specification renders comparable results for both a frequentist and a Bayesian approach that utilizes noninformative priors; this situation is similar to the matching case described by Kass and Wasserman (1996). Similarly, the semivariogram and ICAR approaches render comparable results, which have noticeably larger standard errors for their parameter estimates than are obtained with their SF counterparts.

Table 6.2 Comparative parameter estimates for a LMM quadratic gradient description of LN($\overline{\text{elev}}$ + 17.5) across Puerto Rico, by municipality

	SAS semivariogram (Bessel) model		SAS SF		GeoBUGS-ICAR (100 weeded replications)		WinBUGS-SF	
Parameter	Estimate	se	Estimate	se	Estimate	se	Estimate	se
b_0	6.1906	0.2873	6.1101	0.05467	6.5175	0.1683	6.1101	0.0611
b_{u^2}	−0.5048	0.1219	−0.3881	0.03148	−0.7507	0.1525	−0.3878	0.0347
b_{uv}	−0.2229	0.1227	−0.2939	0.03018	−0.2031	0.1006	−0.2920	0.0300
b_{v^2}	−0.5314	0.1247	−0.5193	0.03217	−0.5683	0.0607	−0.5190	0.0365
var	0.0055	0.0192	0	–	0.0049	0.0068	0.0305	0.0240
varure	0.2856	0.0912	0.0001	–	0.0047	0.0066	0.0318	0.0248
varssre	0.7205	0.2209	0.0282	–	0.4854	0.0925	0.0301	–

NOTE: varure denotes the variance of the unstructured, and varssre denotes the variance of the spatially structured, random effects
U denotes the east-west geocoding coordinate
V denotes the north-south geocoding coordinate

6.4 Generalized Linear Mixed Models

Although LMMs can be implemented as a special case of GLMMs, these latter specifications enable non-normal probability models to be employed in an analysis, too. Consider the 1973/74 geographic distribution of sugar cane harvest area density (SC74/A) across Puerto Rico, by municipality. This ratio of areas can be treated as an empirical probability [i.e., the probability of randomly selecting a cuerdas (0.9712 acres) in a municipality and having it yielding sugar cane]. As such, the adjusted log-odds ratio transformation for municipality i given by $LN\left(\frac{SC74_i/A_i + 0.01}{1 - 0.47 - (SC74_i/A_i + 0.01)}\right)$ results in values that more closely align with a normal frequency distribution (i.e., S-W increases from 0.79 to 0.91); but, the relatively large number of 0s (e.g., 18 of 73) prevents even this transformed variable from achieving an ideal alignment.[2] This transformation moves the zero values slightly above 0, and shrinks the range of values by nearly half in order to better center the set of empirical probabilities (i.e., transforming the distribution to one better resembling a binomial with $p = 0.5$, which tends to better mimic a bell-shaped curve). In other words, a normal approximation essentially fails in this case for the full set of data, although not for the nonzero subset of data. A better model specification would be to employ a binomial distribution, since $100 \times SC74/A$ is a percentage (when both the numerator and the denominator are measured in the same units, such as cuerdas).

The raw harvest density yields MC = 0.395 and GR = 0.477; the normal approximation yields MC = 0.519 and GR = 0.469. Consequently, moderate positive SA is detected in these data. The nature of sugar cane production technology tended to restrict it to the flatter, coastal lowlands; accordingly, one ecological covariate for predicting it is elevation—the variable addressed in the preceding section. A conventional binomial SF generalized linear model description of these data, where the eigenvectors were selected with a stepwise logistic regression procedure, renders

$$LN[\hat{p}_i/(1 - \hat{p}_i)] = -1.6164 - 0.0082\overline{elev} + 4.1761\mathbf{E}_1 + 3.2220\mathbf{E}_4,$$

which is accompanied by considerable overdispersion (the estimated scale parameter is 28.7), a pseudo-R^2 value of roughly 0.60, a SF (i.e., $4.1761\mathbf{E}_1 + 3.2220\mathbf{E}_4$) with MC = 1.005 and GR = 0.139 (representing markedly strong positive SA), and residuals for which MC = 0.087 and GR = 0.728. Of note is that much of the small amount of the detected residual SA is attributable to the municipalities having 0 sugar cane production (e.g., MC decreases to 0.016), even though an indicator variable (I_0) differentiating these from the other municipalities is a nonsignificant covariate (e.g., this covariate accounts for only about 2.5% of the variation in the empirical probabilities, and is not retained after selection by the stepwise regression procedure used to construct the preceding SF).

---

[2] The municipalities with no sugar cane harvest comprise the San Juan metropolitan region and the interior highlands.

Employing SAS PROC NLMIXED, and constraining 0 harvest to be exactly 0, yields

$$\hat{\mathbf{p}}_i = (1 - I_{0,i}) \frac{\exp\left[-1.2867 + \hat{\xi} - 0.0111\overline{elev} + 3.064\boldsymbol{E}_1 + 3.2182\boldsymbol{E}_4\right]}{1 + \exp\left[-1.2867 + \hat{\xi}_1 10.0111\overline{elev} + 3.0646\boldsymbol{E}_1 + 3.2182\boldsymbol{E}4\right]},$$

where exp denotes the base of the natural logarithm, and $\hat{\xi}_i$ denotes the random effects term for municipality i. Of note is that the common practice of centering the covariates strictly for estimation purposes is employed here; by construction, the eigenvectors have a mean of zero. Also of note is that the SAS quadrature algorithm used to estimate the random effects had difficulty converging. The revised SF (i.e., $3.0646\mathbf{E}_1 + 3.2182\mathbf{E}_4$) has MC = 0.967 and GR = 0.158, and hence still represents markedly strong positive SA. A map portraying the geographic distribution of this estimated random effects term appears in Fig. 6.2a. Descriptive statistics for this estimated random effects variable are reported in Table 6.3. Moreover, this estimated random effects term has reasonably good, but not ideal, statistical properties. The spatially structured random effects obtained by adding the SF and this random effects term has MC = 0.356 and GR = 0.739, levels of SA similar to those

Table 6.3 Summary measures for the estimated SF GLMM random effects term

	SAS NLMIXED		WinBUGS (100 weeded replications)			
	(SF)		SF		ICAR	
statistic	Estimate	Standard error	Estimate	Standard error	Estimate	Standard error
b_0	−1.2867	0.2624	−1.3114	0.2852	−1.534	0.2419
$b_{elev - \overline{elev}}$	−0.0111	0.0013	−0.0110	0.0014	−0.0100	0.0013
b_{E_1}	3.0646	1.1559	3.0600	1.3632	***	
b_{E_4}	3.2182	1.3433	3.0116	1.4256	***	
$\hat{\mu}_{\xi}$	0.0015		0.0054		0.0066	
$\hat{\sigma}^2_{\xi}$	0.7045		0.7144		0.3727	
$\hat{\sigma}^2_{\xi+SS}$	0.9787		0.9783		0.9583	
P(S-W)	<0.0001		< 0.0001		< 0.0001	
MC_{ss}	0.967		0.975		0.787	
GR_{ss}	0.158		0.154		0.177	
MC_{ξ}	0.119		0.132		0.036	
GR_{ξ}	1.045		1.000		1.129	
$MC_{SS+\hat{\xi}}$	0.356		0.357		0.388	
$GR_{SS+\hat{\xi}}$	0.739		0.739		0.696	
$r_{\hat{\xi},\overline{elev}}$	0.001		0.001		0.011	
$r_{\hat{\xi},E_1}$	−0.001		−0.009		***	
$r_{\hat{\xi},E_4}$	0.001		0.022		***	

NOTE: ss denotes “spatially structured”

measured in p; its geographic distribution appears in Fig. 6.2b. Finally, p and $\hat{\mathbf{p}}$ are almost identical (e.g., perfect agreement of at least the first three digits to the right of the decimal point), with their residuals having MC = 0.119 and GR = 0.693; the inconsistency here primarily is attributable to the presence of two extreme outliers, one at each end of the distribution of values (see Fig. 6.3). In other words, the random effects term essentially is equivalent to the residual term for the conventional binomial model specification (because repeated measures are not included).

Implementation of this SF model in WinBUGS resulted in the software being unstable (e.g., repeatedly crashing); a successful execution required 840,000 MCMC iterations, of which 20,000 were used as a burn-in period, with every 8,000th outcome retained, yielding 100 values for estimation purposes; of note is that 16 iterations failed during generation of the obtained MCMC chain. As in the LMM analysis, normal priors with sizeable variance were placed on all regression coefficient parameters; gamma distribution priors were posited for the variance terms. The resulting MCMC graphics are very similar to those appearing in Table 6.4. A map portraying the geographic distribution of the arithmetic mean of this estimated random effects term appears in Fig. 6.5c. Descriptive statistics for this average estimated random effects variable also are reported in Table 6.3. As before, the associated, p and $\overline{\mathrm{p}}$ are almost identical, with their residuals having MC = 0.023 and GR = 0.706; again the random effects term essentially is equivalent to the residual term for the conventional binomial model specification. This estimated random effects term has reasonably good, but not ideal, statistical properties, deviating more from a bell-shaped curve but having a mean closer to 0 and slightly less correlation with covariates than its SAS PROC NLMIXED counterpart. Nevertheless, Fig. 6.6 reveals that it is very similar to its SAS counterpart. In both cases, the spatially structured random effects contain roughly the same level of SA.

Successful execution of an implementation of an ICAR model in GeoBUGS required 860,000 MCMC iterations, of which 40,000 were used as a burn-in period, with every 8,000th outcome retained, yielding 100 values for estimation purposes; of note is that 15 iterations failed during this estimation exercise, and that slight but statistically significant first-order serial correlation remained in the two random effects variance estimates at this point in the MCMC chain (i.e., the sample collected thus far). Again, normal priors with sizeable variance were placed on all regression coefficient parameters; gamma distribution priors were posited for the variance terms. Once more, the resulting MCMC graphics are very similar to those appearing in Table 6.4. A map portraying the geographic distribution of the arithmetic mean of this estimated random effects term appears in Fig. 6.5e. Descriptive statistics for this average estimated random effects variable also are reported in Table 6.3. Again, the associated, p and $\overline{\mathrm{p}}$ are almost identical, with their residuals having MC = –0.065 and GR = 0.831; as before, the random effects term essentially is equivalent to the residual term for the conventional binomial model specification. This estimated random effects term has reasonably good, but not ideal, statistical properties, more closely resembling its WinBUGS-generated SF than its SAS PROC NLMIXED counterpart (see Fig. 6.6). In all three cases, the spatially structured random effects contains roughly the same level of SA. More precisely for the ICAR model, the

Table 6.4 Individual GLMM estimation results for each Puerto Rican sugar cane crop year

		Raw per-centages		# 0 s	Point-in-time estimation				
Crop year	Individual SF eigen-vector #s	MC	GR		-a	$b_{\overline{elev} - \overline{\overline{elev}}}$	$\hat{\mu}_\xi$	$\hat{\sigma}^2_\xi$	P(S-W)
1958/59	4,6,15,22, 24	0.320	0.638	0	0.5650	0.0070	0.0014	1.0594	<0.0001
1959/60	1,4,15,24	0.333	0.617	0	0.6324	0.0072	0.0016	1.1546	<0.0001
1960/61	1,4,6,15,24	0.358	0.587	0	0.6210	0.0074	0.0018	1.1981	<0.0001
1961/62	1,4,5,8,12,15,22,24	0.387	0.540	1	0.5639	0.0083	0.0018	1.0412	0.0002
1962/63	1,4,,,8,12,	0.416	0.518	2	0.7528	0.0075	0.0017	0.9814	<0.0001
1963/64	15,,24	0.423	0.514	2	0.8084	0.0075	0.0020	1.0766	<0.0001
1964/65	1,3,4,5,6,8,12,15,22,24	0.461	0.472	2	0.8503	0.0078	0.0022	1.1196	<0.0001
1965/66	1,4,6,24	0.484	0.458	3	1.1959	0.0064	0.0020	1.0135	0.0055
1966/67		0.490	0.454	4	1.3786	0.0060	0.0021	1.1138	0.0027
1967/68		0.474	0.434	6	1.7354	0.0050	0.0018	1.0856	0.0017
1968/69	1,3,4,5,6,8, 12,15,24	0.539	0.401	7	1.7028	0.0078	0.0031	0.8307	0.0572
1969/70	1,4	0.474	0.443	10	1.3680	0.0079	0.0017	1.4294	<0.0001
1970/71		0.430	0.466	14	1.2420	0.0093	0.0015	0.6644	0.0001
1971/72		0.432	0.513	17	1.0903	0.0103	0.0016	0.5211	0.0065
1972/73	none	0.429	0.482	18	1.5514	0.0088	0.0016	1.0744	<0.0001
1973/74	1,4	0.395	0.477	18	1.2867	0.0111	0.0015	0.7045	<0.0001

a denotes the intercept term.
Underlined eigenvector numbers denote eigenvectors that become nonsignificant, at the 10% level, in the presence of a random effects term.
Bold crop years denote the set of points in time used to construct a space-time dataset.

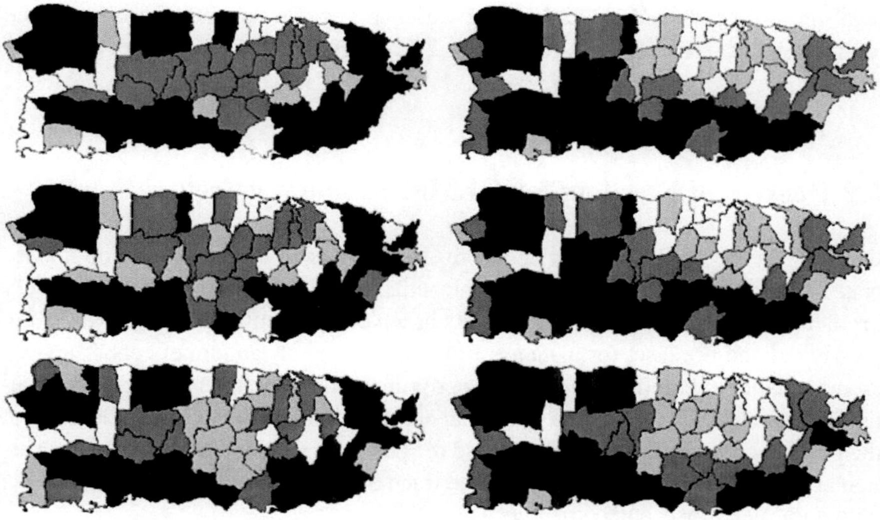

Fig. 6.5 Quantile maps of geographic distributions of unstructured (US; *left*) and spatially structured (SS; *right*) random effects for a binomial model. Darkness of gray scale is directly proportional to the magnitude of random effects values. *Top left* (**a**): from SAS. *Top right* (**b**): SS random effects from SAS based upon a SF model. *Middle left* (**c**): average US random effects from WinBUGS. *Middle right* (**d**): average SS random effects from WinBUGS based upon a SF model. *Bottom left* (**e**): average US random effects from WinBUGS. *Bottom right* (**f**): average SS random effects from WinBUGS based upon an ICAR model

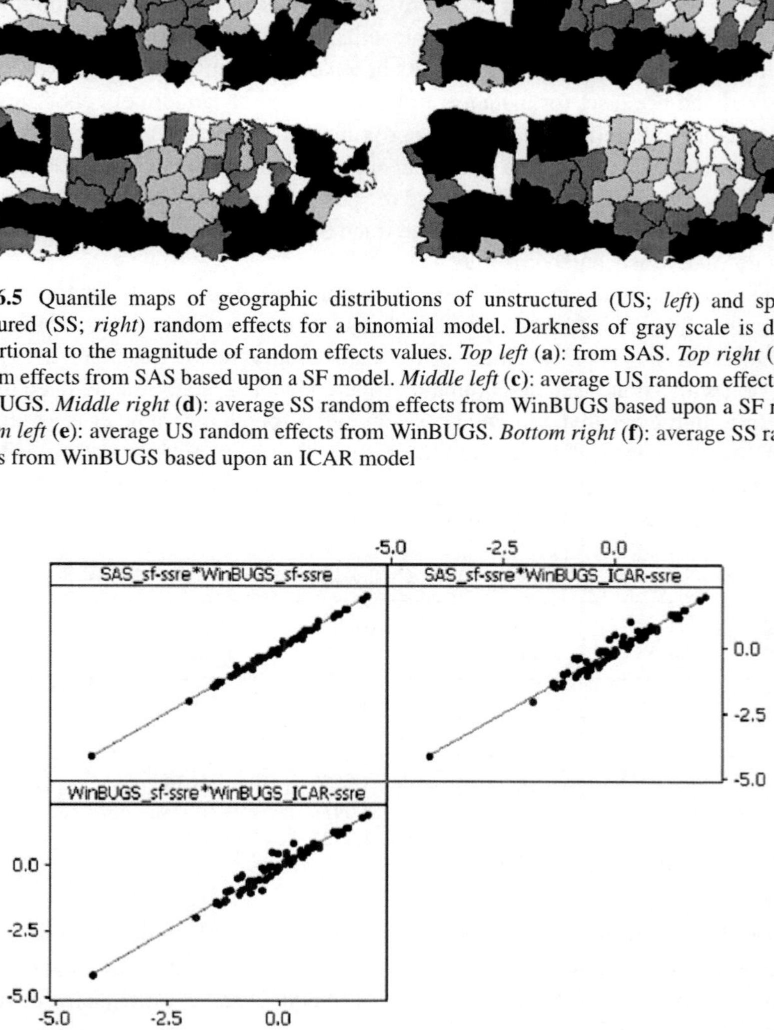

Fig. 6.6 Scatterplot of the SAS and mean WinBUGS estimated spatially structured random effects term. *Top left* (**a**): the SAS versus WinBUGS results based upon SF models. *Top right* (**b**): the SAS SF model results versus the WinBUGS ICAR model results. *Bottom left* (**c**): the WinBUGS SF versus ICAR model results

level of SA is reflected by the relative variance allocated to the ICAR-structured random effects term, which is roughly 48.2%, suggesting the presence of moderate positive SA.

6.5 Degrees of Freedom for GLMM Random Effects

One controversy in the literature concerns the number of degrees of freedom associated with a random effects term. Spiegelhalter et al. (2002) address this very problem for complex hierarchical models in which the number of parameters is not clearly defined because, for instance, of the presences of random effects. They use an information theoretic argument to approximate the effective number of parameters in a model, which for their Bayesian specification is equivalent to the trace of the product of the Fisher information and the posterior covariance matrices; this particular approximation is equivalent to the trace of the 'hat' matrix for linear models with a normally distributed error term.

Here a similar argument is proposed. A random effects term accounts for overdispersion in a generalized linear model. For example, a Poisson regression with overdispersion can be respecified as a negative binomial regression—a Poisson regression with a nonconstant, gamma-distributed mean—or a Poisson mixed model regression. In the negative binomial case, an overdispersion parameter is estimated, causing the deviance statistic to drop to nearly 1. Reformulating the sugar cane production analysis in terms of a Poisson random variable approximation (with an area offset variable) renders the following estimation results:

Statistic	Poisson	Negative binomial	Poisson with random effects
b_0	−1.7901	−0.4865	−1.0923
$b_{\overline{elev}}$	−0.0076	−0.0154	−0.0200
b_{E_1}	3.8218	2.9250	7.9217
b_{E_4}	2.8974	5.9670	10.0280
$\hat{\eta}$	0	2.6978	0
deviance	854.1165	1.2402	0.1982

Quasi-likelihood techniques can be employed with the initial Poisson model specification, resulting in a scale parameter of 27.9607 that can be used to adjust for overdispersion. Meanwhile, the 68 degrees of freedom used to estimate the deviance in the Poisson GLMM are too many, because the random effects almost always has more than 1 degree of freedom associated with it. If the negative binomial estimation results are sound, then this single degree of freedom needs to be increased to 58 (i.e., solve $\frac{13.6779}{73-4-k} = 1.2402$ for k); if the deviance statistic is unknown, then its theoretical value of 1 could be used (as is the practice in quasi-likelihood estimation), resulting here in an estimate of 55 degrees of freedom associated with the random effects term. In other words, only 11–14 degrees of freedom remain after estimation of a GLMM.

Returning to the binomial model specification employed in the preceding section, the equation to solve becomes $\frac{1.4683}{73-4-k} = 1$, implying that the number of degrees of freedom associated with the random effects term coupled with the constraint for $p = 0$ cases is roughly 67.5; in other words, only about 1 degree of freedom remains. This result is corroborated to some degree by the near-perfect predictions obtained by including a random effects term. Because 18 municipalities have a value of 0, the resulting effective degrees of freedom for the random effects term becomes $67.5 - 18 = 49.5$, which is in keeping with the preceding Poisson regression specification results (of note is that a binomial specification has more constraints than a Poisson specification, and hence should have fewer degrees of freedom).

The WinBUGS SF model counterpart to the SF SAS results is furnished by $p_D = 54.8$. This value suggests that the average random effects term has 50.8 degrees of freedom associated with its estimation. This value is of the same order of magnitude as the preceding one, with part of the difference being attributable to the use of a Bayesian analysis.

Finally, the WinBUGS ICAR model counterpart to the two SF results is furnished by $p_D = 54.9$. This value suggests that the average random effects term has 52.9 degrees of freedom associated with its estimation; the ICAR model is accounting for the 2 degrees of freedom associated with the two SF eigenvector coefficient estimates. Again, this value is in keeping with the preceding two random effects degrees of freedom estimates.

Consequently, all three formulations render essentially the same number of degrees of freedom for the estimated random effects term based upon a binomial model specification.

6.6 Extensions to Space-Time Data Sets

One way to increase the number of degrees of freedom in the presence of SA, as well as to move a random effects estimate away from being essentially the model residuals, is to extend a single map into a space-time series of maps all based on the same surface partitioning (see, for example, Lahiri, 1996). For the Puerto Rican sugar cane data example, annual harvest results by municipality are available for crop years 1958/59 to 1973/74. Sixteen points in time are insufficient for properly estimating an ARIMA time series model for any location. SF models allow SA to be described for each separate year (see Table 6.4). Meanwhile, a random effects term can be employed to account for the serial correlation in the space-time series under study here. This term is constant through time, and as such is sensible only for relatively short time series data. The island-wide average level of sugar cane harvest

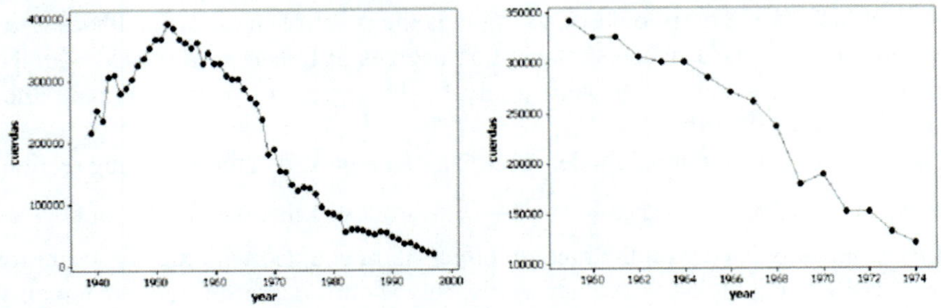

Fig. 6.7 Time series plots of cuerdas of harvested sugar cane for the island of Puerto Rico. *Left* (**a**): annual crop year totals for 1938/39–1996/97. *Right* (**b**): annual crop year totals for 1958/59–1973/74

is nonconstant (see Fig. 6.7), suggesting that at least part of the fixed effects terms in a model specification needs to be year-specific.

According to Table 6.5, the same set of eigenvectors (i.e., $\mathbf{E}_1$, $\mathbf{E}_4$, $\mathbf{E}_6$, and $\mathbf{E}_{24}$) accounts for SA for the crop years 1965/66 through 1967/68. A model specification for this very short space-time series should include year-specific intercept terms, elevation covariate coefficients, and SFs. Estimation results for this dataset appear in Table 6.4. Now the single random effects term that relates to all three years no longer is or can be nearly equivalent to the model's residuals, which vary from year to year for a given location; pseudo-R^2 values have moved only slightly away from 1. Coefficients for the mean elevation covariate and eigenvectors $\mathbf{E}_1$ and $\mathbf{E}_{24}$ change very little across the three years, suggesting that they could be represented by a single temporal parameter. In contrast, the intercept and coefficients for eigenvectors $\mathbf{E}_4$ and $\mathbf{E}_6$ change noticeably from year to year, suggesting that they need to be represented as year-specific effects. Estimation results for this reduced model appear in Table 6.6.

Table 6.5 Space-time GLMM estimation results for Puerto Rican sugar cane crop years 1965/66-1967/68 when all fixed effects are year-specific

	Crop year 1965/1966		Crop year 1966/1967		Crop year 1967/1968	
Statistic	Estimate	Standard error	Estimate	Standard error	Estimate	Standard error
b_0	−1.2291	0.2336	−1.3122	0.2336	−1.4520	0.2336
$b_{\overline{\text{elev}}-\overline{\overline{\text{elev}}}}$	−0.0065	0.0009	−0.0064	0.0009	−0.0065	0.0009
b_{E_1}	4.5040	1.1684	4.5785	1.1684	4.9226	1.1684
b_{E_4}	4.9713	1.2432	5.3372	1.2432	5.8053	1.2432
b_{E_6}	−4.5091	1.1620	−4.8209	1.1620	−5.0203	1.1621
$b_{E_{24}}$	−4.0290	1.0994	−3.9657	1.0994	−4.0285	1.0995
Pseudo-R^2	0.9950		0.9976		0.9929	
$MC_{SS+\hat{\xi}}$	0.449		0.467		0.493	
$GR_{SS+\hat{\xi}}$	0.580		0.562		0.537	
$MC_{\text{residuals}}$	0.019		0.042		0.009	
$GR_{\text{residuals}}$	0.916		0.839		0.808	

Table 6.6 Space-time GLMM estimation results for Puerto Rican sugar cane crop years 1965/66–1967/68 when only some fixed effects are year-specific

Statistic	Crop year 1965/1966 estimate	Crop year 1966/1967 estimate	Crop year 1967/1968 estimate	Standard error
b_0	−1.2440	−1.3041	−1.4444	0.2331
$b_{\overline{elev}-\overline{\overline{elev}}}$		−0.0064		0.0009
b_{E_1}		1.5506		0.3887
b_{E_4}	4.9603	5.3684	5.7511	1.2409
b_{E_6}	−4.5313	−4.7943	−5.0193	1.1600
$b_{E_{24}}$		−1.3330		0.3658
Pseudo-R^2	0.9953	0.9975	0.9936	
$MC_{SS+\hat{\xi}}$	0.449	0.468	0.493	
$GR_{SS+\hat{\xi}}$	0.579	0.561	0.537	
$MC_{residuals}$	0.037	0.070	0.021	
$GR_{residuals}$	0.924	0.802	0.801	

Table 6.7 Space-time GLMM estimation results for Puerto Rican sugar cane crop years 1965/66-1967/68 when all fixed effects are year-specific

Statistic	Crop year 1965/1966		Crop year 1966/1967		Crop year 1967/1968	
	Estimate	Standard error	Estimate	Standard error	Estimate	Standard error
b_0	−1.2291	0.2336	−1.3122	0.2336	−1.4520	0.2336
$b_{\overline{elev}-\overline{\overline{elev}}}$	−0.0065	0.0009	−0.0064	0.0009	−0.0065	0.0009
b_{E_1}	4.5040	1.1684	4.5785	1.1684	4.9226	1.1684
b_{E_4}	4.9713	1.2432	5.3372	1.2432	5.8053	1.2432
b_{E_6}	−4.5091	1.1620	−4.8209	1.1620	−5.0203	1.1621
$b_{E_{24}}$	−4.0290	1.0994	−3.9657	1.0994	−4.0285	1.0995
Pseudo-R^2	0.9950		0.9976		0.9929	
$MC_{SS+\hat{\xi}}$	0.449		0.467		0.493	
$GR_{SS+\hat{\xi}}$	0.580		0.562		0.537	
$MC_{residuals}$	0.019		0.042		0.009	
$GR_{residuals}$	0.916		0.839		0.808	

The single random effects term associated with estimation results reported in Table 6.7 has a mean of 0.0013, a variance of 1.1418, conforms poorly to a bell-shaped curve [P(S–W) = 0.0091], and may well contain some SA (MC = 0.142, GR = 0.969); this increase in SA is accompanied by a decrease in the residual SA for each of the three years. This random effects term is uncorrelated with either mean elevation (r = 0.001) or any of the four eigenvectors used to construct SFs (r lies between –0.001 and 0.001). Each of the three spatially structured random effects reflects approximately the same level of SA displayed by its corresponding sugar cane area harvest percentage. Maps of these variates appear in Fig. 6.8. Overall, the estimated random effects term has a mixture of good and bad properties.

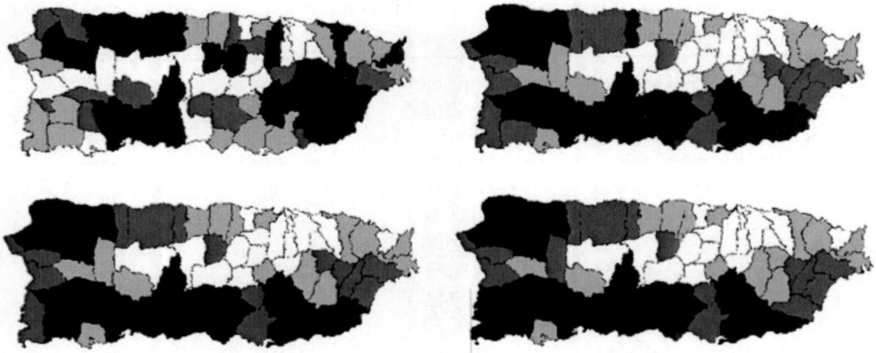

Fig. 6.8 Quantile maps of the geographic distributions of binomial model unstructured (US) and spatially structured (SS) random effects from SAS by year, for 1966, 1967, and 1968. Darkness of gray scale is directly proportional to the magnitude of random effects values. *Top left* (**a**): random effects. *Top right* (**b**): SS random effects based upon a SF model for crop year 1965/66. *Bottom left* (**c**): SS random effects based upon a SF model for crop year 1966/67. *Bottom right* (**d**): SS random effects based upon a SF model for crop year 1967/68

6.7 Discussion and Implications

Comparisons between three common specifications of spatial structuring—namely, semivariogram, spatial autoregressive and SF models—for a random effects term in mixed statistical models reveal that all three perform in an equivalent fashion. Matching Bayesian model priors with their implicit frequentist counterparts yields estimation results from both approaches that are essentially the same. Furthermore, making use of spatially structured random effects tends to furnish an alternative to quasi-likelihood estimation techniques for binomial probability model specifications, as well as to a negative binomial substitution for Poisson probability model specifications.

Semivariogram models offer a geostatistical theoretical basis and have been implemented in SAS for LMMs. A spatial statistics practitioner with the necessary computer programming skills can employ WinBUGS in order to utilize them with GLMMs. Spatial autoregressive modeling also offers a theoretical basis for spatial structuring, and is available in GeoBUGS, but would be very difficult to trick SAS into doing. Meanwhile, spatial filtering, which can be derived from spatial autoregressive model specifications, tends to be more exploratory in nature (being akin to principal components analysis), can be implemented in either SAS or WinBUGS for either LMMs or GLMMs, and can be easily extended to space-time datasets with either of these software packages.

The illustrative Puerto Rico sugar cane examples presented here tend to have a random effects term that virtually equates to the corresponding LMM/GLMM residual variate. But this finding is not always the case, as is highlighted by the extension of a GLMM specification to a space-time sugar cane dataset. In addition, all of the estimated random effects terms for the various Puerto Rico examples presented here tend to be non-normal.

Finally, once a random effects term has been estimated with a frequentist approach, using it when calculating a deviance statistic allows its number of degrees of freedom to be approximated for GLMMs. Although n values are estimated, because they are correlated, the resulting number of degrees of freedom is less than n. This particular finding should help spatial statistics practitioners better understand the cost of employing a statistical mixed model.

Chapter 7
Spatial Filter Versus Conventional Spatial Model Specifications: Some Comparisons

7.1 Introduction

Spatial statistical analysis of geographically distributed counts data has been widely undertaken for many years, with initial analyses involving log-Gaussian approximations because only the normal probability model was first adapted in an implementable form (Ripley, 1990, pp. 9–10) to handle spatial autocorrelation (SA) effects (i.e., similar values tend to cluster on a map, indicating positive self-correlation among observations). In more recent years, linear regression techniques have given way to generalized linear model techniques that account for non-normality (e.g., logistic and Poisson regression), as well as geographic dependence. In very recent years, both linear and generalized linear models have been supplemented with hierarchical Bayesian models, in part to deal with geographic regions having small counts. The objective of this chapter is to furnish a comparison of this variety of principal techniques—both frequentist and Bayesian—available for map analysis with the newly formulated spatial filtering approach.

7.1.1 Background

Over the years multiple regression based upon a normal probability model has been one of the most frequently used statistical methods for undertaking map analysis. More recently the work of McCullagh and Nelder (1983, 1989) has popularized the success applied statisticians have experienced in devising user-friendly implementations of probability models beyond that for the normal curve. Meanwhile, but with a time lag, Wrigley (1985, 2002) helped popularize the use of spatial auto-versions of these models in geographic analyses. "The central role of the Poisson distribution with respect to the analysis of counts is analogous to the position of

---

This material is based upon work supported by the National Science Foundation under Grant No. BCS-9905213. Any opinions, findings, and conclusions or recommendations expressed in this material are those of the author and do not necessarily reflect the views of the National Science Foundation.

D.A. Griffith, J.H.P. Paelinck, *Non-standard Spatial Statistics and Spatial Econometrics*, Advances in Geographic Information Science 1,
DOI 10.1007/978-3-642-16043-1_7, © Springer-Verlag Berlin Heidelberg 2011

the normal distribution in the context of models for continuous data" (Upton and Fingleton, 1989, p. 71). But initial development of an auto-Poisson model proved to be a failure, with this particular model specification being unable to capture the near-universal case of positive SA (Besag, 1974). Circumventing this restriction has been achieved in several different ways. First is the use of an auto-log-Gaussian approximation (e.g., Cressie, 1991), or a Winsorized auto-Poisson specification (Kaiser and Cressie, 1997). Second is the use of spatial filtering model specifications (Getis and Griffith, 2002). Third is the use of hierarchical generalized linear models (HGLM[1]; e.g., see Lee and Nelder, 2001), which can be implemented with software such as GeoBUGS (background discussion is furnished in Casella, 1985; Casella and George, 1992), the add-on spatial statistical module to WinBUGS, the BUGS implementation of Bayesian models. The primary goal of this chapter is to compare these principal approaches to the analysis of georeferenced Poisson random variables.

Comparisons are illustrated with the famous geocoded (by district) Scottish lip cancer data reported by Clayton and Kaldor (1987, pp. 676–677).[2] These data comprise: cases geographically aggregated by district, offset[3] expected values computed on the basis of age and sex compositions of district populations, the number of males at risk (reported in Cressie, 1991, p. 537), and the percentage of each district's outdoor labor force employed in agriculture, fishing, or forestry. Choropleth maps (i.e., thematic maps in which areas are shaded or patterned according to attribute measurements in order to portray their geographic distributions) portraying the geographic distribution of standardized mortality rates (SMRs) computed with these data appear in Fig. 7.1; corresponding boxplots and histograms appear in Fig. 7.2.

Fig. 7.1 Geographic distribution of SMR variates by quartiles: the lowest quartile is denoted by white, the 2nd lowest by *light grey*, the 2nd highest by *dark grey*, and the highest by *black*. *Left*: (**a**) O_i/E_i. *Middle*: (**b**) $(O_i + 0.5)/(E_i + 0.5)$. *Right*: (**c**) $LN[(O_i + 0.5)/(E_i + 0.5)]$

[1] A HGLM is a GLM (e.g., Poisson, binomial, gamma) with multiple levels. The lowest level posits the probability model for individual observations. Higher levels posit probability models for parameters (e.g., prior distributions).

[2] A careful inspection of these data from multiple sources reveals published discrepancies: Cressie (1999), Waller and Gotway (2004) and GeoBUGS enumerate correct lists of geographic neighbors; in contrast, Clayton and Kaldor (1987), Breslow and Clayton (1993), Stern and Cressie (1999)—vis-à-vis Cressie and Guo (1987)—and Lee and Nelder (2001) have the lists of neighbors for Annandale and Tweeddale switched.

[3] An offset variable is one whose regression coefficient is known to be, and hence is set equal to, 1.

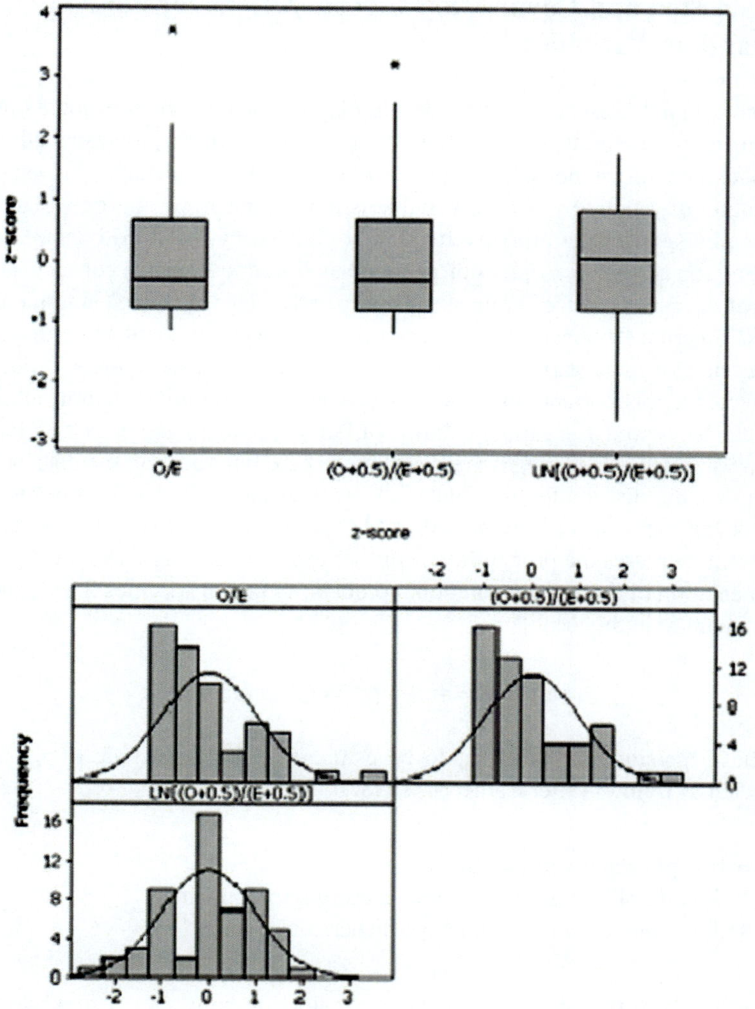

Fig. 7.2 Z-score versions of the three measures of SMR: O_i/E_i, $(O_i + 0.5)/(E_i + 0.5)$, and LN$[(O_i + 0.5)/(E_i + 0.5)]$. *Left* (**a**) boxplots. *Right* (**b**): histograms with normal curves superimposed

Hill et al. (1999) furnish an interesting comparison of focused score tests and Bayesian hierarchical modeling for detecting spatial disease clustering using these data. Furthermore, Stern and Cressie (1999) as well as the GeoBUGS tutorial (Thomas et al., 2004) enumerate these data, but with a markedly different geographic connectivity structure than is used here; in keeping with Clayton and Kaldor (1987), district neighbors separated by water—which are not included in some of the other sources—are included here, resulting in the presence of an additional 30 district-neighbor linkages.

7.2 Variation and Covariation Considerations for Poisson Random Variables

The conventional Poisson random variable may be used to describe counts for the occurrence of rare events, such as the selection of points in a region as locations for some phenomenon, or the number of cases of some event occurring in a given place. One feature of a Poisson random variable is that its mean, μ, and its variance are equal (equidispersion), a property frequently violated by real world data. "Failure of the Poisson assumption of equidispersion has similar qualitative consequences to failure of the assumption of homoskedasticity" associated with the Gaussian distribution (Cameron, 1998, p. 77). One standard way of accommodating overdispersion (the presence of more variation than is expected for a Poisson random variable) is by replacing a Poisson random variable with a negative binomial random variable—which can be viewed as a gamma mixture of Poisson random variables (i.e., Poisson random variables whose means are distributed according to a gamma distribution). In doing so, the distribution of counts is viewed as either (1) having missing variables for the mean specification, and/or (2) being dependent (i.e., the occurrence of an event increases the probability of further events occurring). The most popular implementation of the negative binomial probability model specifies the variance as being quadratic in the mean, or

$$\text{Variance} = \mu + \eta\mu^2 = (1 + \eta\mu)\mu,$$

with the dispersion parameter, η, to be estimated. The magnitude of η may be interpreted as follows (after Cameron, 1998, p. 79):

$\eta = 0$ implies no overdispersion;
$\eta \approx 1/\mu$ implies a modest degree of overdispersion; and,
$\eta \approx 2/\mu$ implies considerable overdispersion.

In other words, if $0 \leq \eta < 0.5/\mu$ (i.e., the midpoint between the first two target values), a map analyst may consider overdispersion detected in georeferenced data to be inconsequential, with little to be gained by replacing a Poisson with a negative binomial probability model specification. Other indices of overdispersion include a statistically significant $\hat{\eta}$ for a negative binomial model, or a Pearson's deviance statistic that is close to or exceeds 2. McCullagh and Nelder (1989, p. 125) suggest that a researcher always should exercise caution and assume overdispersion—doing so costs only 1 degree of freedom for the single additional parameter estimate.

The presence of overdispersion can be attributable to various sources, only one of which is spatial correlation amongst the geographic distribution of counts (see Griffith and Haining, 2006). It can result from heterogeneity latent in grouped data. Because model variance is a function of a model's mean specification, it can arise from missing predictors and/or overlooked non-linear/interaction relationships. As

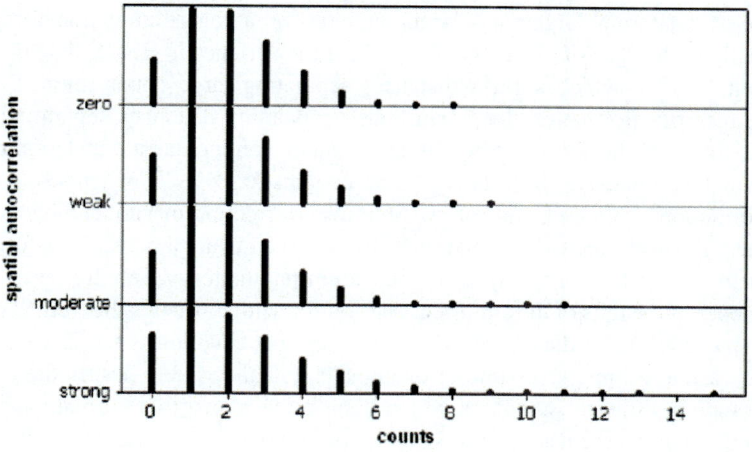

Fig. 7.3 Dotplot of Poisson counts with increasing levels of positive spatial autocorrelation

with the normal probability model, the presence of outliers can introduce extra-Poisson variation. And, it can result from the Poisson probability model being a poor descriptor of a given set of observed counts. With regard to positive SA, which frequently is detected in geographically distributed counts, the clustering of similar values on a map tends to result in the appearance of excessive zeroes, as well as some relatively extreme large counts (i.e., a centripetal force type of effect). In other words, as positive SA increases, extra-Poisson variation is accompanied by a frequency distribution whose form increasingly resembles a descretized negative exponential curve (see Fig. 7.3).

For the Scottish lip cancer data, the raw counts yield $0.5/\hat{\mu} = 0.5/9.5716 \approx 0.0522 << \hat{\eta} = 0.5456$. A negative binomial (i.e., Poisson-gamma) regression based upon only the age-sex-adjusted expected values yields $\hat{\eta} = 0.4895 >> 0.2501$, the maximum value for $0.5/\hat{\mu}_i$. Inclusion of the outdoor labor force covariate reduces this estimate to $\hat{\eta} = 0.3989 >> 0.2810$, the new maximum value for $0.5/\hat{\mu}_i$. In other words, these Scottish lip cancer data exhibit considerable overdispersion.

7.2.1 Heterogeneity in Counts Data

As noted in the preceding discussion, the mean and variance of a Poisson random variable are perfectly positively correlated. One way of stabilizing this non-constant variance is to apply a transformation so that the re-expressed random variable approximately conforms to a normal distribution, which has zero correlation between its mean and variance, and constant variance. This result commonly is achieved by applying a square root (Kuehl, 1994, p. 118) or its finely tuned Freeman-Tukey counterpart (Cressie, 1991, p. 540)—which increasingly shrinks distances separating larger and larger counts—or a logarithmic transformation, especially in the case of overdispersion, where the variance is greater than the mean (Bartlett, 1947)—which shrinks distances separating large counts more than the square root transformation does, while also expanding distances separating small counts. Either of these two types of transformation can include a translation parameter, which compensates for relatively small counts (e.g., $\bar{y} < 3$) when a square root transformation is applied, and for possible over-correction of variance heterogeneity when a logarithmic transformation is applied (i.e., using a power transformation exponent of 0 rather than 0.5). This translation parameter essentially better aligns one or both of the tails of the empirical distribution with those of a theoretical normal distribution. A useful diagnostic for monitoring this situation is the Levene statistic (i.e., a non-normailty assuming diagnostic statistic used to assess the equality of variance in different samples), which is far more robust to a normality assumption violation than the Bartlett test statistic for homogeneity of variance, although it assumes a continuous random variable. Meanwhile, the overdispersion discussed in the preceding section pertains to a Poisson random variable whose variance exceeds its mean. In either case, the nonconstant variance of interest is for attribute counts, most likely arising from geographic aggregations of large numbers of Bernoulli random variables with extremely small success probabilities, varying factors generating the rare events of interest across (sets of) individuals, and/or differences in the number of individuals in each geographic aggregation.[4] A simple heuristic diagnostic for this situation is to test equality of variance after categorizing data as being above/below the sample mean or median number of counts.

Of note is that standardization of an incidence is done by calculating the ratio of observed to expected values in order to remove effects of differences in confounding variables (e.g., age, sex, race/ethnicity). This standardization is based upon a weighted average calculated by decomposing some general reference population into sub-populations (i.e., the standard or benchmark), and enables meaningful direct comparisons to be made amongst various aggregations of individuals under study. The purpose of this adjustment is to account for that heterogeneity arising from various mixtures of sub-populations within each aggregation. In essence, the expected value, say E_i for areal unit i, functions as a covariate whose regression coefficient is set to 1; accordingly, it is labeled an offset variable in the generalized linear modeling literature. But standardization fails to adjust for variation due to the range of values in areal unit denominators: small base values result in more variability than large base values.

The district variable $Y_i = LN\left(\frac{O_i+0.5}{E_i+0.5}\right)$, where O_i denotes the observed counts for district i, for the Scottish lip cancer data conforms closely to a normal distribution. But the range of O_i values (from 0 to 39) potentially could continue to induce nonconstant variance across districts; here this degree of variability seems to be only

[4]Regardless of the context, regional aggregates with small base populations tend to yield imprecise standardized ratios, whereas regional aggregates with large base populations almost always yield significant results.

problematic for the two districts having 0 cases. Setting these two districts aside results in $1 + \frac{1}{O_i}$ accounting for only about 10% of the variation in the location-to-location variance. Conceptually speaking, transforming a variable to better align with a normal distribution essentially eliminates correlation between its mean and variance, and effectively induces constant attribute variance. The transformed values, Y, approximately display this property, which can be detected by comparing the variances of Y above and below its mean and its median; these data splits respectively reduce the Levene homogeneity of variance test statistic from 11.67 to 0.61 ($p = 0.436$), and from 20.81 to 3.80 ($p = 0.056$).

Because of the spatial nature of georeferenced data, geographic variance heterogeneity also is of concern when undertaking a map analysis. Sometimes this data feature is cast as anisotropy; other times it is cast as geographic landscape heterogeneity. One simple diagnostic for this latter situation is to test equality of variance (by computing the Levene statistic) after regionally grouping data according to, say, the four quadrants of the plane[5], or perhaps some set of natural regions—because sample variance is a function of (n–1), the number of arbitrary geographic regions used here will depend upon the total number of areal units, n, contained in a dataset. Classifying the 56 districts into the four quadrants of the plane (based on georeference coordinate medians) yields a Levene test statistic of 3.65 ($p = 0.018$) for $\frac{O_i+0.5}{E_i+0.5}$, which reduces to 0.38 ($p = 0.769$) for Y.

Yet another source of nonconstant geographic variance arises from the use of data aggregated by regions forming an irregular surface partitioning (i.e., irregular lattice data). Because sums of values surrounding a location are used to evaluate SA effects, Besag et al. (1991) comment that varying numbers of entries in these sums need to be accounted for. This particular source of variability is problematic with regard to the conditional autoregressive (CAR) model, whose inverse covariance structure matrix often is given by $(\mathbf{I} - \rho\mathbf{C})$, where $\mathbf{I}$ is the n-by-n identity matrix, ρ is the SA parameter, and the n-by-n matrix $\mathbf{C}$ is binary and has a cell entry $c_{ij} = 1$ if areal units (e.g., the 56 districts of Scotland) i and j are neighbors, and $c_{ij} = 0$ otherwise; often two areal units are considered to be neighbors if they share a common boundary. The term $\sum_{j=1}^{n} c_{ij} = n_i$ counts the number of entries in each of the i (= 1, 2, . . ., n) sums. This particular specification results in a diagonal variance matrix $\mathbf{D}$, where the d_{ii} entries of this matrix are the n_i. Premultiplying matrix $\mathbf{C}$ by matrix $\mathbf{D}^{-1}$ yields matrix $\mathbf{W}$, the row-standardized version of the 0–1 binary geographic configuration matrix $\mathbf{C}$. One appealing feature of this specification is that the autoregressive term $\mathbf{WY}$ renders averages of neighboring values, which stabilizes the geographic variance arising from summing unequal numbers of neighboring values (i.e., the case of an irregular surface partitioning). But the spatial inverse covariance matrix must be symmetric, requiring the specification here to be $(\mathbf{I} - \rho\, \mathbf{D}^{-1}\mathbf{C})\mathbf{D}^{-1}$. In other words,

[5]For the Scottish lip cancer data, latitude and longitude geo-coordinates were retrieved from Waller and Gotway (see http://www.sph.emory.edu/~lwaller/ch9index.htm), and then refined with an ArcView script.

the nonconstant geographic variance incorporated here relates to the inverse of the number of neighbors of each areal unit (i.e., $\frac{\sigma^2}{n_i}$), and must be included in order to satisfy a mathematical requirement of the multivariate normal probability model. Of note is that this nonconstant variance specification is employed by Breslow and Clayton (1993, p. 21), Hill et al. (1999, p. 105), and Stern and Cressie (1999, p. 66), but not by Clayton and Kaldor (1987, p. 678), Cressie (1989, p. 545), or Lee and Nelder (2001, p. 14). One advantage of this specification is that $0 \leq \hat{\rho} \leq 1$ for the case of positive SA. And, for a regular surface tessellation, for which the number of neighbors is approximately constant, this geographic variance heterogeneity all but disappears (although some remains along the edges of a landscape).

The simultaneous autoregressive (SAR) model furnishes an alternative specification that frequently is employed with the auto-Gaussian model. Its spatial covariance structure matrix usually is given by $[(\mathbf{I} - \rho\mathbf{C}\mathbf{D}^{-1})(\mathbf{I} - \rho\mathbf{D}^{-1}\mathbf{C})]^{-1} = [(\mathbf{I} - \rho\mathbf{W}^{T})(\mathbf{I} - \rho\mathbf{W})]^{-1}$, where T denotes matrix transpose, and the resulting matrix is symmetric. This specification also deals with averages of neighboring values and restricts positive values of the autoregressive parameter to the more intuitively interpretable range of $0 \leq \hat{\rho} \leq 1$.

7.2.2 *Spatial Autocorrelation in Poisson Random Variables*

Including the expected values as an offset variable in Poisson regression is equivalent to dealing with a standardized mortality ratio (SMR), which, for example, equals the observed divided by the expected number of lip cancer cases for each district (i.e., O_i/E_i, for district i). Here the SMRs fail to conform to a normal frequency distribution [Shapiro-Wilk, S-W = 0.887; the probability of S-W, P(S-W) < 0.0001], and exhibit moderate positive SA[6]: the Moran Coefficient (MC) = 0.5391 and the Geary Ratio (GR) = 0.2946.

A Box-Cox type of power transformation (Chinn, 1996)—where in keeping with a Poisson random variable the exponent[7] is assumed to be 0—of the SMRs results in

[6]The MC is a covariation-based measure that is similar to a Pearson product-moment correlation coefficient, and has an approximate range of $(1/\lambda_n, 1/\lambda_2)$, where λ_2 and λ_n respectively are the second largest and smallest eigenvalues of matrix **C**. Its expected value is –1/(n–1). The GR is a paired comparisons type of index, is inversely related to the MC, has an approximate range of (0, 2), and has an expected value of 1.

[7]The non-zero exponent functional form is $Y_N = \alpha + \beta[(O_i + \delta)/(E_i + \delta)]^{\gamma}$, which here yields rounded-off parameter estimates of $\hat{\delta} = 0.10$ and $\hat{\gamma} = 0.33$ [RESS = 1.48×10^{-2}; P(S-W) = 0.726]. Setting $\delta = 0.5$ yields $\hat{\gamma} = -0.10$ (RESS = 1.71×10^{-2}), which is very close to 0. Setting $\delta = 0.5$ and executing Friendly's SAS macro boxcox[1].sas (www.math.yorku.ca/SCS/sasmac/boxcox.html) yields $\hat{\gamma} = 0.20$ (RESS = 2.56×10^{-2}), which also is close to 0; setting $\delta = 0.1$ yields $\hat{\gamma} = 0.31$, which essentially is the same result obtained with the quantile equation. As an aside, the Freeman-Tukey transformation (Cressie, 1991, p. 540) furnishes an inferior result with its translation parameter of 1 [P(S-W) = 0.061]; its optimal translation parameter estimate also is 0.5, which modestly improves its performance here [P(S-W) = 0.131].

the set of Y values conforming closely to a normal distribution [P(S-W) = 0.196].[8] This re-expression of the SMRs was computed by constructing normal scores, say Y_N, for the ratio O_i/E_i, using Blom's formula (1958), and then estimating the three-parameter nonlinear regression quantile equation (using SAS PROC NLIN)

$$Y_N = \alpha + \beta \, LN\left(\frac{O_i+\delta}{E_i+\delta}\right); \text{ relative error sum of squares (RESS)} = 3.71 \times 10^{-2}. \tag{7.1}$$

The estimate calculated for δ is 0.4602, which then was rounded off to 0.5. This procedure is consistent with arguments presented by Yeo and Johnson (2000, pp. 954–955). The estimated translation parameter, 0.5, is necessary here because some observed counts equal 0; serendipitously, it equals the commonly used bias-corrected translation parameter (Snedecor and Cochran, 1967, pp.497, 502–503). The detected SA changes little when this transformation is applied: MC = 0.4965 and GR = 0.4383. Meanwhile, the Box-Tidwell linearization transformation identified for the outdoor labor percentage covariate (X_1) is LN(X_1 + 1.2). Employing these transformations in a bivariate regression analysis increases the adj-R^2 from 0.210 (Y regressed on X_1) to 0.278 [Y regressed on LN(X_1 + 1.2)], suggesting that their use is worthwhile. Furthermore, inclusion of the covariate LN(X_1 + 1.2) reduces SA in the residuals: MC = 0.2687 and GR = 0.7212. But this reduction is at the expense of normality [P(S-W) = 0.003].

Presumably the non-normality complication that materializes relates to nonconstant variance (see Sect. 7.2.1). The variable $LN\left(\frac{O_i}{E_i}\right)$ relates to a non-constant variance that is proportional to $1 + \frac{1}{O_i}$ (e.g., see Haining, 1990, pp. 365–366). In the particular lip cancer example explored here, the variance appears to be proportional to $\frac{1}{3}\left(1+\frac{5}{O_i+0.5}\right)$—adj-$R^2 = 0.457$. Second, the residual-versus-predicted plot depicts a funnel-shaped scatter of points. Diagnostic evidence suggests that these complications may be attributable in part to the presences of the two 0-counts districts.

These diagnostic findings imply that results generated by a simple log-normal approximation may suffer from specification error. Possible alternatives are a weighted log-normal approximation, a Poisson regression, or accounting for SA. But the presence of non-zero SA and overdispersion indicates that a simple Poisson regression specification is insufficient for describing the Scottish lip cancer data.

---

[8] A translation parameter is added to both the numerator and the denominator because E_i is based upon a sum of the O_is (i.e., the sums of the E_is and the O_is are equal). In the simple case of each regional expected value being calculated with a landscape-wide rate, for example: $P_i\frac{\sum_{i=1}^{N}(O_i+\delta)}{\sum_{i=1}^{N}P_i} = P_i\frac{\sum_{i=1}^{N}O_i}{\sum_{i=1}^{N}P_i} + \frac{P_i}{\sum_{i=1}^{N}P_i}N\delta \overset{P_i \to P_i}{\longrightarrow} \delta \Rightarrow \frac{O_i+\delta}{E_i+\delta}$, for regional "base populations" P_i in the ith of N areal units.

7.2.3 Spatial Autocorrelation-induced Correlation Inflation

The correlation between Y and LN(X_1 + 1.2) appears to be moderate and positive. But the correlation between two georeferenced variables, X and Y, can be distorted by latent SA, an effect recognized and explored by Clifford et al. (1989), Dutilleul (1993), and Haining (1991). Additional understanding of this effect can be gained through spatial filtering, which involves regressing each variable on a set of synthetic variates representing distinct map patterns that accounts for SA. Griffith (2000a) develops one form of spatial filtering whose synthetic variates are the set of n eigenvectors extracted from matrix $(\mathbf{I} - \mathbf{11}^T/n)\mathbf{C}(\mathbf{I} - \mathbf{11}^T/n)$, the matrix appearing in the numerator of the MC, where **1** is an n-by-1 vector of ones. This procedure is similar to executing a principal components analysis in which the covariance matrix is given by $(\mathbf{I} - \mathbf{11}^T/n)\mathbf{C}(\mathbf{I} - \mathbf{11}^T/n)$. But rather than using the resulting eigenvectors to construct linear combinations of attribute variables, the eigenvectors themselves (instead of principal components scores) are the desired synthetic variates, each containing n elements, one for each areal unit (e.g., Scottish district). The sequential construction of these eigenfunctions enables the extreme values of the MC to be established (de Jong et al., 1984); in other words, this procedure should not be followed by an axis rotation like is done in factor analysis. The extracted eigenvector $\frac{1}{\sqrt{n}}\mathbf{1}$ relates to the mean response, and the remaining (n–1) extracted eigenvectors relate to distinct map patterns characterizing latent SA—whose MCs are given by standardizing their corresponding eigenvalues (see Tiefelsdorf and Boots, 1995)—that can materialize with matrix **C**. Furthermore, for a given geographic landscape surface partitioning, the eigenvectors represent a fixed effect in that matrix $(\mathbf{I} - \mathbf{11}^T/n)\mathbf{C}(\mathbf{I} - \mathbf{11}^T/n)$ does not, and hence they do not, change from one attribute variable to another. Theory for this type of spatial filtering is presented in Griffith (2003).

One difficulty associated with this eigenfunction decomposition is that n eigenvectors are extracted from matrix $(\mathbf{I} - \mathbf{11}^T/n)\mathbf{C}(\mathbf{I} - \mathbf{11}^T/n)$. Restricting attention to only those eigenvectors describing substantive positive (e.g., MC > 0.25)[9] SA, when latent SA is positive, further reduces the candidate set. Supervised stepwise selection from the remaining eigenvectors is a useful and effective approach to identifying the subset of eigenvectors that best describes latent SA in a particular georeferenced Poisson variable. This procedure begins with only the intercept included in a regression specification. Next, at each step an eigenvector is considered for addition to the model specification. For the stepwise linear Gaussian model, commonly the eigenvector having the largest partial correlation with variable Y is selected, but only if its corresponding F-ratio achieves or surpasses a prespecified level of significance; this is the criterion used to establish statistical importance of an eigenvector.

[9]A value of 0.25 for the MC tends to relate to about 5% of the variance in Y being attributable to redundant information arising from latent spatial autocorrelation, given a particular areal unit neighborhood configuration.

Meanwhile, in stepwise Poisson regression, the eigenvector that produces the greatest reduction in the log-likelihood function chi-square test statistic is selected, but only if it produces at least a prespecified minimum reduction; as before, this is the criterion used to establish statistical importance of an eigenvector. In each statistical procedure, at each step all eigenvectors previously entered into a spatial filter (SF) equation are reassessed, with the possibility of removal of vectors added at an earlier step. The forward/backward stepwise procedure terminates automatically when some prespecified threshold values (respectively for F-ratios and chi-square statistics) are encountered for entry and removal of all candidate eigenvectors. The ultimate inclusion criterion is determined by the MC value of the residuals, which should indicate an absence of SA. Satisfying this MC condition sometimes requires supervised backward elimination of marginally selected eigenvectors because their inclusion has forced the residual MC value to decrease too far below 0. This final stopping criterion for the linear Gaussian model is relatively easy to implement because MC distributional theory is known for linear regression residuals; a corresponding stopping rule for Poisson regression is far more difficult to implement because of a lack of such distributional theory.

As mentioned previously, because all attribute variables are collected for the same surface partitioning of a given geographic landscape, matrix $(\mathbf{I} - \mathbf{11}^T/n)\mathbf{C}(\mathbf{I} - \mathbf{11}^T/n)$ and its eigenfunctions do not change from one attribute variable to another. Employing this eigenfunction analysis approach for a given geographic landscape, then, a bivariate correlation coefficient (denoted by r_{XY}) can be decomposed into the following components: (1) a set of eigenvectors that strongly correlates with both X and Y, and hence is common to the SFs of both X and Y (denoted by $\mathbf{E}_c$); (2) a set of eigenvectors that correlates strongly with X but very weakly with Y, and hence is unique to the SF for X (denoted by $\mathbf{E}_{uX}$); (3) a set of eigenvectors that correlates strongly with Y but very weakly with X, and hence is unique to the SF for Y (denoted by $\mathbf{E}_{uY}$); and, the regression residuals for the full SFs (respectively denoted by e_X and e_Y, for variables X and Y). For the Scottish lip cancer data example,

$$\mathbf{Y} = 0.03391 + 3.2090\mathbf{E}_1 - 2.4032\mathbf{E}_2 - 1.4216\mathbf{E}_9 + 1.6042\mathbf{E}_3 + 1.7929\mathbf{E}_{19} + \mathbf{e}_Y, \tag{7.2a}$$

and

$$\text{LN}(\mathbf{X}_1 + 1.2) = 1.95341 + 1.8809\mathbf{E}_1 - 3.5301\mathbf{E}_2 - 2.3159\mathbf{E}_9 - 1.3491\mathbf{E}_4 + 1.1572\mathbf{E}_{34} + \mathbf{e}_X. \tag{7.2b}$$

The set of common eigenvectors (see Fig. 7.4, for the Scottish lip cancer data example) will inflate the correlation between X and Y. Because the eigenvectors of matrix $(\mathbf{I} - \mathbf{11}^T/n)\mathbf{C}(\mathbf{I} - \mathbf{11}^T/n)$ are mutually orthogonal as well as mutually uncorrelated (see Griffith, 2003), those appearing in only one of the two SFs will deflate the correlation between X and Y. And, the correlation between residuals from the full SF models represents pure attribute correlation. This is the correlation coefficient that is being inflated or deflated.

Fig. 7.4 Spatial filter map patterns (i.e., eigenvectors $\mathbf{E}_1$, $\mathbf{E}_2$ and $\mathbf{E}_9$) common to Y, LN(X_1 + 1.2), and LN(X_2/area – 22.2)

For the Scottish lip cancer data, the various correlation coefficients are as follows:

raw data: $r_{XY} = 0.53919$, which is the correlation between LN(X_1 + 1.2) and Y

SF residuals: $r_{e_Y e_X} = 0.16027$, which is the correlation between [Y – (0.03390 + 3.20900$\mathbf{E}_1$ – 2.40321$\mathbf{E}_2$ + 1.60420$\mathbf{E}_3$ – 1.42158$\mathbf{E}_9$ + 1.79290$\mathbf{E}_{19}$)] and [LN(X_1 + 1.2) – (1.95339 + 1.88087$\mathbf{E}_1$ – 3.53007$\mathbf{E}_2$ –1.34907$\mathbf{E}_4$ – 2.31591$\mathbf{E}_9$ + 1.15719$\mathbf{E}_{34}$)]

linear combinations of common eigenvectors: $r_{\mathbf{E}_c} = 0.90595$, which is the correlation between (3.20900$\mathbf{E}_1$ – 2.40321$\mathbf{E}_2$ – 1.42158$\mathbf{E}_9$) and (1.88087$\mathbf{E}_1$ – 3.53007$\mathbf{E}_2$ – 2.31591$\mathbf{E}_9$)

linear combinations of unique eigenvectors: $r_{\mathbf{E}_{uX}\mathbf{E}_{uY}} = 0$, which is the correlation between (–1.34907$\mathbf{E}_4$ + 1.15719$\mathbf{E}_{34}$) and (1.60420$\mathbf{E}_3$ + 1.79290$\mathbf{E}_{19}$)

cross-correlations of residuals and linear combinations of unique eigenvectors: $r_{\mathbf{E}_{uY} e_X} = 0.28620$ and $r_{\mathbf{E}_{uX} e_Y} = -0.04224$, which are the correlations between, respectively, (1.60420$\mathbf{E}_3$ + 1.79290$\mathbf{E}_{19}$) and [LN(X_1 + 1.2) – (1.95339 + 1.88087$\mathbf{E}_1$ – 3.53007$\mathbf{E}_2$ –1.34907$\mathbf{E}_4$ – 2.31591$\mathbf{E}_9$ + 1.15719$\mathbf{E}_{34}$)], and (–1.34907$\mathbf{E}_4$ + 1.15719$\mathbf{E}_{34}$) and [Y – (0.03390 + 3.20900$\mathbf{E}_1$ – 2.40321$\mathbf{E}_2$ + 1.60420$\mathbf{E}_3$ – 1.42158$\mathbf{E}_9$ + 1.79290$\mathbf{E}_{19}$)].

Identification of common and unique eigenvectors for the two Scottish lip cancer data SFs, as well as coefficients for the various linear combinations, appear in Table 7.1. The bivariate correlations reported in Table 7.1 confirm that only those eigenvectors having the strongest correlations with Y and with LN(X_1 + 1.2) appear in the preceding SFs; eigenvectors unique to a SF have a relatively strong correlation with their attribute variable, and a relatively weak correlation with the other attribute variable. These results reveal that the pure attribute correlation between percentage outdoor labor and the SMR is only 0.16027, with the net effect of SA being an inflation of this correlation to 0.53919. Using results appearing in Table 7.2, the detailed calculation of coefficient 0.53919 is given by

$$\begin{aligned}&0.16027\sqrt{(1-0.57314)\times(1-0.51160)}+0.90595\sqrt{0.43423\times 0.44569}\\&\quad+0\sqrt{0.13890\times 0.06591}+0.28620\sqrt{0.13890\times(1-0.51160)}\\&\quad+(-0.04224)\sqrt{(1-0.57314)\times 0.06591}.\end{aligned}$$

These terms reveal that the common eigenvectors dramatically inflate the correlation coefficient, the unique eigenvectors only modestly deflate the correlation coefficient, and the cross-correlations add a very modest inflation.

Table 7.1 Eigenvector spatial filter regression coefficients for transformed standard mortality ratio (SMR) and percentage outdoor labor (X_1), for Scottish lip cancer ($n = 56$)

Eigenvector	$LN[(O_i + 0.5)/(E_i + 0.5)]$	$LN(X_1 + 1.2)$
None (intercept)	0.03390	1.95339
Common eigenvectors		
$\mathbf{E}_1$	3.20900	1.88087
$\mathbf{E}_2$	–2.40321	–3.53007
$\mathbf{E}_9$	–1.42158	–2.31591
Unique eigenvectors		
$\mathbf{E}_3$	1.60420	
$\mathbf{E}_{19}$	1.79290	
$\mathbf{E}_4$		–1.34907
$\mathbf{E}_{34}$		1.15719
bivariate correlations coefficients for the eigenvectors and the attribute variables		
E_1	**0.49712**	**0.27168**
E_2	**–0.37229**	**–0.50989**
E_3	**0.24852**	***0.12119***
E_4	***0.11480***	**–0.19486**
E_5	0.06596	–0.01622
E_6	–0.03921	–0.11869
E_9	**–0.22022**	**–0.33451**
E_{12}	–0.04159	0.14997
E_{15}	–0.05538	0.00703
E_{19}	**0.27775**	***0.15996***
E_{24}	–0.05534	–0.13554
E_{28}	0.08643	–0.02157
E_{32}	0.05843	–0.04198
E_{34}	***0.09146***	**0.16715**

NOTE: E_k denotes eigenvector k. NOTE: X denotes the percentage of each district's outdoor labor force employed in agriculture, fishing, or forestry (bold denotes 5% significance, and bold italic denotes 10% significance)

Table 7.2 Eigenvector spatial filter regression results for transformed standard mortality ratio (SMR) and percentage outdoor labor (X_1), for Scottish lip cancer, using a 10% level of significance selection criterion

Feature	$LN[(O_i + 0.5)/(E_i + 0.5)]$	$LN(X_1 + 1.2)$
Common eigenvectors	Unadjusted-$R^2 = 0.43423$	Unadjusted-$R^2 = 0.44569$
Unique eigenvectors	Unadjusted-$R^2 = 0.13890$	Unadjusted-$R^2 = 0.06591$
All selected eigenvectors	Unadjusted-$R^2 = 0.57314$	Unadjusted-$R^2 = 0.51160$
Residual MC	$z_{MC} \approx 0.52$	$z_{MC} \approx 0.09$
P(S-W)	0.949 ($p = 0.019$)	0.983 ($p = 0.590$)
MC for linear combination of eigenvectors	0.907	0.860

The principal finding here is that the relationship between Y and $LN(X_1 + 1.2)$ is weak, rather than moderate, and positive. And, it only may appear to be significant because of an exaggeration of the relationship attributable to the presence of positive

SA [$t = 4.7$, versus $= 1.1$]! In addition, about half of the information contained in the geographic distribution of Scottish lip cancer cases by district is redundant.

7.3 Principal Spatial Statistical Model Specifications

Auto- models are models that have the response variable on both the left- (e.g., Y) and right-hand (e.g., **WY**) sides of an equation. These specifications may be employed when the assumption of independent observations fails to hold. This feature of data commonly occurs in time series, as well as geographic distributions of data values. The principal complication of correlated observations is a loss of efficiency (i.e., statistical precision) for conventional parameter estimators, requiring specially derived, more sophisticated spatial statistical techniques to conduct valid map analyses. In contrast, SF model specifications allow SA to be accounted for with synthetic variates in a fashion that enables conventional statistical techniques to be employed. Two SF models can be posited: (1) a Gaussian-approximation multiple linear regression specification in which Y is the regressand (Griffith, 2000a), and both $LN(X_1 + 1.2)$ and a judiciously selected subset of eigenvectors (i.e., synthetic variates) extracted from matrix $(\mathbf{I} - \mathbf{11}^T/n)\mathbf{C}(\mathbf{I} - \mathbf{11}^T/n)$ are the regressors; and, a Poisson regression specification in which LN(E + 0.5) is an offset variable, $LN(X_1 + 1.2)$ is a covariate, as are judiciously selected eigenvectors (Griffith, 2002). Meanwhile, random-effects HGLM specifications also can be used to deal with non-normal data. One appealing feature of this latter approach is that SA in a non-Gaussian (e.g., Poisson) georeferenced random variable can be captured without having to derive an explicit multivariate generalization of its distributional form.

7.3.1 The Log-normal Approximation

A SAR model—whose auto- prefix indicates that Y appears on both sides of its equation—was estimated with the Scottish lip cancer data, yielding, for $\mathbf{Y}$, $\mathbf{X}^* = LN(X_1 + 1.2)$, and spatial covariance structure matrix $[(\mathbf{I} - \rho\mathbf{W})^T(\mathbf{I} - \rho\mathbf{W})]^{-1}$,

$$\hat{\mathbf{Y}} = 0.67202\mathbf{WY} - 0.31447\,(1 - 0.67202)\,\mathbf{1} + 0.20174\,(\mathbf{I} - 0.67202\mathbf{W})\,\mathbf{X}^*. \quad (7.3)$$

For this equation, the pseudo-R^2 is 0.541, and the residuals contain negligible SA (MC = –0.05367, GR = 1.05801), fail to conform to a normal distribution [P(S-W) = 0.0007], and display some nonconstant variance in the weighted residual-versus-predicted plot. To some degree, the two districts having 0 cases appear to be the culprits. Classifying the 56 districts into the four quadrants of the plane yields a Levene test statistic of 1.17 ($p = 0.330$) for the residuals.

One important finding from this analysis is that latent SA is positive in nature, reconfirming that the conventional auto-Poisson model specified by Besag (1974) is inapplicable here. A second important finding is that the regression coefficient for $LN(X_1 + 1.2)$ is not statistically significant, confirming the spatial filtering-based conclusion reported in Sect. 7.2.3. A third important finding is that the SA structure appears to be second-order, rather than first-order. In other words, the estimated SAR model, whose covariance matrix is $[(\mathbf{I}-0.67202\mathbf{W})^T(\mathbf{I}-0.67202\mathbf{W})]^{-1}$, includes the two spatial-lags terms $-1.34404(\mathbf{W}^T + \mathbf{W})$ and $0.67202^2\mathbf{W}^T\mathbf{W} = 0.45161\mathbf{W}^T\mathbf{W}$, suggesting that a covariance matrix of the simple form $(\mathbf{I} - \rho\mathbf{C})^{-1}$ defined in Sect. 7.2.1 would be inadequate. The term $(\mathbf{W}^T + \mathbf{W})$ captures 1st-order spatial dependency effects; the term $\mathbf{W}^T\mathbf{W}$ captures 2nd-order spatial dependency effects.

7.3.2 A Winsorized Auto-Poisson Model

Winsorizing Poisson counts data involves systematically replacing extremely high counts with the value of some cut-off count (after Barnet, 1978). It is a compromise between the infinite sum of a Poisson probability model and utilizing all of the Poisson-type counts information in some dataset while establishing a most extreme acceptable count. One advantage of this approach emphasized by Kaiser and Cressie (1997) is that the Winsorized alternative to the auto-Poisson model is capable of capturing positive SA, which the auto-Poisson specification is unable to do (Besag, 1974). In other words, the probabilities of excessively large counts, whose Poisson probabilities of occurring essentially are 0, are set to exactly 0, allowing positive spatial dependence in counts to be modeled with a distribution exhibiting Poisson-like behavior. Kaiser and Cressie (1997) also show that the Winsorized alternative has a relatively simple mean response specification, and has an expected value that is near that of the regular auto-Poisson version (i.e., these expectations are nearly the same).

The auto- prefix in the Winsorized auto-Poisson specification refers to a term appearing in the mean response specification that is a function of the sum of neighboring observed counts. This is similar to the **WY** term appearing in the SAR model; here the term is $\sum_{j=1}^{n} W_{ij}LN\left(\frac{O_j+0.5}{E_j+0.5}\right) = \sum_{j=1}^{n} W_{ij}Y_j$. Parameter estimation requires Markov chain Monte Carlo (MCMC) maximum likelihood techniques (MCMC-MLE; Gilks et al., 1996; Hubbell et al., 2001) employing a Gibbs sampler and, when necessary, a Metropolis-Hasting algorithm. This procedure frequently is initiated by computing pseudo-likelihood estimates (PLEs) of the parameters with conventional Poisson regression statistical software, where the mean response specification includes the **WY** type of term. These estimates are for the parameters of conditional Poisson mass functions, whereas of interest is the estimation of parameters for joint multivariate Poisson mass functions. For the Scottish lip cancer data, negative binomial probability model estimates of these PLEs are: $\hat{\eta} = 0.0959$, $\hat{\alpha} = -0.4497$, $\hat{\beta}_1 = 0.2551$, and $\hat{\rho} = 0.7500$.

When count data display overdispersion, the Gibbs sampler involves sampling from a Poisson-gamma distribution. The gamma distribution has a shape parameter given by $\frac{1}{\hat{\eta}}$, where $\hat{\eta}$ is the PLE from a negative binomial distribution. If the random sampling outcome is denoted by G_i, for district i, then the log-mean for the Scottish lip cancer situation is given by the expression

$$\mathrm{LN}\,(G_i)\ +\mathrm{LN}(\hat{\eta}) + [\hat{\alpha} + \hat{\beta}_1\,\mathrm{LN}\,(X_{i,1} + 1.2) + \hat{\rho}\sum_{j=1}^{n} w_{ij}y_j + \mathrm{LN}\,(E_i + 0.5)], \quad (7.4)$$

where $\hat{\eta}$, $\hat{\alpha}$, $\hat{\beta}_1$, and $\hat{\rho}$ are the PLEs, and exponentiation of the sum of the second and third terms constitutes the gamma distribution scaling parameter. An initial geographic distribution of Scottish district counts is obtained by randomly sampling from $n = 56$ independent Poisson-gamma distributions with shape parameter $\frac{1}{\hat{\eta}}$ for the gamma distribution, and mean $\mu_i = G_i\frac{1}{\hat{\eta}}\exp\left[\hat{\alpha} + \hat{\beta}_1\,\mathrm{LN}\,(x_{1,i} + 1.2) + \mathrm{LN}\,(E_i + 0.5)\right]$ for the accompanying Poisson distribution; this set of computations allows $\hat{\rho}\sum_{j=1}^{n} w_{ij}y_j$ to be calculated in the second iteration. During each subsequent sampling iteration, each of the 56 Scottish districts is visited in turn, but in a random order, and the district count in question is replaced by sampling from a Winsorized Poisson distribution with mean $G_i\frac{1}{\hat{\eta}}\exp[\hat{\alpha}+\hat{\beta}_1\,\mathrm{LN}\,(x_{1,i} + 1.2) + \hat{\rho}\sum_{j=1}^{n} w_{ij}y_j + \mathrm{LN}\,(E_i + 0.5)]$; the Winsorizing threshold employed here is $3O_{max}$, where O_{max} is the maximum observed count, and the district-by-district expected number of cases were rescaled after each random sampling selection to maintain the equality of observed and expected cases sums. These iterations are repeated until convergence of the triplet of sufficient statistics (i.e., $T_1 = \sum_{i=1}^{n} O_i$ for $\hat{\alpha}$, $T_2 = \sum_{i=1}^{n} O_i L(x_{1,i} + 0.5)$ for $\hat{\beta}_1$, and $T_3 = O_i \sum_{i=1}^{n}\sum_{j=1}^{n} w_{ij}y_j/2$ for $\hat{\rho}$) is attained.

The first 25,000 of 525,000 iterations executed for a given MCMC chain were discarded (i.e., the "burn in" period) to remove transient states toward the equilibrium distribution for $\hat{\alpha}$, $\hat{\beta}_1$, and $\hat{\rho}$. The remaining 500,000 iterations were weeded such that the sufficient statistics for only every 100th simulated map were retained, resulting in chains of length 5,000. Using three independently generated chains of sufficient statistics, constructed time series plots and correlograms confirms independence of retained maps, and after arbitrarily dividing each chain into 125 consecutive groups of 40 map results, ANOVA confirms within-chain convergence. Two-way ANOVA confirms consistency of the three trajectories across the 125 groupings (i.e., between-chain convergence). Summary statistics for these various diagnostics appearing in Tables 7.3 and 7.4[10] indicate that the chains are well behaved and should yield sound estimates (the lowest marginal probability is 0.170;

[10] The Levene test statistic was used to assess homogeneity of variance across groupings because the magnitude of the numbers involved allows them to be treated as though they approximate a continuous random variable. Meanwhile, there is no reason to expect that these sets of numbers conform to normal distributions, eliminating the possibility of using the Bartlett test statistic. The R measure is described in Gelman and Rubin (1992).

	Chain #1		Chain #2		Chain #3	
Parameter	Estimate	Asymptotic standard error	Estimate	Asymptotic standard error	Estimate	Asymptotic standard error
$\hat{\alpha}$	–0.4583	0.0524	–0.4556	0.0465	–0.4631	0.0376
$\hat{\beta}_1$	0.2527	0.0286	0.2515	0.0254	0.2550	0.0201
$\hat{\rho}$	0.8061	0.0622	0.8050	0.0604	0.7985	0.0352

Table 7.3 Diagnostic statistics for the MCMC-generated Winsorized auto-Poisson chains (5000 retained iterations)

		Time series plot slope		AR(1)		ANOVA			
Chain	Sufficient statistic	t	Prob	t	Prob	F	Prob	Levene	Prob
1	T_1	–0.79	0.428	1.08	0.281	0.71	0.994	1.03	0.404
	T_2	–0.83	0.404	0.99	0.324	0.70	0.995	1.06	0.315
	T_3	0.09	0.932	0.11	0.913	0.87	0.851	0.87	0.841
2	T_1	–0.25	0.801	0.91	0.361	0.92	0.723	0.92	0.731
	T_2	–0.35	0.727	0.89	0.376	0.92	0.715	0.90	0.766
	T_3	–1.37	0.170	0.85	0.396	1.03	0.396	0.99	0.521
3	T_1	–0.09	0.931	0.72	0.475	0.98	0.532	0.83	0.921
	T_2	–0.08	0.938	0.85	0.398	0.95	0.633	0.81	0.938
	T_3	–0.17	0.866	–0.69	0.492	0.72	0.992	0.83	0.908

NOTE: prob denotes the probability under the null hypothesis of a zero effect; these are two-tail for the t-statistics

the Gelman-Rubin R indices are greater than but close to 1.2, suggesting that perhaps the chains could be somewhat longer). The respective MCMC-MLEs and their approximate asymptotic standard errors[11] (Huffer and Wu, 1998, p. 513) are as follows:

Parameter	Average	Standard deviation	P(S-W)	PLE standard error
$\hat{\alpha}$	–0.4613	0.0033	0.759	0.1765
$\hat{\beta}_1$	0.2542	0.0017	0.608	0.0853
$\hat{\rho}$	0.8039	0.0056	0.908	0.1168

These results were further corroborated by generating 100 additional independent chains of size 5,000 (35.2 million Winsorizings occurred across these simulations), using the previously outlined MCMC protocol, and computing the MCMC-MLEs for each additional chain. Average results for the 100 chains are included in the following tabulation:

[11]These computations are based upon the Fisher information matrix.

Table 7.4 Diagnostics for 3 MCMC-generated Winsorized auto-Poisson chains, each grouped into 125 batches of 40 weeded iterations (5000 retained iterations)

Two-way ANOVA statistics

Sufficient statistic	Groupings F	Groupings prob	Chains F	Chains prob	Interaction F	Interaction prob	Levene	Prob
T_1	0.79	0.960	0.49	0.610	0.91	0.832	0.92	0.849
T_2	0.77	0.972	0.57	0.563	0.90	0.858	0.92	0.851
T_3	0.81	0.938	0.67	0.514	0.90	0.861	0.90	0.919

R measures

Sufficient statistic	Overall R	$R(k = 40)$ Mean	Standard deviation	Minimum	Maximum	# > 1.2
T_1	1.189	1.296	0.256	0.992	2.147	70
T_2	1.217	1.293	0.253	0.989	2.235	71
T_3	1.248	1.298	0.241	0.989	2.066	73

NOTE: prob denotes the probability under the null hypothesis of a zero effect

These standard deviations mat be viewed as the simulation equivalents[12] to the preceding asymptotic standard errors, and indicate that: relatively little variation occurs across chains, the asymptotic standard errors may be too large (perhaps because $n = 56$), and the PLE standard errors are far too large (which commonly is the case in the presence of positive SA). And, the three individual chain results are in keeping with these averages, as well as the negative binomial pseudo-likelihood parameter estimates and their standard errors computed with the 100 simulated maps from each 525,000th iteration are in keeping with their observed data counterparts.

7.3.3 A Proper CAR Model Specification via GeoBUGS

The following Poisson HGLM with the log link function was estimated using the Scottish lip cancer data and GeoBUGS:

$$\mathrm{LN}(\mu_i) = \mathrm{LN}\,(\mathrm{E}_i + 0.5) + \alpha + \beta_1 \mathrm{LN}\left(\mathrm{X}_{i,1} + 1.2\right) + \nu_i, \tag{7.5}$$

[12] Huffer and Wu (1998, p. 514) note that studying the multivariate behavior of MCMC parameter estimates is a rather complicated and daunting problem, and suggest examining only univariate aspects of the sampling distributions of the individual MCMC estimates (i.e., each parameter estimate separately).

where ν_i denotes unobserved district-specific log-relative risks. The prior distributions attached to this log-mean response equation are:

$$O_i \sim \text{Poisson}(e^{LN(E_i+0.5)+\alpha+\beta_1 LN(X_{i,1}+1.2)+\nu_i}),$$

$\alpha \sim \text{normal}(0, 0.0001)$,
$\beta_1 \sim \text{normal}(0, 0.0001)$,
$\nu_i \sim \text{auto-normal}(\sum_{j=1}^{n} c_{ij}v_j / \sum_{j=1}^{n} c_{ij}, \sigma_\varepsilon^{-2} / \sum_{j=1}^{n} c_{ij})$, with a conditional autoregressive model specification corresponding to a proper multivariate Gaussian distribution with a full-rank covariance matrix $(\mathbf{I} - \rho\mathbf{D}^{-1}\mathbf{C})\mathbf{D}^{-1}$, where matrices **C** and **D** are defined in Sect. 7.2,
$\sigma_\varepsilon^{-2} \sim \text{gamma}(0.5, 0.0005)$, and
$\rho \sim \text{uniform}(1/\lambda_{56}, 1/\lambda_1)$, where λ_{56} and λ_1 respectively are the smallest and largest eigenvalues of matrix $\mathbf{D}^{-1/2}\ \mathbf{C}\mathbf{D}^{-1/2}$,

where ~ denotes "distributed as." This is a hybrid version of specifications contained in the GeoBUGS example and in Lee and Nelder (2001). Variable $LN(X_1 + 1.2)$ is used in place of $X_1/10$ to maintain consistency with other analyses summarized in this chapter. Of note, as before, the translation value of 0.5 added to E_i has been retained to facilitate comparisons. In part, these priors were selected because counts tend to conform to a Poisson distribution, sampling distributions of regression coefficients tend to conform to a bell-shaped curve, inverse variance tends to conform to a gamma distribution, and the SA parameter value is contained in a restricted interval (see Sect. 7.2.1).

Of a total of 40,000 iterations executed for a given GeoBUGS MCMC chain (random number generator seeds: 314159 for chain #1, and 50001 for chain #2; each chain required roughly 30 minutes of execution time), the first 15,000 were discarded to remove transient states toward the equilibrium distribution for $\hat{\alpha}$, $\hat{\beta}_1$, $\hat{\rho}$, and $\hat{\sigma}_\varepsilon^2$, where $\hat{\rho}$ is the estimated CAR spatial autoregression coefficient and $\hat{\sigma}_\varepsilon^2$ is the estimated error variance; these first 15,000 results are the "burn in" period. The remaining 25,000 iterations were weeded such that the parameter estimates for only every hundredth simulated map were retained, resulting in chains of length 250. Using two independently generated chains of parameter estimates, time series plots and correlograms were constructed to confirm independence of retained maps, and after arbitrarily dividing each chain into 10 consecutive groups of 25 map results, ANOVA was performed to confirm within-chain convergence. A two-way ANOVA was used to confirm consistency of the two trajectories across the 10 groupings (i.e., between-chain convergence). Summary statistics for these various diagnostics are similar to those reported in Tables 7.3, 7.4. Overall these diagnostics indicate that the chains are well behaved and should yield sound estimates. The respective BUGS estimates together with their standard errors and normality diagnostics (based on the S-W test statistic probability) are as follows:

These results compare favorably with their MCMC maximum likelihood counterparts obtained with the Winsorized auto-Poisson analysis (see Sect. 7.3.2). The

	Chain #1			Chain #2		
Parameter	Average	Standard error	P(S-W)	Average	Standard error	P(S-W)
$\hat{\alpha}$	–0.6486	0.4995	<0.0001	–0.7187	0.5844	<0.0001
$\hat{\beta}_1$	0.3140	0.0991	0.287	0.3262	0.0888	0.002
$\hat{\rho}$	0.9630	0.0421	<0.0001	0.9551	0.0471	<0.0001
$\hat{\sigma}_\varepsilon^2$	0.4908	0.1984	<0.0001	0.4741	0.2046	<0.0001

value of 0.96 for $\hat{\rho}$ is very close to its upper limit value of 1, similar to the value of 0.175 (whose upper limit is 1/5.70803 = 0.17519) reported by Lee and Nelder (2001, p. 14) based upon $\mathbf{D} = \mathbf{I}$. The precision of these estimates could be improved by increasing the chain lengths, but only at the cost of considerably more computer execution time

The posterior distribution of the 250 estimates of the covariate coefficient, $\hat{\beta}$, appear to conform to a normal distribution, as is indicated by their S-W test statistic probabilities, P(S-W). But the two posterior distributions of the 250 intercept term estimates, $\hat{\alpha}$, fail to conform to a normal distribution, as is indicated by their S-W test statistic probabilities. The 250 estimates of the SA parameter, $\hat{\rho}$, are strongly skewed by being constrained to be less than its upper limit of 1. Meanwhile, similar to a product moment correlation coefficient, the LN$(0.04 + \sqrt{\frac{1+\hat{\rho}}{1-\hat{\rho}}})$ transformed values conform closely to a normal distribution, indicated by respective S-W test statistic probabilities of 0.895 and 0.728 (S-W increases from 0.700 to 0.997 for chain #1, and from 0.718 to 0.996 for chain #2).

One important finding here is that $\hat{\rho}$ is very close to its upper limit, suggesting that perhaps a 2nd-order spatial covariance matrix would be more appropriate (see Sect. 7.2.1), or that an improper CAR specification can be employed. Another important findings is that the percentage of each district′s outdoor labor force employed in agriculture, fishing, or forestry remains a statistically significant covariate.

7.4 Spatial Filter Model Specifications

SF models are models that include synthetic map pattern variates to account for the presence of non-zero SA. These specifications may be employed with conventional statistical theory, usurping spatial dependency effects from model residuals. Because the synthetic variates are landscape specific, spatial filtering is semi-parametric in nature.

7.4.1 The Log-normal Approximation Spatial Filter Model

The log-normal approximation SF specification relates directly to the preceding discussion about inflation and deflation of correlation coefficients (Sect. 7.2.3).

Because a conventional multiple linear regression model is employed, the eigenvectors retain their properties of orthogonality and uncorrelatedness. These two properties can be compromised if a weighted multiple linear regression analysis is undertaken.

Part of the SA latent in Scottish lip cancer cases is captured by SA latent in the outdoor labor percentage covariate. This commonality is made explicit by the preceding correlation decomposition (see Sect. 7.2.1). A stepwise linear regression analysis forcing this covariate to be in the equation results in the selected subset of eigenvectors being from those listed in Table 7.1 (i.e., $\mathbf{E}_1$, $\mathbf{E}_2$, $\mathbf{E}_3$, $\mathbf{E}_4$, and $\mathbf{E}_{19}$). The outdoor labor percentage covariate has a significant regression coefficient, in part because it completely captures latent SA in lip cancer cases represented by eigenvector $\mathbf{E}_9$, and hence removes this eigenvector from the regression equation, and through multicollinearity pulls eigenvector $\mathbf{E}_4$ into the regression equation. With MC = 0.91753, the linear combination of eigenvectors represents strong positive SA. Both the covariate and this linear combination account for roughly 58 per cent of the variance in the transform lip cancer cases. The residual MC is nonsignificant ($z_{MC} = 0.64$). The residuals fail to conform to a normal distribution [P(S-W) = 0.008]. And, the cloud of points for the residual-versus-predicted scatterplot continues to be well-behaved except for one of the two 0-count districts.

7.4.2 A Poisson Spatial Filter Model

Weaknesses of the preceding log-normal approximation, especially those identified with diagnostic statistics, suggest that changing the underlying probability model from a bell-shaped curve may be worthwhile. Griffith (2002a) outlines specification of a SF version of the auto-Poisson model. Adopting this specification is equivalent to maintaining that overdispersion implied by $\hat{\eta} = 0.3989$ (reported in Sect. 7.2) is attributable to the geographically aggregated lip cancer case counts being dependent—i.e., SA. Hence, the following Poisson specification is posited here:

$$\mathrm{LN}(\mu_i) = \mathrm{LN}\left(E_i + 0.5\right) + \alpha + \beta \mathrm{LN}\left(X_{i,1} + 1.2\right) + \sum_{k=1}^{K} \beta_k E_{i,k}, \qquad (7.6)$$

where $E_{i,k}$ is the ith element of eigenvector $\mathbf{E}_k$, and β_k is the GLM parameter associated with eigenvector $\mathbf{E}_k$. The linear combination of eigenvectors, $\sum_{k=1}^{K} \beta_k E_{i,k}$, is the SF. Of note is that the translation parameter, 0.5, added to the expected values, E_i, has been retained here to facilitate comparisons with the SAR model. A researcher most likely would not include this translation value in practice because an expected count cannot be 0 unless the population at risk is of size 0 (a trivial case).

$K = 8$ eigenvectors were selected using the stepwise Poisson regression procedure available in Stata (see Table 7.5). This procedure resulted in selection of

Table 7.5 Spatial filter generalized linear model parameter estimates

	Poisson probability model				negative binomial probability model			
Variable	Parameter estimate	Standard error	χ^2 statistic	Probability of exceeding χ^2	Parameter estimate	Standard error	χ^2 statistic	Probability of exceeding χ^2
Intercept	–0.3296	0.1460	5.10	0.0240	–0.4830	0.1787	7.30	0.0069
LN(X_1+1.2)	0.2074	0.0725	8.19	0.0042	0.2873	0.0859	11.18	0.0008
$\mathbf{E}_1$	2.3397	0.3720	39.55	<0.0001	2.2760	0.4910	21.48	<0.0001
$\mathbf{E}_2$	–1.9282	0.4074	22.40	<0.0001	–1.8397	0.5616	10.73	0.0011
$\mathbf{E}_3$	1.0477	0.3690	8.06	0.0045	0.9264	0.4976	3.47	0.0626
$\mathbf{E}_4$	0.9763	0.3311	8.69	0.0032	0.9727	0.4547	4.58	0.0324
$\mathbf{E}_9$	–0.7188	0.3990	3.25	0.0716				
$\mathbf{E}_{19}$	1.6312	0.3521	21.46	<0.0001	1.5853	0.4850	10.69	0.0011
$\mathbf{E}_{24}$	–0.7183	0.3538	4.12	0.0423				
$\mathbf{E}_{32}$	0.6117	0.3063	3.99	0.0458				
dispersion	0				0.0820	0.0357		

those eigenvectors appearing in Table 7.1, as well as the following additional eigenvector: $\mathbf{E}_{24}$. Unfortunately, this additional eigenvector may be a consequence of losing the orthogonality and uncorrelatedness properties because of the weighting involved in Poisson regression parameter estimation; moreover, collinearity among the weighted eigenvectors causes some difficulty in estimation. The standardized Pearson residuals for counts from this analysis display only trace SA (MC = 0.00508, GR = 0.83780), and predicted counts from this analysis have a pseudo-R^2 of 0.702, while predicted SMR values have a pseudo-R^2 of 0.603. Although detected overdispersion has been reduced (the deviance statistic decreases from 3.64 to 2.06), 23 of the district means fail to satisfy the condition $\hat{\eta} = 0.0820 << \frac{1}{\mu}$. But the decrease in the deviance statistic together with the decrease in $\hat{\eta}$ from 0.2621 to 0.0820 support the notion that much of the overdispersion detected in the preceding analysis is attributable to spatial dependence.

$K = 5$ eigenvectors were selected using the stepwise negative binomial regression procedure available in Stata (see Table 7.5). This procedure resulted in selection of a subset of those Poisson regression eigenvectors appearing in Table 7.5, with $\mathbf{E}_9$, $\mathbf{E}_{24}$, and $\mathbf{E}_{32}$ having statistically nonsignificant coefficients here. This specification yields $\hat{\eta} = 0.0820$, and a deviance statistic of 1.23. Because essentially only the regression coefficient standard errors are altered by this change in specifications, with insignificant eigenvectors not being selected by the stepwise procedure, the pseudo-R^2 values changed little; the one for predicted counts decreases to 0.630, whereas the one for predicted SMR values increases to 0.631. The SA indices also basically remain unchanged (MC = 0.00572, GR = 0.82068).

7.4.3 A Spatial Filter Model Specification via BUGS

The improper (or intrinsic) CAR is one alternative to the proper CAR prior specification for random effects, where estimation of the SA parameter ρ is replaced by setting it equal to 1 and then including a second random effects term with an exchangeable, say normal, prior. Thus, the total random effect for each areal unit is the sum of two terms: a spatially structured and an unstructured random component. Given a $\hat{\rho}$ value very close to 1, this option offers a reasonable alternative for the Scottish lip cancer data. This specification is called a convolution prior (Besag et al., 1991; Mollie, 1996), and is viewed as being more flexible than simply assuming CAR random effects, because results can be partitioned into those due to spatially structured variation, and those due to unstructured over-dispersion. The relative amounts of these two components indicate the comparative importance of each.

In keeping with the tradition of principal components regression analysis, a SF, which first is estimated with a conventional stepwise Poisson regression procedure, also can function as a spatially structured component for random effects, replacing the improper CAR term. Accordingly, its estimated coefficient should be close to 1. One advantage of this approach is that a spatial autoregressive structure is not needed; rather, it is replaced by a composite map pattern component specifying the areal unit means in such a way that spatial structure is introduced into random effects.

The following SF Poisson HGLM with the log link function was estimated using the Scottish lip cancer data and BUGS:

$$\mathrm{LN}(\mu_i) = \mathrm{LN}\left(\mathrm{E_i} + 0.5\right) + \alpha + \beta_1 \mathrm{LN}\left(\mathrm{X_{i,1}} + 1.2\right) + \beta_2 \mathrm{F_i} + \nu_i, \qquad (7.7)$$

where $\mathrm{F_i} = \sum_{k=1}^{K} \beta_k \mathrm{E_{i,k}}$ denotes the SF term for areal unit i. The prior distributions attached to this log-mean response equation are the same as before for α, β_1, and σ_ε^{-2}, and are as follows for the other parameters:

$$\mathrm{O_i} \sim \mathrm{Poisson}(e^{\mathrm{LN(E_i+0.5)}+\alpha+\beta_1\mathrm{LN(X_{i,1}+1.2)}+\beta_2\mathrm{F_i}+\nu_i}),$$

$$\beta_2 \sim \mathrm{normal}\,(0, 0.0001)\,,$$

$$\nu_i \sim \mathrm{normal}(0, \sigma_\varepsilon^{-2}).$$

Of note, as in the preceding section, the translation value of 0.5 added to $\mathrm{E_i}$ has been retained to facilitate comparisons. In part, these priors were selected because counts tend to conform to a Poisson distribution, sampling distributions of regression coefficients tend to conform to a bell-shaped curve, and inverse variance tends to conform to a gamma distribution.

Of a total of 525,000 iterations executed for a given BUGS MCMC chain (seeds: 314159 for chain #1, and 50001 for chain #2; each chain required roughly 8 minutes of execution time), the first 25,000 were discarded to remove transient states toward the equilibrium distribution for $\hat{\alpha}$, $\hat{\beta}_1$, $\hat{\beta}_2$, and $\hat{\sigma}_{\varepsilon}^2$, where $\hat{\beta}_2$ is the estimated SF regression coefficient; these first 25,000 results are the "burn in" period. The remaining 500,000 iterations were weeded such that the parameter estimates for only every hundredth simulated map were retained, resulting in chains of length 5,000. Using two independently generated chains of parameter estimates, time series plots and correlograms were constructed to confirm independence of retained maps, and after arbitrarily dividing each chain into 125 consecutive groups of 40 map results, ANOVA was performed to confirm within-chain convergence. A repeated measures ANOVA was used to confirm consistency of the two trajectories across the 125 groupings (i.e., between-chain convergence; repeated measures were used because chains tended to be nearly identical, regardless of the random number seeds or the initial parameter estimates used). Summary statistics for these various diagnostics are similar to those reported in Tables 7.3 and 7.4. Overall these diagnostics indicate that the chains are well behaved and should yield sound estimates (the most problematic chains are for $\hat{\sigma}_{\varepsilon}^2$). The respective BUGS posterior distribution estimates together with their standard errors and normality diagnostics [based on the Kolmogorov-Smirnov (K-S) test statistic probabilities, P(K-S)] are as follows:

	Chain #1			Chain #2		
Parameter	Average	Standard error	P(K-S)	Average	Standard error	P(K-S)
$\hat{\alpha}$	–0.3735	0.1562	0.023	–0.3735	0.1562	0.025
$\hat{\beta}_1$	0.2213	0.0745	> 0.150	0.2213	0.0744	> 0.150
$\hat{\beta}_2$	0.9898	0.1234	> 0.150	0.9898	0.1235	> 0.150
$\hat{\sigma}_{\varepsilon}^2$	0.0511	0.0353	< 0.010	0.0510	0.0353	< 0.010

These results compare favorably with the reported in Table 7.5 (now the $\hat{\beta}_2$ variance is 0.1371).

7.5 Discussion

This chapter examines a variety of ways to model georeferenced counts data, contrasting conventional specifications with SF specifications. In doing so, it highlights several important issues that arise in this context. Foremost is the use of spatial filtering techniques. Second is the interplay between missing covariates and an accounting for SA in model specifications. And, third is implications for data mapping.

Table 7.6 Cross-validation results for spatial filter model parameter estimates

	Poisson parameter estimates		Negative binomial parameter estimates	
Variable	Average	Standard deviation	Average	Standard deviation
Deviance/df	2.1020	0.0656	–0.4834	0.0331
Intercept	–0.3303	0.0315	0.2874	0.0164
LN(X_1+1.2)	0.2078	0.0161	2.2766	0.0712
$\mathbf{E}_1$	2.3412	0.0693	–1.8398	0.0854
$\mathbf{E}_2$	–1.9306	0.0931	0.9268	0.0701
$\mathbf{E}_3$	1.0448	0.0859	0.9731	0.0619
$\mathbf{E}_4$	0.9744	0.0675	–0.4834	0.0331
$\mathbf{E}_9$	–0.7208	0.0771		
$\mathbf{E}_{19}$	1.6302	0.0691	1.5848	0.0615
$\mathbf{E}_{24}$	–0.7164	0.0816		
$\mathbf{E}_{32}$	0.6113	0.0655		
Dispersion	0		0.0816	0.0055

7.5.1 Cross-validation Results for the Poisson Spatial Filter Model

The SF model results reported in Sect. 7.3.4 were subjected to a cross-validation analysis (Table 7.6). In other words, leaving out one Scottish district at a time, in turn, the Poisson SF model was estimated with data for the remaining 55 districts. This analysis yielded an average pseudo-R^2 of 0.704, with a standard error of 0.011. It also produced a mean squared prediction error of 35.9, which compares favorably with the model's predicted mean squared error of 21.2. The correlation between the predicted and observed values is moderate-to-strong ($r = 0.736$).

The maximum likelihood parameter estimates and the average cross-validation parameter estimates are very similar, implying unbiasedness. In addition, cross-validation standard deviations are relatively small, all being less than 0.1. The cross-validation ratio of the deviance statistic to the number of degrees of freedom (i.e., deviance/df) also is very similar to its full model estimated value (i.e., 2.1020). In both cases the value is very close to 2, which is on the threshold of problematic overdispersion (Hardin, 2001, p. 115). Similar findings were obtained for the negative binomial model.

7.5.2 A Simulation Experiment Based Upon the Poisson Spatial Filter Model

Mean responses for area-specific Poisson random variables are given by Eq. (7.6). These n values can be used to simulate n independent drawings from a Poisson distribution with the means (which automatically equal the corresponding variances) defined by Eq. (7.6). Spatial autocorrelation effects captured by Eq. (7.6) are transferred to the simulated data, resulting in preservation of the SF map pattern. Once

Table 7.7 Simulation experiment results for spatial filter model parameter estimates

	Poisson parameter estimates		Negative binomial parameter estimates	
Variable	Average	Standard deviation	Average	Standard deviation
intercept	–0.3394	0.1469	–0.4938	0.1250
$LN(X_1+1.2)$	0.2073	0.0727	0.2864	0.0621
$\mathbf{E}_1$	2.3477	0.3746	2.2880	0.3535
$\mathbf{E}_2$	–1.9353	0.4078	–1.8454	0.4072
$\mathbf{E}_3$	1.0466	0.3652	0.9188	0.3838
$\mathbf{E}_4$	0.9778	0.3297	0.9746	0.3286
$\mathbf{E}_9$	–0.7226	0.4049		
$\mathbf{E}_{19}$	1.6371	0.3593	1.6045	0.3734
$\mathbf{E}_{24}$	–0.7091	0.3541		
$\mathbf{E}_{32}$	0.5993	0.3110		
dispersion	0		0.0018	0.0058

NOTE: the simulation experiment consisted of 10,000 replications; all negative binomial dispersion parameters less than 0 were replaced with 0

these simulated samples have been drawn, Poisson SF models can be estimated with each of them, allowing sampling distributions for each of the model parameters to be simulated. The outcome of this exercise allows assessment of both the parameter estimates and their accompanying standard errors computed with the observed data.

The first simulation experiment undertaken here consisted of 10,000 drawings of 56 sets of values from independent Poisson distributions. The accompanying 10,000 parameter estimation results are summarized in Table 7.7. The most noticeable discrepancies, albeit modest ones, between the observed data SF results reported in Table 7.5 and the means and standard deviations calculated with this simulation exercise (see Table 7.7) are for eigenvector $\mathbf{E}_{32}$'s coefficient, and for the variability of eigenvector $\mathbf{E}_{19}$'s coefficient. But none of these differences exceed roughly 2%, suggesting that the simpler Poisson model should be preferable to the negative binomial model. In other words, an assumption of 56 independent Poisson random variables whose means (and hence variances) are given by Eq. (7.6) appears reasonable. The second simulation experiment replicated the first one, but with random sampling from a Poisson-gamma distribution. Although average parameter estimates are roughly the same as their empirical counterparts, simulation standard errors for this case (see Tables 7.6 and 7.7) do not agree very well with the empirical results; this inconsistency may be attributable to $n = 56$ being too small for agreement with asymptotic results.

This simulation approach has several advantages over the MCMC approach associated with a Winsorized auto-Poisson model. First, SA is embedded via the SF parameterization, resulting in a much simpler simulation procedure that is void of convergence issues and capable of quickly generating massively large numbers of maps. Second, estimation results are directly comparable with those obtained for generalized linear models. And, third, estimation does not depend upon pseudo-likelihood estimation results. One disadvantage is that the SF simulation depends upon the selected set of eigenvectors that is determined with the original sample data, which may well result in some specification error.

7.5.3 Impacts of Incorporating Additional Information

One appealing feature of map analysis is that SA latent in a georeferenced response variable can be exploited to compensate for variables missing from a model specification. In doing so, dealing with redundancies allows considerable information invisible to traditional statistical techniques to be extracted from georeferenced data. Moreover, sums of surrounding nearby values, or the structure of the configuration of georeferenced values, furnish useful surrogate information for a statistical analysis. As additional variables are measured and included in a model specification, SA contained in a response variable begins to be accounted for by SA contained in these covariates (see Sect. 7.2.3). This transfer of map pattern effects from the left-hand to the right-hand side of an equation reduces the residual redundancies that can be exploited.

This situation can be illustrated with the number of males at risk (X_2) from the Scottish lip cancer data analysis reported by Cressie (1991, p. 537), whose Box-Tidwell transformation[13] was determined to be $LN(\frac{X_2}{area} - 22.2)$. Employing a spatial filtering decomposition [see Eqs. (7.2a) and (7.2b)], SA latent in this second covariate accounts for roughly half of its variability, and can be described by the six eigenvectors $\mathbf{E}_1$, $\mathbf{E}_2$, $\mathbf{E}_9$, $\mathbf{E}_{12}$, $\mathbf{E}_{15}$, $\mathbf{E}_{19}$, and $\mathbf{E}_{34}$, whose linear combination represents moderate-to-strong positive SA (MC = 0.76089). The first three of these eigenvectors are common to the SFs describing the transformed variables Y and $LN(X_1 + 1.2)$. In addition, vector $\mathbf{E}_{19}$ is common to Y, and vector $\mathbf{E}_{34}$ is common to $LN(X_1 + 1.2)$; $\mathbf{E}_{12}$ and $\mathbf{E}_{15}$ are unique to $LN(X_2/area - 22.2)$. The attribute correlation structure for these three georeferenced variables may be summarized as follows:

Y	1	0.160	–0.213
$LN(X_1 + 1.2)$	0.539	1	–0.210
$LN(X_2/area - 22.2)$	–0.540	–0.441	1
	Y	$LN(X_1 + 1.2)$	$LN(X_2/area - 22.2)$

The lower left-hand triangle of this tabulation contains the unadjusted (for SA) pairwise correlation coefficients; the upper right-hand triangle contains the spatially filtered correlation coefficients (see Sect. 7.2.3). These results reveal that the natures

[13] Cressie employs the transformation $\sqrt{\frac{\#\text{ of lip cancer cases}}{\#\text{ of males at risk}}} + \sqrt{\frac{\#\text{ of lip cancer cases}+1}{\#\text{ of males at risk}}}$.

of the attribute relationships are not changed by the presence of spatial dependencies, but that all coefficients are substantially inflated (from roughly 210% to 337%).

Analysis of the following stepwise results for the SAR and negative binomial SF models illustrates the compensatory feature of SA:

Variable added to a model	SAR model specification		Negative binomial spatial filter model specification		
	$\hat{\rho}$	Pseudo-R^2	Eigenvector #s	$\hat{\eta}$	Pseudo-R^2
none/offset	NA		none	0.4895	0.237
spatial autocorrelation	0.7199	0.541	1–4, 9, 19, 24, 32	0.0772	0.660
LN(X_1 + 1.2)	0.6720	0.549	1–4, 9, 19, 32	0.0689	0.663
LN(X_2/area – 22.2)	0.6493	0.549	1–4, 9, 19, 32, 34	0.0529	0.689

These results illustrate that including SA in a model specification can have a big impact, accounting for roughly 50% of the variance in the SAR model specification, and roughly 42% in the negative binomial SF model specification; overdispersion is dramatically reduced for this latter model specification, too. Addition of the first covariate only slightly increases the percentage of variance accounted for, while slightly reducing the role of the SA component—this is the trade-off, which partly is determined by the number of eigenvectors common to the response variable and covariate. Addition of the second covariate again only slightly increases the percentage of variance accounted for, while further reducing the magnitude of the SA component.

One important result emphasized here is that considerable insight into the geographic distribution of Scottish lip cancer incidence (i.e., the Box-Cox transformed ratio of observed to expected counts) can be gained simply by exploiting its latent SA. Another is that different model specifications yield different parameter estimates for the same covariates, mostly because of varying distributional assumptions.

7.5.4 Implications for Data Mapping

Identifying outlier areal units is a primary interest here: cases poorly accounted for by a model (i.e., either markedly exceed or markedly less than their expected values), while accounting for redundant information arising from the presence of SA. Studentized residuals for the SAR model, and Pearson residuals for the various Poisson model specifications, may be used to statistically establishing local deviations of Y values from their global map means. Each of these residuals has constant variance, and because multiple testing is involved, may be compared with a t-distribution[14] having n-p-1 degrees of freedom using the most liberal Bonferroni-adjusted critical region of size 0.05/n for each tail. The p+1 values for the respective Scottish lip cancer models are: 3 for the SAR, 4 for the Winsorized auto-Poisson-gamma, 8 for the Poisson-gamma SF, 19 for the SF BUGS, and 27 for the PCAR[15] BUGS.[16]

Based upon count residuals, areal units may be classified as *hot spots* (i.e., places where predicted counts are significantly less than their observed counterparts), *cold spots* (i.e., places where predicted counts are significantly greater than their observed counterparts), or places where predicted counts are as expected given the socio-economic/demographic composition of their inhabitants. Visual outlier identification of Scottish lip cancer counts based upon the preceding criteria [i.e., $(O_i - \hat{O}_i)/\sqrt{\hat{O}_i}$] reveals that a single aberrant district (Banff Buchan), due to its excessively large number of cases, is highlighted by the Poisson SF model. Meanwhile, Annandale is flagged as a cold spot by the proper CAR HGLM. This district is one of the two 0-case districts that introduce noticeable specification error into the log-normal approximation model specification. Because both 0-case districts fail to be detected as outliers here, differentiating them with an indicator variable as coming from a different population than the remaining 54 districts may be ill-advised. The Winsorized auto-Poisson model reveals Aberdeen as a second cold spot outlier. No outliers are identified for the other two model specifications (see Fig. 7.5a). The Bonferroni t-value criterion results in Glasgow and Dundee being added to the list of potential cold spots by both the SAR and the Windsorized auto-Poisson residuals.

Based upon SMR ratios, districts also may be classified as hot spots (i.e., places where predicted SMR values are significantly greater than 1), cold spots (i.e., places where predicted SMR values are significantly less than 1), or places where SMR values are as expected given the age and sex composition of their inhabitants. Visual outlier identification of Scottish lip cancer SMR values based upon the criterion of $\frac{O_i+0.5}{E_i+0.5} / \frac{\hat{O}_i}{E_i+0.5}$ reveals that only the proper CAR HGLM specification uncovers anomalies (see Fig. 7.5b): Clydebank appears as a hot spot, and Annandale and Tweeddale (the two 0-count districts) appear as cold spots. But some of the SMR ratios are substantially more than (by nearly 250%) or less than (by nearly 90%) 1. Furthermore, if $SMR = \widehat{SMR}$, then $SMR/\widehat{SMR} = 1$. But the variance of this ratio is less straightforward to compute. Simulation results for the GeoBUGS proper CAR model, for example, suggest a standard error value of roughly 0.20858–implying that districts whose observed SMRs are greater than 140% of their predicted SMRs

[14]The P(S-W) values for the various models are: 0.922 for the SAR, 0.703 for the Winsorized auto-Poisson, 0.445 for the Poisson spatial filter, 0.067 for the GeoBUGS proper CAR, and 0.575 for the BUGS spatial filter specification.

[15]PCAR denotes the proper conditional autoregressive model specification, which restricts the value of the autoregressive parameter to its feasible parameter space.

[16]Effective degrees of freedom were calculated in BUGS as parameter estimate p_D (see Spiegelhalter et al., 2002).

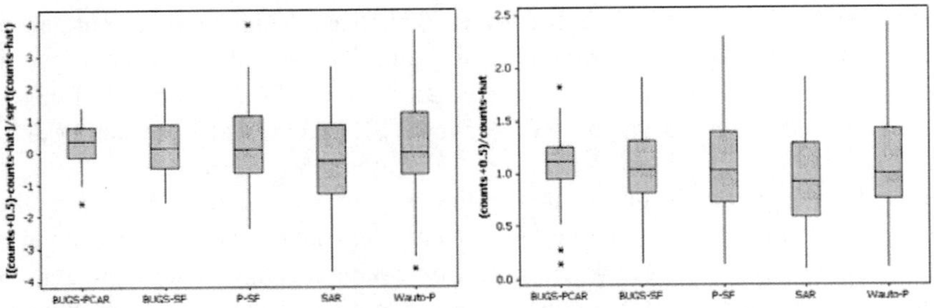

Fig. 7.5 Boxplots of deviations produced by the various model specifications. Auto-P: Winsorized auto-Poisson; BUGS-PCAR: GeoBUGS proper CAR; BUGS-SF: BUGS spatial filter; P-SF: Poisson spatial filter; and, SAR: simultaneous autoregressive. *Left* (**a**) standardized residual counts. *Right* (**b**): ratio of observed to expected SMRs

Fig. 7.6 Geographic distribution of potential *hot spots* (*black*) and *cold spots* (*gray*)

Fig. 7.7 Spatial structure for random effects. The gray quantile groups go from relatively low, negative to relatively high, positive values. *Left* (**a**): the GeoBUGS output based upon a proper CAR specification. *Right* (**b**) the BUGS output based upon a spatial filter specification

constitute potential hot spots, and districts whose ratios of these two values are less than 60% constitute potential cold spots. Because each of the sets of ratio values also conforms to a normal distribution[17], Bonferroni-adjusted t-values again can be used to uncover hot and cold spots. The results suggest that the list of potential hot spots include not just Clydebank (identified here by all but the SAR and Winsorized auto-Poisson models), but also Skye Lochlash (identified here by all but the proper CAR HGLM), NE Fife and Ettrick (identified here by all but the two HGLMs), Berwickshire (identified here by the two SF models), and Banff Buchan, Wigtown and Monklands (identified here only by the Poisson SF model). In addition, the list of potential cold spots also includes Strathkelvin (identified by the SAR and Winsorized auto-Poisson models). The map of potential hot and cold spots for the Scottish lip cancer example appears in Fig. 7.6.

The spatial structure components associated with the HGLM results appear in Fig. 7.7; these two map patterns are very similar, with a commonality exceeding 75 percent. Posterior distributions of SMRs for the two HGLMs, of which at least 53 of the 56 conform to a normal frequency distribution, reveal no hot or cold spots. But both of these analyses involve estimating $n = 56$ random effects terms. These estimated terms produce multiplicative factors for the more conventional types of predicted SMRs.

7.6 Concluding Comments

In conclusion, the SAR model implies the presence of moderate positive SA in the Scottish lip cancer data, while producing the least informative results. Its diagnostics suggest the possibility of two different populations for the SMRs, a data feature unsupported by other analyses. In part, these weaknesses may be attributable to model assumption violations (e.g., non-normal, heteroskedastic residuals). But this type of specification has the important advantage of enabling a decomposition of correlation coefficients into spatial and aspatial terms. In contrast, the Winsorized auto-Poisson specification replaces a log-normal with a Poisson distribution assumption, which results in a slight increase in its descriptive power (see Table 7.8).

The Poisson SF specification furnishes additional improvement in descriptive power. For the Scottish lip cancer SMRs, its diagnostics suggest that the violation of equidispersion primarily is attributable to latent SA. For the single covariate specification, its parameter estimates are very similar to those for the SAR model, and its analytical estimation results appear to be robust. Meanwhile, HGLMs yield appealing results based upon a Poisson distribution of case counts and a set of spatially structured normal priors. The proper CAR specification shares an auto-normal

[17]The P(S-W) values for the various models are: 0.401 for the SAR, 0.289 for the Winsorized auto-Poisson, 0.464 for the Poisson spatial filter, 0.092 for the GeoBUGS proper CAR, and 0.926 for the BUGS spatial filter specification.

Table 7.8 Bivariate regression results for observed and model-predicted SMRs

	Back-transformed SAR		Winsorized auto-Poisson		Poisson spatial filter		HGLM			
							Proper CAR		Spatial filter	
Statistic	Estimate	Prob	Estimate	Prob	Estimate	Prob	Estimate	Prob	Estimate	Prob
Intercept	2.7387	0.009	2.8468	0.004	1.1747	0.219	0.0352	0.932	–0.2201	0.714
Slope	0.5915	<0.0001	0.6681	<0.0001	0.8773	<0.0001	1.0480	<0.0001	1.0411	<0.0001
Pseudo- R^2	0.601	***	0.644	***	0.702	***	0.943	***	0.886	***

specification with the SAR and Winsorized auto-Poisson models, but only for its n random effects terms. The SF specification shares a mean response specification with the Poisson SF model. The Poisson SF specification requires numerically intensive eigenfunction calculations, whereas the Winsorized auto- model and HGLMs require numerically intensive MCMC calculations. The former can be computed with many standard software packages, while the latter requires specialized software packages (e.g., GeoBUGS). And, the SF and proper CAR specifications capture SA through their parameters, rather than through their observed values. Furthermore, similar to inclusion of a SA term in a model specification, a random effects term represents missing covariates that need to be uncovered in order to have a more

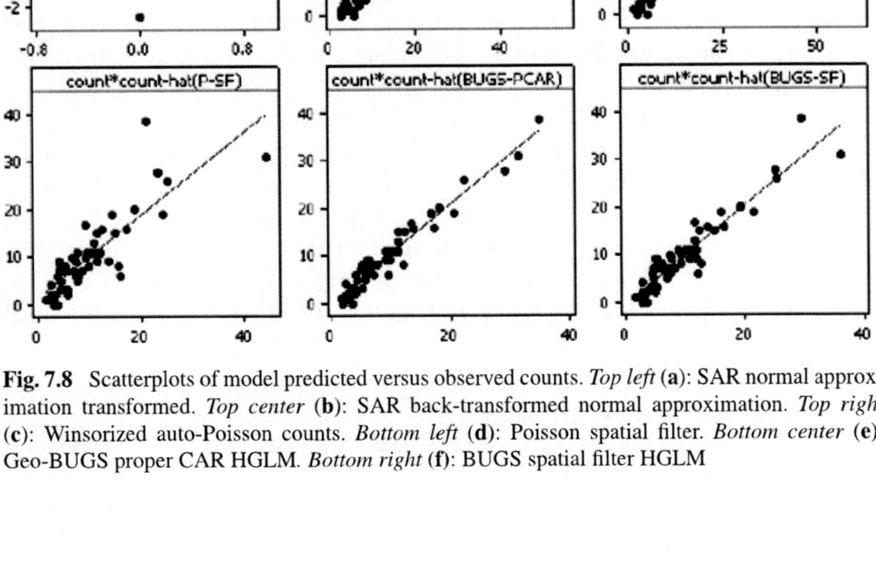

Fig. 7.8 Scatterplots of model predicted versus observed counts. *Top left* (**a**): SAR normal approximation transformed. *Top center* (**b**): SAR back-transformed normal approximation. *Top right* (**c**): Winsorized auto-Poisson counts. *Bottom left* (**d**): Poisson spatial filter. *Bottom center* (**e**): Geo-BUGS proper CAR HGLM. *Bottom right* (**f**): BUGS spatial filter HGLM

complete understanding of predicted SMRs, an understanding that would support better delineation of hot and cold spots.

Although the six models fail to agree on which districts are hot spots and which are cold spots, they do highlight some potential districts for these two categories. Furthermore, Table 7.8 and Fig. 7.8 summarize bivariate regression results obtained by regressing observed counts on predicted counts. Overall, predicted counts generated by HGLMs align best with the observed counts, with the proper CAR specification outperforming all other specifications in terms of its pseudo-R^2 value, but at a cost of considerably more degrees of freedom. Alignment for predicted counts generated by the SAR and Poisson SF models are comparable. The anticipated values for the intercept and slope regression coefficients are, respectively, –0.5 [based upon the quantity (O_i + 0.5) used throughout] and 1. The HGLM SF model specification renders bivariate regression parameter estimates most similar to these two values.

Findings reported in this chapter imply that a log-normal approximation is least desirable, a Poisson SF specification offers some useful insights, and a HGLM furnishes useful predictions, when undertaking data mapping. In addition, this case study should motivate a fuller appreciation of the variety of models researchers unfamiliar with spatial filtering can employ in their analyses.

Chapter 8
The Role of Spatial Autocorrelation in Prioritizing Sites Within a Geographic Landscape

Superfund program legislation—primarily the U.S. Comprehensive Environmental Response, Compensation and Liability Act—and its public health motivations catapulted environmental contamination issues into the forefront of society's concerns. One outcome was a report by the U.S. National Research Council (NRC, 1994) examining principal methods considered or actually employed by federal and state government agencies to prioritize the remediation of hazardous waste sites. The emphasis was on between-site variation among locations, initially overlooking within-site variation for locations. The purpose of this paper is to extend more recent work on prioritizing the remediation of subregions within a given hazardous waste site, emphasizing within-site variation for locations. These extensions are illustrated with a case study of the Murray superfund site.

8.1 Introduction: The Problem

Because an enormous amount of money and people-years of effort are needed to complete the necessary environmental restoration targeted by superfund legislation, prioritizing schemes need to identify those sites in greatest need of remediation, followed by a determination of the extent to which a selected site needs to be remediated. The environmental evaluation involved consists of three stages (after NRC, 1994, p. 66): (1) identification of environmental landscapes and concomitant biomarkers indexing the risk of exposure; (2) estimation of the sources and magnitudes of contamination; and, (3) determination of appropriate remedial actions (e.g., soil removal, groundwater treatment). Heavy metal contaminants posing significant potential threats to human health, due to their known or suspected toxicity and their abundance at superfund sites, that have been identified by the USEPA's Office of Solid Waste and Emergency Response and the Agency for Toxic Substances and Disease Registry, with some being highlighted by the Centers for Disease Control (CDC, 2001), include: arsenic (As), barium (Ba), cadmium (Cd), chromium (Cr), copper (Cu), mercury (Hg), nickel (Ni), lead (Pb), and zinc (Zn). Both As and Pb are analyzed in the case study presented in this paper. As is naturally present in groundwater, and sometimes is a residue of industrial production; As is a poison that

D.A. Griffith, J.H.P. Paelinck, *Non-standard Spatial Statistics and Spatial Econometrics*, Advances in Geographic Information Science 1,
DOI 10.1007/978-3-642-16043-1_8,

is linked to, among other diseases, cancer and diabetes. Pb is a naturally occurring, ubiquitous element that human activities geographically concentrate in the environment far beyond its natural background level; Pb is a poison that is linked to neurological and developmental illnesses, especially in children.

The second prioritization stage involves the collection of soil, water and/or air samples—called extent of contamination samples—whose pollution contents are measured. If within-site subregions are to be identified, in order to help determine the extent to which remediation should be undertaken, then samples must be geocoded. Frequently the implemented geographic sampling design is poor, in that some subregions (e.g., hot spots—concentrations of excessively high levels of a pollutant) are oversampled while other subregions are undersampled. This outcome occurs mainly because the initial objective of sampling often is to find out which toxic materials are present, and to ascertain the site-wide extent of contamination. A subregion in which high levels of contamination are detected tends to be oversampled in order to verify the clustering of high levels. But budget constraints result in other subregions of a site being more sparsely sampled, sometimes causing their evaluations to be based upon too few samples, or even no samples when the wrong locations have been sampled.

Once measures of a contaminant have been made, the relative level of the contaminant can be established. EPA bases its exposure assessment guidelines on the upper 95% confidence limit (UCL) calculated using the mean and standard deviation of contaminant concentration computed with a site′s sample measures (Bowers et al., 1996). This criterion could suggest that a site should receive a low priority score for remediation, when in fact some subregions of the site should be assigned a high priority score. Or, this criterion could suggest that a site receive a high priority score, when not every subregion of the site is severely contaminated. Subregional assessment is further complicated by the presence of spatial autocorrelation (SA) in the sample data; nearby samples contain redundant contamination information, which in turn impacts upon the UCL that is calculated.

The research problem addressed here asks:

(1) What is the correct UCL calculation? and
(2) What method should be used to identify high priority subregions of a site?

Formulating answers to these two questions requires the use of both spatial statistics and geographic information systems (GISs). These answers are illustrated here in terms of the Murray superfund site.

8.2 The Murray Superfund Site: Part I

In all, 253 geocoded aggregated surface (0–2") soil samples—a number of nearly adjacent soil samples, whose assay results are pooled for a composite measure, and then tagged with a common georeferencing coordinate—were collected in a

Fig. 8.1 Location of soil samples in the Murray superfund site. *Left* (**a**): division of the site into the four quadrants of the plane. *Right* (**b**): division of the site into the smelter parcel and residential neighborhoods, and the Thiessen polygon surface partitioning based upon soil sample locations

0.5 square mile area of Murray, Utah, and their concentrations of As and Pb measured. Of these, 173 were collected in an abandoned lead smelting facility superfund site, and 80 were collected in two of its adjacent residential neighborhoods located along the western and southern borders of the smelter site. Airborne emissions and placement of waste slag from the smelting process polluted this area. Sample Pb concentrations range from 37 parts per million (ppm) to 33,000 ppm. Sample As concentrations range from 5 ppm to 7,700 ppm. Besides differentiating geographic variability between the smelter site and its two adjacent neighborhoods, geographic variability also can be analyzed in terms of the four quadrants of the plane, which in counter-clockwise rotation respectively contain 63, 57, 68 and 65 soil sample locations. The geographic configuration of the sample points can be articulated with Thiessen polygons. These various features of this geographic landscape are portrayed in Fig. 8.1.

8.2.1 State-of-the-Art Practice

A considerable amount of effort has been devoted to handling the log-normal nature of most contamination measures—transforming a set of contamination measures by replacing them with their logarithm values results in a sample that more closely mimics a normal frequency distribution. The key analytical benefit here is reducing specification error attributable to wrongly assuming a normal distribution probability model for inferential purposes, one that does not characterize the raw data. The key communication complication here is the ability to discuss the UCL, which is based upon the normal probability model, in terms of the original measurements. Consequently, substantial effort has been expended on how to calculate accurate back-transformations (see Armstrong, 1992; Bowers et al., 1996). But whether the UCL is expressed in logarithm or raw-data measurement terms, it is severely limited when its calculation fails to accommodate SA that is latent in data.

In recognizing geographic pattern, several studies promote the use of spatial analysis for identifying high priority subregions of a contaminated site. Ginevan and Splitstone (1997) outline how kriging can be used to generalize a contamination surface from a set of sample points. Burmaster and Thompson (1997) outline the use of Thiessen polygons, with specific reference to incorporating spatial pattern of contamination into the UCL calculation; more specifically, they calculate a weighted average whose weights are the inverse areas of the Thiessen polygons.

The state-of-the-art practice illustrated by these researchers is to exploit SA in order to construct generalized contour maps, but otherwise to overlook SA, although not necessarily outcomes of the geographic configuration of sample data, in order to calculate the UCL. The methodology outlined in this paper corrects this second deficiency, incorporating SA into the UCL calculation through the use of a spatial simultaneous autoregressive (SAR) model specification, marries it to kriging based upon a semivariogram model that is consistent with the SAR model, and extends assessment to a bivariate situation. This extension satisfies Burmaster and Thompson's (1997) requirement of preserving the individual spatial patterns of, as well as the correlation between, two contaminant concentrations.

8.2.2 A Spatial Methodology: Stage 1, Spatial Sampling Data Collection and Preprocessing

The spatial methodology involves steps ranging from sample selection to identification of remediation regions. Sampling should be undertaken with two goals in mind. First, a site needs to be adequately covered. Second, pollution hot spots need to be verified. Stehman and Overton (1996) outline how to implement a hexagonal tessellation stratified random sample. This design ensures adequate coverage across a study site. It suggests that the first nearest neighbor statistic should be around 2, indicating a strong tendency for sample locations to be uniformly spaced; random selection within a hexagon avoids the sample being geographically systematic, and prevents this statistic from equaling its maximum (approximately 2.14). Often regions surrounding sample locations revealing high levels of a pollutant then are intensively sampled, in order to verify the existence of a hot spot. This second stage of the sampling process will further reduce the nearest neighbor statistic. Both of these stages would be well served by a model-informed sampling strategy that involves estimation of the nature and degree of latent SA in the geographic distribution of the pollutant. As sample intensity increases, SA tends to increase. As SA increases, total sample size should decrease, in order to minimize the collection of redundant information. An equilibrium between these two opposing trends is desirable.

The second step is to identify a variable transformation that converts the pollution measures into values that closely mimic a bell-shaped frequency distribution. Most all sample pollution measures exhibit a log-normal type of distribution (Gilbert, 1987; Millard and Neerchal, 2001), or empirically a frequency distribution where changing each data value to its natural logarithmic counterpart yields a set of values that conforms to a normal distribution. This frequency distribution tends to describe pollution measures well because they are bounded below at 0 and usually are strongly positively skewed. But a heavy metal such as Pb occurs naturally in all soils, implying that its lower bound may differ from zero, requiring a threshold parameter to be included in the log-normal distribution specification. Pollution is deposited in a geographic landscape by point source human activities, such as Pb emissions dispersing from the smoke stack of a smelter. Relatively small amounts are deposited in most locations, while increasingly larger amounts are deposited in fewer and fewer locations (perhaps near the smoke stacks). If the process depositing pollution is repetitive, then with some stochastic fluctuation (e.g., wind pattern change), each layer of pollution has approximately the same geographic distribution, resulting in new deposit amounts being proportional to existing deposit amounts at each location. Thus, the cumulative effect of many layers of small deposits is multiplicative, resulting in a log-normal distribution, and a transformed variable of the form

$$LN\,(\text{pollution concentration measure } + \delta)\,, \tag{8.1}$$

where LN denotes the natural logarithm, and δ is a translation parameter at least accounting for the naturally occurring background level of a pollutant.

Often real-world data, especially if they are georeferenced, contain considerable heterogeneity. This heterogeneity frequently is related to the magnitude of a measure. Equation (8.1) is equivalent to a Box-Cox power transformation with an exponent of zero. This zero exponent transforms positively skewed frequency distributions into ones that are more symmetric; it moves the left-hand frequency bump to the right, and squashes this bump downward, which forces the two tails to inflate. With regard to the raw measures, relatively speaking, this transformation shrinks very large values, magnifies very small values, and preserves intermediate values. Including the translation parameter, δ, primarily impacts upon one or both tails, modifying their inflation so that it better corresponds to that of a bell-shaped curve. At least some additional data heterogeneity can be accounted for by allowing δ to vary by the size of measures, or

$$LN\left[\text{pollution concentration measure} + \delta_0 + \delta_1 \left(\frac{r_1}{n+1}\right)^{\gamma_1} \left(\frac{r_2}{n+1}\right)^{\gamma_2}\right], \tag{8.2}$$

where r_1 and r_2 respectively are the ascending and descending rankings of the n pollution concentration measures, δ_0 is a translation parameter constant, δ_1 is a constant of proportionality, and γ_1 and γ_2 are exponents attached to the relative rankings. Equation (8.1) is the special case of $\delta_1 = 0$. The nonconstant translation parameter should have values contained within the range of the data, and should result in a closer alignment of the empirical and theoretical cumulative frequency distributions basically by stretching one or both of the tails of the empirical distribution. Additional heterogeneity can be accounted for by allowing the exponent to vary by the size of measures, or

$$\left[\text{pollution concentration measure} + \delta_0 + \delta_1 \left(\frac{r_1}{n+1}\right)^{\gamma_1} \left(\frac{r_2}{n+1}\right)^{\gamma_2}\right]^{\delta_2+\delta_3\left(\frac{r_1}{n+1}\right)^{\gamma_3}\left(\frac{r_2}{n+1}\right)^{\gamma_4}}, \tag{8.3}$$

where the terms of $\delta_2 + \delta_3 \left(\frac{r_1}{n+1}\right)^{\gamma_3} \left(\frac{r_2}{n+1}\right)^{\gamma_4}$ are defined in a similar fashion to those for Eq. (8.2). Equation (8.3) will tend to better align both the tails as well as the center of the empirical frequency distribution. Equations (8.1) and (8.2) are special case of $\delta_2 = \delta_3 = 0$. Equation (8.3) could have $\delta_1 = 0$, hence capturing heterogeneity solely with a nonconstant exponent. Equation (8.3) is suggested when the translation parameter values of Eqs. (8.1) and/or (8.2) fall outside the interval $(-y_{min}, y_{max})$, where y_{min} and y_{max} denote the extreme values of Y.

The third step is to krig values—spatial interpolation—and to produce the necessary quantities to calculate UCLs. Statistical analyses engaged in during this step should be nearly void of specification error, given the accommodation of assumptions of normality, constant variance, and observation independence. For remediation purposes, the important consideration is avoiding specification error. But for communication purposes, the important consideration is expressing decision criteria in understandable quantitative terms. Hence, this is the step in which a back-transformation could be calculated for communication purposes. The fourth, and final, step is to demarcate remediation subregions of a site. These third and fourth steps are spelled out in more detail in the ensuing sections of this paper.

8.3 The Murray Superfund Site: Part II

Griffith (2002b) reports a 1st nearest neighbor statistic of 0.06208 for the Murray site, indicating that the sample locations are highly clustered. Visual inspection of the maps in Fig. 8.1 suggests subregions that are under- or unsampled, subregions that are oversampled, and some apparent sampling transect.

Equation (8.2) was calibrated for both As and Pb. Normal distribution quantile plots appear in Fig. 8.2 and show the evolution of the transformed values. Both pollutants begin with the hooked quantile plot typifying untransformed log-normal data, and achieve their greatest conformity gains merely by being subjected to a simple logarithmic transformation. Inclusion of a constant translation parameter primarily better aligns the lower tails of the empirical cumulative frequency distributions with their theoretical normal cumulative frequency distribution counterpart. Capturing heterogeneity by letting the translation parameter vary with data value order ranking essentially aligns all but the largest two As values, and all of the

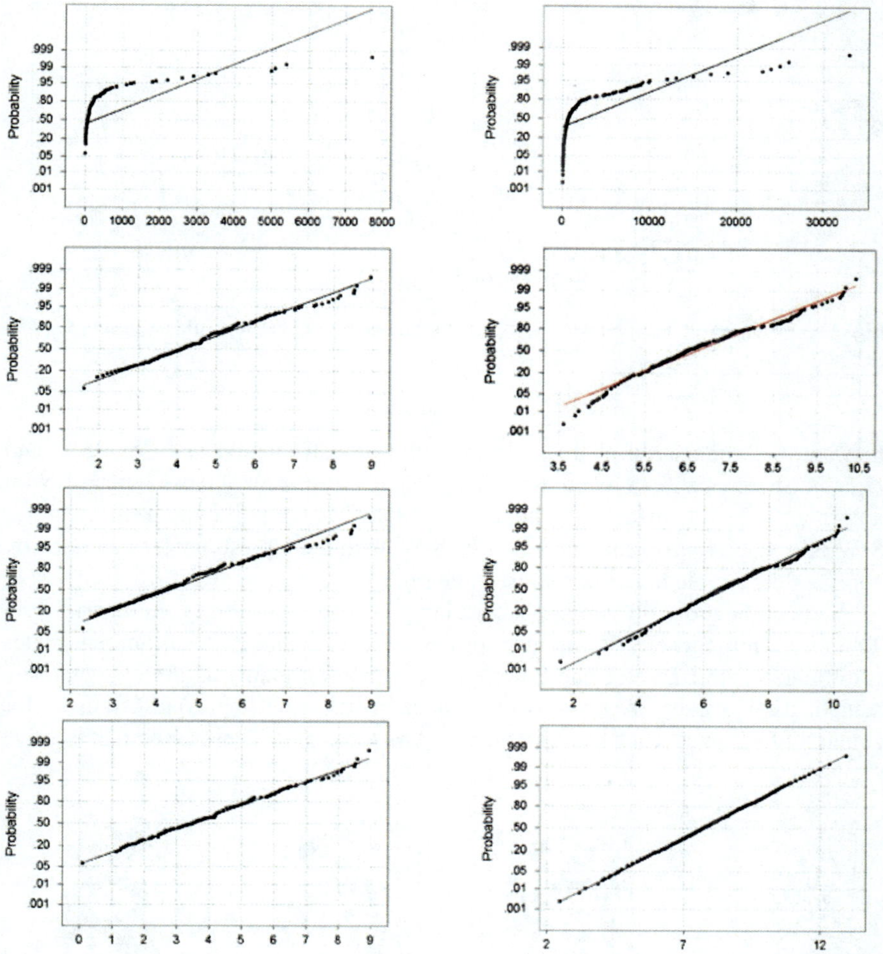

Fig. 8.2 Evolution of the pollutant data transformation quantile plots. *Left* (**a**): from top to bottom, raw As values, logarithmic As values, Eq. (8.1) As values, and Eq. (8.2) As values. *Right* (**b**): from top to bottom, raw Pb values, logarithmic Pb values, Eq. (8.1) Pb values, and Eq. (8.2) Pb values

Pb values. These results are corroborated by boxplots for these sequential transformations, which appear in Fig. 8.3. These latter graphics reveal that the frequency distribution bumps spread out from lower values toward high values, the highest values shrink toward the lower values, and improved symmetry emerges. Of note is that the As analysis is complicated by the presence of 32 measures occurring at the detection level of 5 ppm.

Geographic distributions of the relative transformation effects appear in Fig. 8.4. Both for As and Pb, conspicuous clusters of raw values are replaced by swaths of relatively high values that, for the most part, differentiate between the smelter site and the residential neighborhoods. Again, little difference is visually detectable

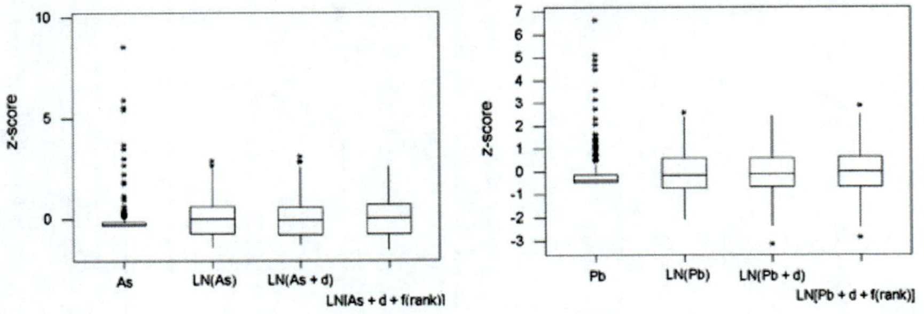

Fig. 8.3 evolution of the pollutant data transformation box plots: *top* (**a**): as results. *bottom* (**b**): pb results

between the application of a simple logarithm transformation and Eqs. (8.1) and (8.2). A Shapiro-Wilk (S-W) statistic indexing of conformity of these measures with a normal frequency distribution appears in Table 8.1; the null hypothesis value for S-W is 1. Each transformation increases S-W, with the largest increase attained by applying the simple logarithm transformation.

A quantification of geographic variability heterogeneity is summarized in Table 8.1. Homogeneity of variance for the various As and Pb measurement scales is evaluated with Bartlett's and Levene's (i.e., a non-normailty assuming diagnostic statistic used to assess the equality of variance in different samples) test statistics for equality of variance; each has a null hypothesis value of 0. These assessments are in

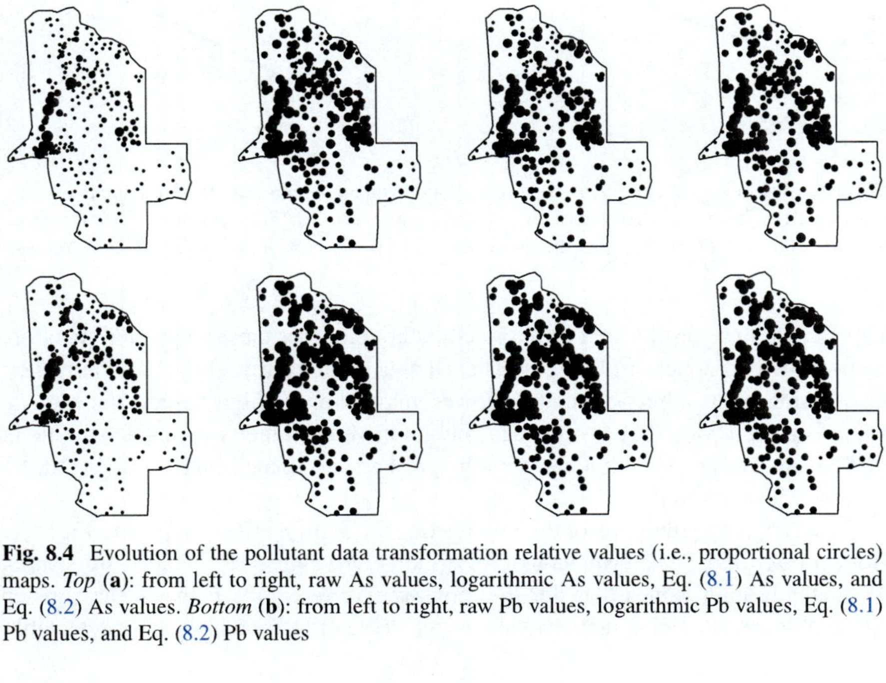

Fig. 8.4 Evolution of the pollutant data transformation relative values (i.e., proportional circles) maps. *Top* (**a**): from left to right, raw As values, logarithmic As values, Eq. (8.1) As values, and Eq. (8.2) As values. *Bottom* (**b**): from left to right, raw Pb values, logarithmic Pb values, Eq. (8.1) Pb values, and Eq. (8.2) Pb values

Table 8.1 Sequential construction of the variable transformations

	As ($5 \leq$ As $\leq 7{,}700$)				Pb ($37 \leq$ Pb $\leq 33{,}000$)		
Variable	Bartlett	Levene	S-W	R	Bartlett	Levene	S-W
Raw y	221.660***	9.017***	0.348***	0.589	34.161***	16.192***	0.523***
	221.161***	2.785***			168.109***	6.316***	
LN(y)	2.731***	23.600***	0.964***	0.740	3.005***	30.413***	0.976***
	9.011**	2.871**			28.067***	10.408***	
LN(y+δ)	3.201***	27.837***	0.970***	0.739	2.887***	26.912***	0.990*
	10.602**	2.607*			30.298***	10.297***	
LN[y+δ+f(r)]	2.465***	22.912***	0.972***	0.748	2.521***	19.819***	0.999
	7.860**	3.024**			28.361***	9.205***	
SAR residuals	1.842***	12.561***	0.995	0.706	2.236***	20.022***	0.995
	13.280***	6.098***			31.520***	13.304***	
Filter residuals	1.404	5.050**	0.996	0.688	1.442**	8.065***	0.991
	11.414**	3.550**			29.165***	9.702***	

NOTE 1: ***, **, * denote a significant difference from the null hypothesis value (0 for the Bartlett and Levene, and 1 for the S-W statistics) at, respectively, the 1, 5% and 10% level

NOTE 2: the first row Bartlett and Levene statistics test variance differences between the smelter site and neighboring residential neighborhoods

NOTE 3: the second row Bartlett and Levene statistics test variance differences between the four quadrants of the plane

NOTE 4: no evidence was found to support the presence of a heterogeneous transformation exponent

terms of both the smelter site/residential regions and the four quadrants of the plane (see Fig. 8.1). The Eq. (8.2) values display considerably less heterogeneity than do the raw values, and basically less than the simple logarithmically transformed values. But nonconstant geographic variance does persist, even though its magnitude is substantially less.

8.3.1 A Spatial Methodology: Stage 2, Spatial Statistics for Calculating UCLs

A spatial SAR model was fitted to the transformed data. A suitable surface tessellation for this purpose can be constructed using Thiessen polygons (see Fig. 8.1). The configuration of points depicted by this surface partitioning can be represented by a standard binary 0–1 geographic weights matrix, say **C**, where $c_{ij} = 1$ if two distinct points i and j share a Thiessen polygon boundary, and $c_{ij} = 0$ otherwise. The SAR model results allow the SA adjusted calculation of a mean, a standard error, and a t-statistic.

Calculation of a UCL requires an estimate of the mean, an estimate of the variance, and the number of degrees of freedom. The simplest, pure SAR model may be written as

$$\mathbf{Y} = \mu\,(1 - \rho)\,\mathbf{1} + \rho\mathbf{W}\mathbf{Y} + \boldsymbol{\varepsilon}\,, \qquad (8.4)$$

where **Y** is an n-by-1 vector of georeferenced values, **1** is an n-by-1 vector of ones, **W** is the row-standardized version of matrix **C**, μ is the mean of Y, ρ is a SA parameter, $\mu(1-\rho)$ is the mean of $(\mathbf{Y} - \rho\mathbf{WY})$, and $\boldsymbol{\varepsilon}$ is an n-by-1 independent and normally distributed, constant variance random error vector. An estimate of the mean, corrected for the presence of SA, is given by $\hat{\mu}$ obtained with Eq. (8.4), which actually is the conditional mean of Y given $\mathbf{W}_i\mathbf{Y}$ (the average of surrounding values of Y for observation i). This interpretation is based upon two features of Eq. (8.4). First, if $\rho = 0$, then SA is absent and μ is calculated with independent observation values. Second, if $\mathbf{W}_i\mathbf{Y} = 0$, then the average of the surrounding values is 0. Although this second interpretation is weakened when 0 lies outside the range of the data, conceptually it is sensible; here the transformed As minimum is close to 0, equaling 0.1, while the transformed Pb minimum of 2.5 relates to the minimum value inflated by two-thirds via the translation parameter. While gathering additional sample data that include 0 would strengthen this latter interpretation of $\hat{\mu}$, such a data collection exercise often is impractical, if not impossible.

Meanwhile, the variance estimate corrected for the presence of SA is given by

$$\hat{\sigma}^2 = (\mathbf{Y} - \hat{\mu}\mathbf{1})^{\mathrm{T}}(\mathbf{I} - \hat{\rho}\mathbf{W})^{\mathrm{T}}(\mathbf{I} - \hat{\rho}\mathbf{W})(\mathbf{Y} - \hat{\mu}\mathbf{1})/(\mathrm{n} - 2), \tag{8.5}$$

where **I** denotes the identity matrix, T denotes the matrix transpose operation, and division is by (n–2) because both μ and ρ are estimated. Because positive SA inflates the variance, the quantity yielded by Eq. (8.5) will tend to be less than its conventional counterpart of $s^2 = (\mathbf{Y} - \hat{\mu}\mathbf{1})^{\mathrm{T}}(\mathbf{Y} - \hat{\mu}\mathbf{1})/(\mathrm{n} - 1)$; the variance inflation factor here is given by $TR\{[(\mathbf{I} - \hat{\rho}\mathbf{W})^{\mathrm{T}}(\mathbf{I} - \hat{\rho}\mathbf{W})]^{-1}\}/n$, where TR denotes the matrix trace operator. This variance inflation plays a critical role in determining the effective sample size—the number of independent observations to which a set of georeferenced observations are equivalent—say n^*. In the presence of SA, as the distance between two sample locations decreases, their respective attribute values become increasingly similar, and their information content becomes increasingly redundant. Overlooking this redundant information introduces specification error into a data analysis. The purpose of calculating quantities like Eq. (8.5), using equations like (8.4), is to adjust for or remove impacts of the redundant information.

Next, consider the variance of the sampling distribution of the sample mean of variable Y, $\bar{y}$, when the variance of Y is unknown, which is given by

$$\{\mathbf{1}^{\mathrm{T}}[(\mathbf{I} - \hat{\rho}\mathbf{W})^{\mathrm{T}}(\mathbf{I} - \hat{\rho}\mathbf{W})]^{-1}\mathbf{1}/n\}\hat{\sigma}^2/\mathrm{n}, \tag{8.6}$$

and which reduces to the conventional $\hat{\sigma}^2/\mathrm{n}$ when $\rho = 0$. Rewriting Eq. (8.6) in terms of s^2 renders the following estimate of effective sample size;

$$\mathrm{n}^* = \mathrm{n}\ TR\{[(\mathbf{I}-\hat{\rho}\mathbf{W})^{\mathrm{T}}(\mathbf{I}-\hat{\rho}\mathbf{W})]^{-1}\}\ /\mathbf{1}^{\mathrm{T}}[(\mathrm{I} - \hat{\rho}\mathbf{W})^{\mathrm{T}}(\mathrm{I} - \hat{\rho}\mathbf{W})]^{-1}\mathbf{1}, \tag{8.7}$$

which reduces to n when $\rho = 0$, and asymptotically converges on 1 as ρ approaches 1 (see Griffith and Zhang, 1999). Equation (8.7) allows determination of the appropriate t-statistic, which has n^*–2 degrees of freedom.

Finally, normal curve theory states that the 95% UCL is given by

$$\bar{y} + t_{n-1,0.95}\frac{s}{\sqrt{n}}$$

which here translates into

$$\mathbf{1}^T\mathbf{Y}/n + \mathbf{t}_{n^*-2,0.95}\left(\{\mathbf{1}^T[(\mathbf{I} - \hat{\rho}\mathbf{W})^T(\mathbf{I} - \hat{\rho}\mathbf{W})]^{-1}1/n\}\times \right.$$
$$\left.[(\mathbf{Y} - \hat{\mu}\mathbf{1})^T(\mathbf{I} - \hat{\rho}\mathbf{W})^T(\mathbf{I} - \hat{\rho}\mathbf{W})(\mathbf{Y} - \hat{\mu}\mathbf{1})/(n-2)]\right)^{1/2}/\sqrt{n},$$

or

$$\mathbf{1}^T\mathbf{Y}/n + \mathbf{t}_{n^*-2,0.95}\left(\mathrm{TR}\{[(\mathbf{I} - \hat{\rho}\mathbf{W})^T(\mathbf{I} - \hat{\rho}\mathbf{W})]^{-1}\mathbf{1}/n\}\times \right.$$
$$\left.[(\mathbf{Y} - \hat{\mu}\mathbf{1})^T(\mathbf{I} - \hat{\rho}\mathbf{W})^T(\mathbf{I} - \hat{\rho}\mathbf{W})(\mathbf{Y} - \hat{\mu}\mathbf{1})/(n-2)]\right)^{1/2}/\sqrt{n^*}$$

As an aside, $\mathbf{1}^T[(\mathbf{I} - \hat{\rho}\mathbf{W})^T(\mathbf{I} - \hat{\rho}\mathbf{W})]^{-1}\mathbf{1}/n\} \approx e^{0.95\hat{\rho}/(1-0.91\hat{\rho})}$, $0 \leq \hat{\rho} < 1$, which allows a quick, easier calculation of these expressions. Ignoring impacts of SA on the sampling distribution of $\bar{y}$ results in use of the incorrect expression

$$\mathbf{1}^T\mathbf{Y}/n + \mathbf{t}_{n-2,0.95}\left(\mathrm{TR}\{[(\mathbf{I} - \hat{\rho}\mathbf{W})^T(\mathbf{I} - \hat{\rho}\mathbf{W})]^{-1}\mathbf{1}/n\}\times \right.$$
$$\left.([(\mathbf{Y} - \hat{\mu}\mathbf{1})^T(\mathbf{I} - \hat{\rho}\mathbf{W})^T(\mathbf{I} - \hat{\rho}\mathbf{W})(\mathbf{Y} - \hat{\mu}\mathbf{1})/(n-2)])\right)^{1/2}/\sqrt{n}.$$

This first expression renders UCL boundary values greater than or equal to (when $\rho = 0$) those calculated with this second expression. These are the equations used to calculate entries in Table 8.3.

8.4 The Murray Superfund Site: Part III

Results of fitting Eq. (8.4) to both the As and the Pb data appear in Table 8.2. In both cases a moderate level of positive SA is detected, with roughly a fifth of the variance in variable Y accounted for by variable $\mathbf{W}_i\mathbf{Y}$. Residual normality and variance homogeneity results appear in Table 8.1, and show close conformity with a normal distribution, but with the continued persistence of nonconstant geographic variance. The traditional predicted-versus-residual plots appear in Fig. 8.5, and suggest that, other than for the As = 5 complication, conspicuous deviations from conventional variance homogeneity are absent.

Table 8.2 Simultaneous autoregressive (SAR) modelling results

As				Pb			
		Residual				Residual	
$\hat{\rho}$	Adj-R^2	MC	GR	$\hat{\rho}$	Adj-R^2MC	GR	
0.532	0.245	–0.030	1.045	0.494	0.208	–0.034	1.085

Table 8.3 Quantities used to calculate, and the resulting, UCLs

	As		Pb	
Statistic	Uncorrected	Corrected	Uncorrected	Corrected
$\hat{\mu}$	3.46316	3.46316	7.69417	7.69417
standard error of $\hat{\mu}$	0.13344	0.25459	0.11539	0.20791
n*	253	68.9	253	77.6
Df	252	66.9	252	75.6
t-statistic for 0.95 level	1.6509	1.6680	1.6509	1.6653
UCL	3.68346	3.88782	7.88467	8.04040

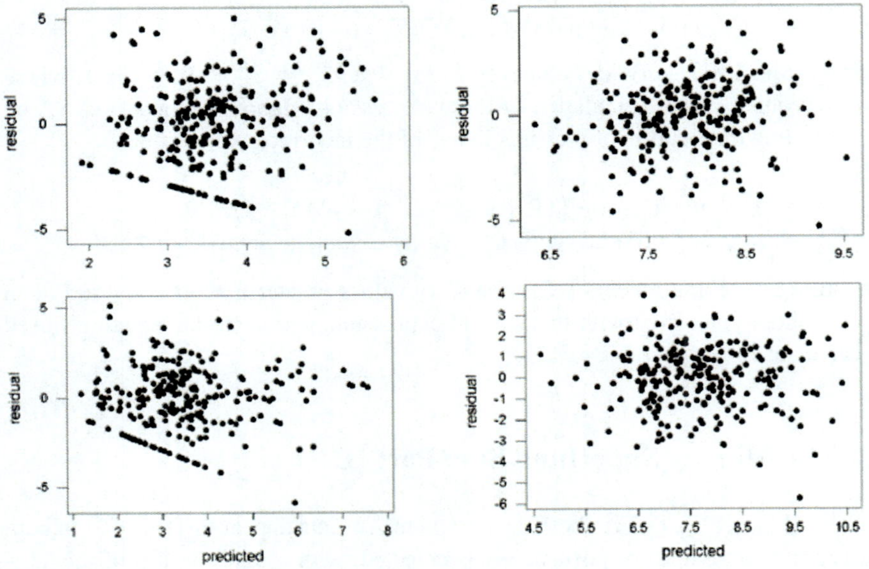

Fig. 8.5 Conventional homogeneity of variance scatter plots. *Left* (**a**): top, for spatial SAR model describing As; bottom, for spatial filter model describing As. *Right* (**a**): top, for spatial SAR model describing Pb; bottom, for spatial filter model describing Pb

The UCL results appear in Table 8.3. Variance inflation results in both the uncorrected means and their standard errors as larger than they should be, consequences that are compensatory to some degree since the mean is divided by the standard error. The presence of a moderate degree of positive SA results in effective sample sizes that are less than a third of n. This result has only a very modest impact upon the correct t-statistic, though, partially because a t-statistic converges upon a normal variate z-score as n goes to infinity; the only marked discrepancies are for values of n or n* very close to 1. The overall outcome is a UCL that expands by 2–6%. In other words, some subregions of the Murray superfund site would be misclassified as not being high priority remediation locations when in fact they are.

Geographic impacts of the changes in these UCLs include a shrinkage in area by about 8.3% of the As, and by about 14.2% of the Pb, subregions that qualify for remediation in the site. When SA is overlooked, roughly 38.5% of the Murray superfund site qualifies for remediation of As contamination, whereas roughly 35.9% of the site qualifies for remediation of Pb contamination. Respectively each of these percentages decreases to 35.3% and 30.8% once SA effects are taken into account. The marginal areas vulnerable to misclassification are located along the borders of the subregions identified with classical statistics.

8.4.1 A Spatial Methodology: Stage 3, Prioritizing Subregions for Remediation

The third step of the spatial methodology is to krig values produced by the most appropriate transformation equation [i.e., (8.1), (8.2), or (8.3)]. The semivariogram model selected for this spatial interpolation exercise needs to be consistent with the model selected for the spatial autoregressive analysis. Griffith and Layne (1999) argue that the SAR and Bessel function geostatistical semivariogram models conceptually and numerically are closely linked. This pair of models is used here to krig the As and Pb surfaces, and to compute the As and Pb UCLs.

The following Bessel function semivariogram model [Eq. (3.8), Sect. 3.2] was used to interpolate both the As and the Pb surfaces; the effective range is approximately 4r, where r denotes the range parameter. The graph of Eq. (3.8) displays a cusp in the neighborhood of $\overline{\mathrm{d}} = 0$, a characteristic of a second-order SA mechanism that also is captured by the spatial SAR model. Equation (3.8) is used to estimate the covariance (say, using matrix notation, $\mathbf{S}_{\mathrm{om}}$) between sample point pollutant measures and unsampled point pollutant measures, which are the ones to be interpolated. The m interpolated values are given by

$$\hat{\mathbf{Y}}_{\mathrm{m}} = \hat{\mu}\mathbf{1}_{\mathrm{m}} + \mathbf{S}_{\mathrm{mo}}^{\mathrm{T}}\mathbf{S}_{\mathrm{oo}}^{-1}(\mathbf{Y}_{\mathrm{o}} - \hat{\mu}\mathbf{1}_{\mathrm{o}}), \tag{8.8}$$

where the subscript m denotes values to be interpolated, the subscript o denotes observed sample values, and $\mathbf{S}_{\mathrm{oo}}$ denotes the variance-covariance matrix for observed sample values, the measures to which Eq. (3.8) is fitted (see Griffith and Layne, 1999). In effect, Eq. (8.8) spreads the information content in a sample over a map, much like spreading icing over the top of a cake. If SA does not exist in variable Y, then $\mathbf{S}_{\mathrm{oo}} = \mathbf{I}, \mathbf{S}_{\mathrm{om}} = \mathbf{0}$, and $\hat{\mathbf{Y}}_{\mathrm{m}} = \hat{\mu}\mathbf{1}_{\mathrm{m}}$; the conventional maximum likelihood estimate of a univariate missing value is the mean of the observed values.

8.5 The Murray Superfund Site: Part IV

Restricting attention to point pairs within a 1000-foot radius, Eq. (3.8) estimation results are as follows:

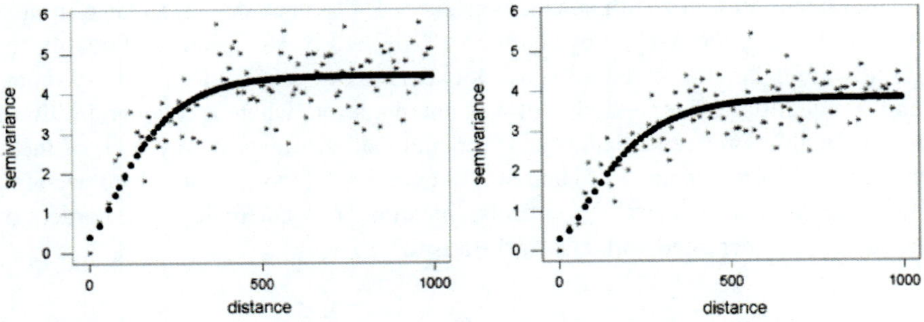

Fig. 8.6 Observed and Bessel function fitted semivariogram plots. *Top* (**a**): As results. *Bottom* (**b**): Pb results

$$\text{As}: \gamma(\bar{d}) = 0.4119 + 4.0426\left[1 - \left(\frac{\bar{d}}{109}\right)K_1\left(\frac{\bar{d}}{109}\right)\right], \text{RESS} = 0.448,$$

and

$$\text{Pb}: \gamma(\bar{d}) = 0.2995 + 3.5355\left[1 - \left(\frac{\bar{d}}{122}\right)K_1\left(\frac{\bar{d}}{122}\right)\right], \text{RESS} = 0.381,$$

where RESS denotes the relative error sum of squares (the error sum of squares divided by the total sum of squares adjusted for the mean). The semivariogram plots for these two situations appear in Fig. 8.6.

The fourth, and sometimes final, spatial methodology step is to demarcate remediation subregions of a site using both the kriged surface and the UCL of the adjusted mean. While a back-transformation can be used to compute the UCL in terms of the original pollutant measurement scale, the mapping exercise can and should retain precision by being done in the transformed variable space. The UCLs reported in Table 8.3 have been applied to the interpolation results based upon Eqs. (3.8) and (8.8). As is expected, the visually detectable swaths appearing in Fig. 8.4 reflect the high priority remediation subregions demarcated in Fig. 8.7. About half of the smelter site is ranked as high priority for remediation, as is much of the immediately adjacent residential neighborhoods, both for As and for Pb. Common to these subregions is a large portion of the western residential neighborhood, the southwest quadrant of the smelter site, and the northwester corner of the southern residential neighborhood. An additional feature of the remediation maps is the scattered set of isolated point UCLs. These locations signify subregions that are prime candidates for subsequent intensive sampling, but only when they are based upon the UCL adjusted for SA.

8.5.1 A Spatial Methodology: Stage 4, Covariation of Contaminants and Joint Pollutant Analyses

When contamination by more than a single pollutant is of concern, several additional aspects of the remediation prioritizing task arise. Foremost are covariations among

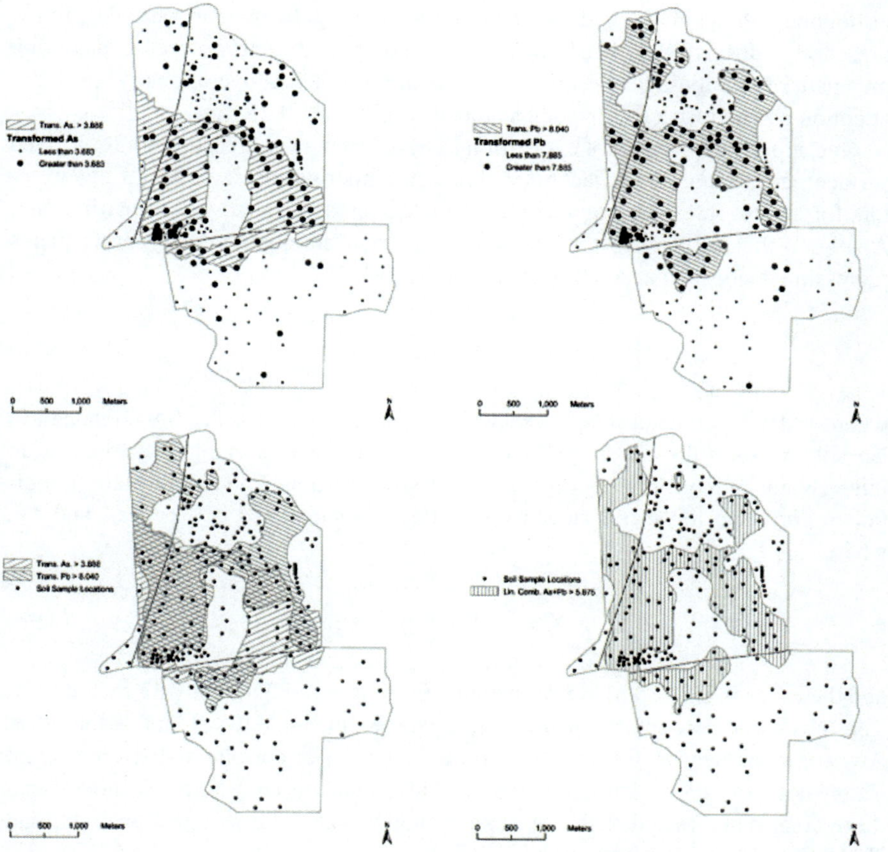

Fig. 8.7 Remediation subregions based on 95% UCLs. *Top, left* (**a**): As results. *Top, right* (**b**): Pb results. *Bottom, left* (**c**): map overlay of As and Pb results. *Bottom, right* (**d**): joint As and Pb results

pollutants. In a bivariate case, the focus is on correlation between the two pollutants as well as the SA contained in each pollutant.

Linear correlation measures are impacted upon by the log-normal nature of pollution data. Hence, correlations calculated with raw data values often do not accurately capture actual covariations. The more informative correlations are those calculated with Box-Cox transformed data values.

Meanwhile, SA also can disguise attribute covariations. Removing SA, either by dealing with the residuals of an SAR model, or the residuals from a spatial filter model, corrects for spatial dependency effects. Spatial filtering can be based upon the eigenfunctions of the numerator of the Moran Coefficient (MC) index of SA (see Griffith, 2000a), given by expression (5.8) [see Sect. 5.5]. Tiefelsdorf and Boots (1995) show that all of the eigenvalues of matrix expression (5.8) relate to specific MC values. The eigenvectors of expression (5.8) may be treated as synthetic variates, and interpreted in the context of map pattern as described in Sect. 6.2. Hence these n eigenvectors describe the full range of all possible mutually orthogonal map patterns, and may be interpreted as synthetic map variables. In the presence of positive SA, then, an analysis can employ those eigenvectors depicting map patterns exhibiting consequential levels of positive SA; operationally speaking, attention can be restricted to eigenvectors having MC ≥ 0.25, say.

One appealing property of expression (5.8) is that matrix **C** is constant for a given surface partitioning and adjacency definition, rendering the same set of eigenvectors for all attributes geographically distributed across a given surface partitioning. Another is that the eigenvectors can be used in a conventional, ordinary least squares regression analysis to account for SA. In other words,

$$\mathbf{Y} = \alpha_Y \mathbf{1} + \mathbf{E}_k \boldsymbol{\beta} + \boldsymbol{\varepsilon}_Y, \tag{8.9a}$$

where k denotes the subset of eigenvectors that accounts for the SA contained in variable Y, α_Y is the conditional mean intercept term for variable Y, and $\boldsymbol{\varepsilon}_Y$ is an independent random error term associated with variable Y. The correlation coefficient corrected for spatial dependency effects is calculated between $\boldsymbol{\varepsilon}_X$ and $\boldsymbol{\varepsilon}_Y$, where

$$\mathbf{X} = \alpha_X \mathbf{1} + \mathbf{E}_h \boldsymbol{\beta} + \boldsymbol{\varepsilon}_X, \tag{8.9b}$$

and the terms in Eq. (8.9b) are defined like those in Eq. (8.9a), but with regard to X. The subset of eigenvectors contained in $\mathbf{E}_h$ and $\mathbf{E}_k$ most likely will not be the same. Any common eigenvectors will tend to inflate the linear correlation between X and Y; any non-common eigenvectors will tend to deflate this correlation. Of note is that these eigenvectors capture the separate X and Y map patterns that Burmaster and Thompson require to be preserved.

Finally, the joint treatment of X and Y require adjustments to the individual UCLs. Now two sources of redundant information exist: correlation between variables, and SA within each variable. Dutilleul (1993) updates the Richardson-Clifford discussion about how SA impacts upon the correlation coefficient. Extending his discussion reveals that covariation also has a variance inflation factor similar to that presented in Eq. (8.6), with this factor largely being compensated for by the individual variable variance inflation factors when a correlation coefficient is computed. Moreover, spatial dependency impacts upon a correlation coefficient increase as the correlation moves away from zero, and decrease again as the correlation approaches ± 1. If the correlation between X and Y is zero, then Eqs. (8.9a) and (8.9b) would contain no common eigenvectors; if the correlation between X and Y is ± 1, then Eqs. (8.9a) and (8.9b) would contain exactly the same set of eigenvectors. Meanwhile, constructing a weighted average of X and Y, say [wX + (1–w)Y] for $0 \leq w \leq 1$, yields as the sampling distribution variance for $w\bar{x} + (1 - w)\bar{y}$

$$\frac{w^2\sigma_X^2 + (1 - w)^2\sigma_Y^2 + 2w(1 - w)\rho_{XY}\sigma_X\sigma_Y}{n},$$

where σ_X^2 and σ_Y^2 respectively denote the variance of variables X and Y, ρ_{XY} denotes the product moment correlation between variables X and Y, and the term $2w(1-w)\rho_{XY}\sigma_X\sigma_Y$ adjusts for the presence of redundant attribute information in the bivariate georeferenced dataset.

In this bivariate case, effective sample size becomes a weighted average of the individual pollutant effective sample sizes that is adjusted for the correlation between X and Y. The numerator of Eq. (8.7) becomes

$$\frac{w^2\hat{\sigma}_X^2+(1-w)^2\hat{\sigma}_Y^2+2w(1-w)\hat{\rho}_{XY}\hat{\sigma}_X\hat{\sigma}_Y}{w^2\hat{\sigma}_X^2+(1-w)^2\hat{\sigma}_Y^2}, \tag{8.10a}$$

times

$$w^2\hat{\sigma}_X^2 TR\{[(\mathbf{I}-\hat{\rho}_X\mathbf{W})^T(\mathbf{I}-\hat{\rho}_X\mathbf{W})]^{-1}\}+(1-w)^2\hat{\sigma}_Y^2\{[(\mathbf{I}-\hat{\rho}_Y\mathbf{W})^T(\mathbf{I}-\hat{\rho}_Y\mathbf{W})]^{-1}\}, \tag{8.10b}$$

and the denominator of Eq. (8.7) becomes

$$w^2\hat{\sigma}_X^2\mathbf{1}^T[(\mathbf{I}-\hat{\rho}_X\mathbf{W})^T(\mathbf{I}-\hat{\rho}_X\mathbf{W})]^{-1}\mathbf{1}+(\mathbf{1}-w)^2\hat{\sigma}_Y^2\mathbf{1}^T[(\mathbf{I}-\hat{\rho}_Y\mathbf{W})^T(\mathbf{I}-\hat{\rho}_Y\mathbf{W})]^{-1}\mathbf{1}$$

$$+\,2w(1-w)\hat{\rho}_{XY}\hat{\sigma}_X\hat{\sigma}_Y\mathbf{1}^T[(I-\hat{\rho}_X\mathbf{W})^T(I-\hat{\rho}_Y\mathbf{W})]^{-1}\mathbf{1}. \tag{8.10c}$$

where $\hat{\rho}_X$ is the SA parameter estimate for variable X, $\hat{\rho}_Y$ is the SA parameter estimate for variable Y, and the sample statistics are $s_X^2=\hat{\sigma}_X^2$, $s_Y^2=\hat{\sigma}_Y^2$, and $r_{XY}=\hat{\rho}_{XY}$. Therefore, n^* equals n times expression (8.10a) times expression (8.10b) divided by expression (8.10c). If $\hat{\rho}_X=\hat{\rho}_Y=0$, then this product of the three expressions reduces to n. If $w=0$, $w=1$ or $\hat{\rho}_X=\hat{\rho}_Y$, then this product of the three expressions reduces to Eq. (8.7). In other words, the bivariate effective sample size is a weighted average of the individual univariate effective sample sizes (i.e., it must be contained in the interval defined by them). And, as $\hat{\rho}_X$ and $\hat{\rho}_Y$approach 1, n^* approaches 1. The weighting is determined by both the relative variances of X and Y and the weights used in constructing a linear combination of X and Y, and is impacted little by the value of r_{XY}.

Therefore, Eqs. (8.9a) and (8.9b) can be used to compute $\hat{\rho}_X$, $\hat{\rho}_Y$, r_{XY}, s_Y^2 and s_X^2, followed by expressions (8.10a)–(8.10c) being used to compute n^*. Variables X and Y can be averaged if contaminants X and Y are considered to be equally important for remediation prioritizing, rendering the statistic $\frac{1}{2}(\overline{X}+\overline{Y})$ and the need to construct a UCL for $\frac{1}{2}(\mu_X+\mu_Y)$, where μ_X and μ_Y respectively denote the means for variables X and Y. Results from this analysis identify the high priority subregions within a site in terms of X and Y jointly, a demarcation that may well differ from that identified by simply doing a map overlay of the UCL of X and the UCL of Y.

8.6 The Murray Superfund Site: Part V

Correlations between As and Pb are reported in Table 8.1. The logarithmic transformation markedly increases the linear correlation estimate from 0.589 to 0.740. Addition of the constant translation parameter, and then the heterogeneity translation parameters only slightly changes this result. Adjusting for the presence of SA in both As and Pb reduces the correlation to roughly 0.706. In addition, given $\hat{\rho}_X = 0.532$ and $\hat{\rho}_Y = 0.494$ (see Table 8.2), where As and Pb respectively were arbitrarily linked to X and Y, the effective sample size is $n^* = 72.5$, which is contained in the interval [68.9, 77.6].

Table 8.4 summarizes stepwise regression results for Eqs. (8.9a) and (8.b). Fourteen eigenvectors account for approximately 30 percent of the variance in As and in Pb; a graphical portrayal of these equations appears in Fig. 8.8. Eight of these eigenvectors are common to Eqs. (8.a) and (8.b); their map patterns appear in Fig. 8.9. Equations (8.9a) and (8.9b) furnish a modestly better data description than does the spatial SAR model specified in equation (8.4). The residuals produced by both Eqs. (8.4) and (8.9) appear to contain only trace levels of SA. Equations (8.9a) and (8.9b) suggest a slightly weaker correlation between X and Y than is obtained through the use of Eq. (8.4).

For a bivariate analysis, assuming equal importance of Pb and As for prioritizing (i.e., w = 0.5), normal curve theory states that the 95% UCL is needed for

$$[\mathbf{1}^T\mathbf{X}/n + \mathbf{1}^T\mathbf{Y}/n]/2,$$

Table 8.4 Stepwise spatial filter modeling results

	As					Pb				
				Residual					Residual	
Step	Eigen-vector	Coefficient probability	Adj-R^2	MC	GR	Eigen-vector	Coefficient probability	Adj-R^2	MC	GR
0	***	***	0	0.321	0.692	***	***	0	0.255	0.772
1	**3**	0.000	0.083	0.221	0.776	**3**	0.000	0.059	0.203	0.826
2	**1**	0.000	0.129	0.172	0.817	**12**	0.000	0.111	0.161	0.866
3	**28**	0.000	0.161	0.152	0.851	**28**	0.000	0.158	0.132	0.919
4	**12**	0.004	0.182	0.130	0.868	**20**	0.000	0.204	0.093	0.974
5	**31**	0.005	0.202	0.117	0.877	**17**	0.006	0.222	0.074	0.990
6	10	0.006	0.221	0.095	0.897	**6**	0.010	0.238	0.052	1.002
7	**20**	0.014	0.235	0.079	0.917	8	0.013	0.253	0.031	1.018
8	4	0.019	0.249	0.059	0.942	**31**	0.027	0.264	0.020	1.029
9	17	0.039	0.258	0.047	0.951	**2**	0.039	0.273	0.002	1.037
10	6	0.045	0.267	0.032	0.960	**1**	0.049	0.281	−0.014	1.050
11	35	0.045	0.276	0.023	0.970	**26**	0.060	0.288	−0.024	1.060
12	11	0.047	0.284	0.009	0.983	**34**	0.062	0.295	−0.033	1.064
13	33	0.082	0.290	0.002	0.992	**22**	0.072	0.302	−0.043	1.075
14	7	0.091	0.296	−0.011	1.004	**24**	0.091	0.307	−0.051	1.085

NOTE: common eigenvectors are highlighted with bold numbers

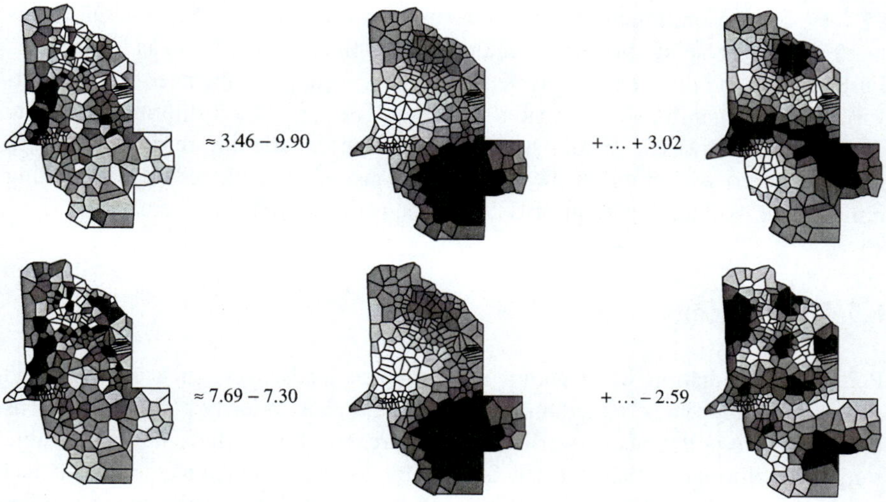

Fig. 8.8 Choropleth maps portraying the spatial filtering equation. *Top* (**a**): the As map together with eigenvector maps of E_3 and E_7. *Bottom* (**b**): the Pb map together with eigenvector maps of E_3 and E_{24}

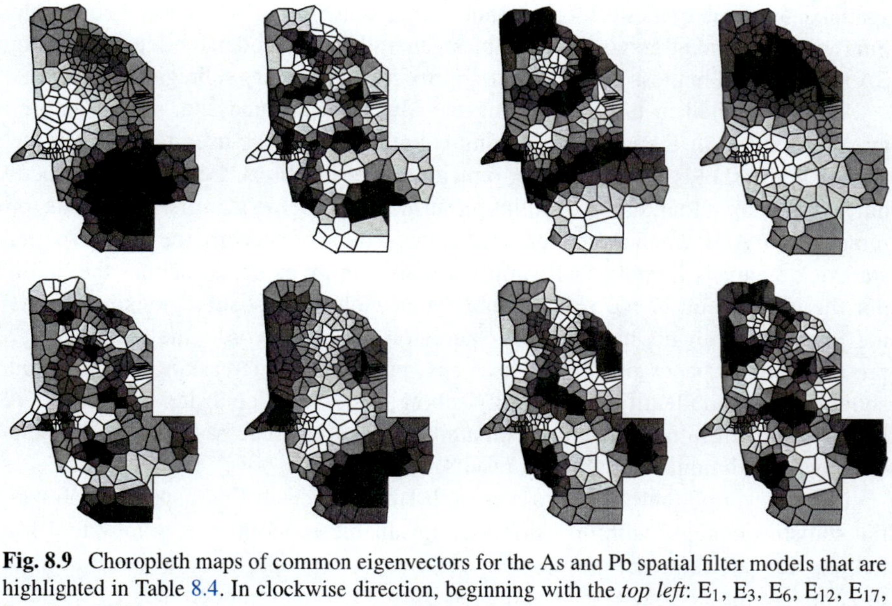

Fig. 8.9 Choropleth maps of common eigenvectors for the As and Pb spatial filter models that are highlighted in Table 8.4. In clockwise direction, beginning with the *top left*: E_1, E_3, E_6, E_{12}, E_{17}, E_{20}, E_{28}, and E_{31}

using the accompanying t-statistic of $t_{n*-4,0.95}$. The UCL value here increases from 5.77074 (the calculation result when ignoring latent SA) to 5.87521, an increase of nearly 2%.

The high priority subregions for remediation appear in Fig. 8.7. Figure 8.7d identifies those parts of the site whose joint As and Pb contamination meets the joint

95% UCL criterion. Figure 8.7c is the result of overlaying Figs. 8.7a, b. Of note is that Fig. 8.7d is not simply the union or intersection of Figs. 8.7a, b, as is Fig. 8.7c. Furthermore, the differences between Figs. 8.7c, d supports the need to do multivariate rather than univariate spatial analyses. The cost of substituting univariate overlay for multivariate spatial statistics would be undertaking remediation work on lower priority locations at the expense of consuming resources for undertaking remediation work on higher priority locations, perhaps even elsewhere.

8.7 Implications

Pollution remediation work within contaminated landscapes, such as superfund sites, may compromise remediation of only the highest priority polluted locations if SA latent in pollutants is overlooked. Researchers could believe they have more statistical information than actually is available to them, as well as more statistical precision than exists when calculating confidence intervals for demarcating subregions for remediation. The same is true if considerable heterogeneity is overlooked. In other words, the methodology presented in this paper furnishes an answer to the question asking what the correct UCL calculation is. In doing so, it highlights that spatial heterogeneity merits more attention when drawing a model-based geographic inference, the size of geographic samples generally is misunderstood, and ignoring SA reveals good but less efficient first-approximation priority subregions.

As is illustrated in the paper using the Murray superfund site, spatial autoregressive models or their spatial filtering counterparts can be used to establish the statistical thresholds for prioritizing remediation of locations. Geostatistical procedures can be used to interpolate pollution surfaces in order to identify subregions for remediation. And, when more than one heavy metal is of concern, the proper spatial statistical analysis is more than simply a map overlay exercise; neither the union nor the intersection of individual contaminant high priority subregions represents the joint contaminants high priority subregion. In other words, the methodology presented in this paper also furnishes an answer to the question asking what method should be used to identify high priority subregions of a polluted site. In doing so, it highlights that results for multiple contaminants should not be based simply on map overlays of individual contaminant results.

The most important finding of research summarized in this paper is that spatial statistics coupled with GIS offers an invaluable economic geography tool for helping allocate the enormous amount of money and people-years of effort needed to complete the necessary environmental restoration being undertaken by modern society. In other words, accounting for SA makes a difference!

Acknowledgment Part of this research was funded by the National Science Foundation, research grant #BCS-9905213. The superfund site data were made available by Dr. Philip E. Goodrum, Research Scientist, and Mr. William C. Thayer, Syracuse Research Corporation; Dr. Susan Griffin, U.S. E.P.A. Region 8, furnished the original Murray superfund site data. Help with preparation of the graphics by Mr. Michael Schramm, Department of Geography, Syracuse University, also is gratefully acknowledged.

Chapter 9
General Conclusions: Spatial Statistics

The original version of our *Annals in Regional Science* paper enumerates a number of topics that serve as focal points for the frontiers of spatial statistics and spatial econometrics. This first part of the book addresses some of these topics, which are loosely connected, in considerably more detail:

1. The ecological fallacy: Chap. 2
2. spatially adjusted statistical techniques, and quantifying spatial autocorrelation: Chap. 3
3. exploratory spatial data analysis: Chaps. 4 and 5
4. Bayesian hierarchical models: Chap. 6
5. auto-model specification (normal, Poisson, binomial), and spatial structure as a covariate (spatial filtering): Chap. 7
6. sampling network structure: design-based inference: Chap. 8

This selected list reflects research preferences of one of the authors, rather than some rank ordering of importance.

Considerable work still needs to be undertaken about the ecological fallacy. Two important aspects of this problem highlighted in Chap. 2 are: (1) georeferenced data are messy—standard statistical model and technique assumptions are not justified, and (2) sometimes only geographic aggregates can be treated. In this first case, many relationships are non-linear, which prevents them from being transferred from individuals to aggregates of individuals in a simple way. This is a critical feature that interacts with mixtures of non-identical observations, creating heterogeneity and excessive variation for geographic random variables. Spatial autocorrelation accounts for only part of this total excess variation. In this second case, rates, for example, require aggregates of individuals, as do variables such as the rural-urban dichotomy.

Seminal work establishing linkages between spatial autoregressive and geostatistical models is interesting and illuminating. This articulation needs to be extended to space-time contexts, as well as to inclusion of other model specifications such as spatial filtering and geographically varying coefficients. Autocorrelation is the key concept in these clusters of research; accordingly, spatial autocorrelation is fundamental, too. Many space-time datasets are dirty because they contain missing values

D.A. Griffith, J.H.P. Paelinck, *Non-standard Spatial Statistics and Spatial Econometrics*, Advances in Geographic Information Science 1,
DOI 10.1007/978-3-642-16043-1_9, © Springer-Verlag Berlin Heidelberg 2011

(often in addition to unusual values). Using both spatially and temporally redundant information latent in a dataset allows imputations to be calculated for such missing values. This theme constitutes one of the principal problems needing solved by spatial analysis; an urgent need exists for procedures that compute extremely accurate and precise imputations.

Two facets of exploratory spatial data analysis that merit attention are a better understanding of frequency distributions constructed with georeferenced data, and correlations between georeferenced random variables. Frequently spatial scientists inspect histograms as a first step in data analysis, often finding that these graphs fail to closely align with any of the numerous existing ideal frequency distributions. Chap. 4 furnishes basic insights into why this occurs. But a mathematical statistics theoretical basis needs to be established for the intuition and numerical demonstrations appearing in that chapter. Meanwhile, spatial scientists need to recognize that correlation coefficients can be dramatically altered by latent spatial autocorrelation; depending upon prevailing spatial patterns, these coefficients can be inflated toward 1 or -1, or they can be deflated toward 0. In other words, a correlation coefficient for a pair of georeferenced random variables cannot be taken at face value!

Contemporary statistical methodology allows spatial scientists to approximate impacts of unmeasured (i.e., latent) variables and/or measurement error by including a random effects term in a model specification, acknowledging that georeferenced data are noisy (i.e., contain considerable variability). This spatial statistical topic is at the forefront of the subdiscipline today. Estimates of these impacts can be obtained with Bayesian techniques, allowing analysis of a single geographic distribution (positing prior distributions furnishes the necessary ancillary information), or with frequentist techniques when repeated measures (i.e., multiple geographic distributions, which furnish ancillary information as repeated measures) are available. Such random effects almost always have a spatially structured component, which relates to the spatial autocorrelation displayed by a georeferenced random variable. Spatial structuring can be captured with an autoregressive model (e.g., the conditional autoregressive model used in Bayesian map analysis), or by a spatial filter (i.e., regressing a random effects terms on a set of eigenvectors to separate them into a geographically varying mean response and a random error term).

These preceding discussions raise the question of relationships between spatial filtering and conventional spatial statistical models, which is the topic of Chap. 7. Spatial filtering offers the advantage of allowing a spatial scientist to work within the context of conventional statistical technology. It is consistent with statistical specifications associated with Bayesian map analysis: it represents spatial autocorrelation as a feature of model parameters, rather than correlated response variable values; as such, it casts a model intercept as an observation-specific surrogate for unobserved variables by expressing it as a spatially structured random deviation from some global intercept. This conceptualization posits that empirical probabilities are correct, while simple model parameters are not. In contrast, an auto-model posits that simple model parameters are correct, while empirical probabilities are conditional on other observations. Consequently, direct dependency between values of a response variable is replaced by the incorporation of spatial autocorrelation into prior parameter distributions, in Bayesian analysis, or a random effects intercept term, in frequentist analysis.

Finally, as can be surmised from impacts of spatial autocorrelation on histograms (e.g., Chap. 4) and correlation coefficients (e.g., Chap. 5), spatial autocorrelation affects prioritizing, say, polluted sites for remediation, based upon unusual values (e.g., hot spots)—an attribute of dirty data. Any rankings of sets of georeferenced objects (e.g., the rank size distribution of city sizes) suffer from this same corruption. This feature of georeferenced data has been recognized for decades, but little work has been produced while at the same time increasingly more sets of georeferenced objects have been ranked, some on an annual basis.

In conclusion, our *Annals of Regional Science* paper emphasizes a sizeable number of non-standard spatial statistics topics, some of which are treated in more depth in this book. The comprehensive treatments presented here initialize a quest to suscitate interest in the methodologies exposed and possible further applications of these methodologies.

Part II
Non-standard Spatial Econometrics

"Solem orientem plures adorant quam occidentem..."
More people admire the rising rather than the setting sun...

Plutarch

Chapter 10
Introduction: Spatial Econometrics

In spatial econometrics, various topics have their own importance: specification, estimation and testing are the main building blocks of a spatial econometric model.

An economist should attach special importance to the specification stage; experience has taught that the functioning of spatial economies follows a complex pattern, and that is the pattern that should be adequately modeled.

In this part, a certain number of working papers are brought together, most of which have been presented and commented on at special sessions of international conferences, sessions devoted to spatial econometrics and statistics.

Apparently they stand loose from each other; but, a common thread links them all together, to wit the necessity of discovering the complex bindings—static or dynamic—of spatial economic units. Hopefully this may stimulate fresh thinking about this very important aspect of spatial econometric modeling.

Some non-standard specifications have not been included, as they have been or are going to be published (e.g. Coutrot et al. 2009; Griffith and Paelinck, 2009).

D.A. Griffith, J.H.P. Paelinck, *Non-standard Spatial Statistics and Spatial Econometrics*, Advances in Geographic Information Science 1,
DOI 10.1007/978-3-642-16043-1_10,

Chapter 11
A Mixed Linear-Logarithmic Specification for Lotka-Volterra Models with Endogenously Generated SDLS-Variables

In Arbia and Paelinck (2003a, b), a Lotka-Volterra model (LVM) is applied to the convergence-divergence problem of European regions in terms of incomes per capita. As the latter have to be non-negative, a double logarithmic version may be substituted for the original specification, a modification that removes at least part of the non-linearity of LVMs; this chapter introduces this non-linearity again. Discussion begins with a general section on LVMs, to go on with a mixed linear-logarithmic specification, of which the positivity of the (possible) equilibrium solution is proved, and for which a (sufficient) stability condition is derived. Section 11.3 presents an application of the model to the classical four macro-regions in which the Netherlands is subdivided.

11.1 Lotka-Volterra Models

In this section, generalized Lotka-Volterra models are introduced, and examples given of some applications, including estimation aspects of the latter.

11.1.1 A General Specification

A generalized LVM can be written in matrix-vector notation as

$$\dot{\mathbf{u}} = \hat{\mathbf{u}}\ (\mathbf{A}\,\mathbf{u}\ +\ \mathbf{a})\,, \tag{11.1}$$

where **u** is a column-vector of (endogenous) variables, $\hat{\mathbf{u}}$ its diagonal matrix version, **A** a square matrix, and **a** is a column-vector of fixed coefficients; the •-notation denotes the time derivative, ∂/∂t.

Given equation (11.1), the variables **u** describe a time path that can take all the characteristics of general continuous dynamic processes (e.g., convergence, divergence, limit circles; see Braun, 1975, §4.9; Gandolfo, 1996, in particular §24.4; Peschel and Mende, 1986). What can be said about equation (11.1) to converge to its focus, $-\mathbf{A}^{-1}\,\mathbf{a}$? Constructing a Lyapunov-function (Hahn, 1963)

D.A. Griffith, J.H.P. Paelinck, *Non-standard Spatial Statistics and Spatial Econometrics*, Advances in Geographic Information Science 1,
DOI 10.1007/978-3-642-16043-1_11, © Springer-Verlag Berlin Heidelberg 2011

$$v = (\mathbf{u} + \mathbf{A}^{-1}\mathbf{a})'(\mathbf{u} + \mathbf{A}^{-1}\mathbf{a}) \tag{11.2}$$

gives

$$\dot{v} = 2(\mathbf{u} + \mathbf{A}^{-1}\mathbf{a})'\hat{\mathbf{u}}\,\mathbf{A}(\hat{\mathbf{u}} + \mathbf{A}^{-1}\mathbf{a}) \tag{11.3a}$$

$$= (\mathbf{u} + \mathbf{A}^{-1}\mathbf{a})'\,(\hat{\mathbf{u}}\mathbf{A} + \mathbf{A}\,\hat{\mathbf{u}})\,(\mathbf{u} + \mathbf{A}^{-1}\mathbf{a}). \tag{11.3b}$$

In the purely linear case, if the real parts of **A**'s eigenvalues are negative, v is negative definite (Hahn, 1965, p. 26; La Salle and Lefschetz, 1961, p. 48), and the sufficient conditions for asymptotic stability are satisfied. In the LVM case the problem is more involved; the proof of the above sufficiency conditions still being satisfied is given in Paelinck (1992, pp. 142–143).

11.1.2 Applications

Originally, special versions of the LVM have been applied to the field of biomathematics, i.a., to build so-called "predator-prey" models. An example is the following model (all parameters strictly positive):

$$\dot{x} = x\,(a - b\,y), \tag{11.4a}$$

$$\dot{y} = y(-c + dx). \tag{11.4b}$$

Here x is the prey, developing at a constant rate a, but preyed upon by the predators y; the latter, in the absence of prey animals, fade out at a rate c, but are kept alive by x.

The resulting state diagram in the x-y plane shows a "pseudo-elliptic" closed curve, and the time-explicit graph shows sinusoidal lagged curves of different amplitudes.

The model just described was proposed by Samuelson (1971) as a candidate for dynamic economic analysis, and applied by Dendrinos and Mullaly (1981) to the evolution of urban populations, although no explicit econometric estimation was performed. Before presenting some econometric results, an appropriate estimation method is unfolded here.

The flexibility of the LVM specification is shown by the various time-paths and singular points resulting from various parameter combinations (presence or absence, signs); Braun (1975, pp. 590–599) gives examples of this. For instance, if a term $-ex$ is added to equation (11.4a), and a term $-fy$ to equation (11.4b), both terms representing competition for limited resources, within the prey and the predator group, the solution becomes $[x^o = a/e;\ y^o = 0]$ for $c/d > a/e$.

11.1.3 Simultaneous Dynamic Least Squares (SDLS) Estimation

Consider a (e.g., sectorally, spatially, dynamically) interdependent econometric model (Paelinck, 1996b, §2.1)

$$\mathbf{A\,u} + \mathbf{B\,x} = \boldsymbol{\varepsilon}, \tag{11.5}$$

where **u** is a column-vector of endogenous variables, **x** a column-vector of exogenous ones, **ε** being the usual column-vector of random elements. Equation (11.5) always can be rewritten as

$$\mathbf{y} = \mathbf{Z}\,\boldsymbol{\beta} + \boldsymbol{\varepsilon}, \tag{11.6}$$

where **Z** comprises at the same time endogenous and exogenous variables. The basic idea of SDLS is to minimize the sum of squared deviations between the observed and the *endogenously computed* (shown by caps) values of the endogenous variables, **u;** this leads to

$$\mathbf{u} - \hat{\mathbf{u}} = [\mathbf{u} - (\mathbf{Z} - \hat{\mathbf{Z}}\boldsymbol{\beta}] - \mathbf{Z}\boldsymbol{\beta})], \tag{11.7}$$

and minimizing as said before gives

$$\hat{\boldsymbol{\beta}} = (\mathbf{Z'}\hat{\mathbf{Z}})^{-1}\mathbf{Z\,u}, \tag{11.8}$$

where $\hat{\mathbf{Z}}$ includes the computed values of the endogenous variables; a possible computing process is an iterative one, but Sect. 11.4 presents a specification with endogenously computed **u** values.

The following properties hold (Paelinck, 1990b, p. 7–8):

– $\boldsymbol{\beta}$ is a generalized reduced form estimator;
– if $\boldsymbol{\varepsilon} \sim \mathbf{N}(\mathbf{0}, \sigma^2\mathbf{I})$, then $\boldsymbol{\beta}$ is a maximum likelihood estimator; and,
– $\boldsymbol{\beta}$ is a consistent estimator, and plim $\boldsymbol{\beta\beta'} = \sigma^2(\mathbf{Z'Z})^{-1}$, for homoscedastic $\boldsymbol{\varepsilon}$.

The method has been applied to a two-equation full-parameter LVM process for the city of Rotterdam, The Netherlands, and for the time period 1946–1978 (Paelinck, 1996a, §3), the equations being

$$\Delta'\ln x_t = a + b\,x_{t-1} + c\,y_{t-1}, \tag{11.9a}$$

$$\Delta'\ln y_t = d + e\,x_{t-1} + f\,y_{t-1}, \tag{11.9b}$$

Table 11.1 Parameter values of the Rotterdam application

Parameter	Value	Student's t
a	−0.8798	−7.68
b	0.0711	7.33
c	0.3988	4.22
d	1.0870	9.49
e	−0.0825	−8.51
f	−0.5355	−5.67
a*	0.0362	1.64
d*	0.0538	2.43

where x represents population and y per capita income. Table 11.1 (taken from Paelinck, 1990a) presents the estimation results, which can be given plausible interpretations.

Note also the presence of the two parameters a^* and d^*, which allow the initial values to be shifted optimally by the computed process with respect to the observed initial values. Moreover, the resulting relevant eigenvalues lie between *–1* and *0*, so—abstracting from the discretionarity problem (see Gandolfo, 1996, pp. 411–412)—the process would be asymptotically convergent in terms of population and per capita income. According to the divergence criterion (Gandolfo, 1996, p. 456) the system is possibly anti-dissipative along certain stretches of its time-path, although conservative in its non-trivial singular point.

11.2 Mixed Specification

In this section, a combined linear-logarithmic specification is presented.

11.2.1 Equations

Instead of equation (11.1), now consider

$$\hat{\mathbf{u}}^{-1}\mathbf{u} = \mathbf{A}\,(\mathbf{u} + \ln \mathbf{u}) + \mathbf{a}, \tag{11.10}$$

for which the equilibrium solution (if it exists) is

$$\mathbf{u}^{\mathbf{o}} + \ln \mathbf{u}^{\mathbf{o}} = -\mathbf{A}^{-1}\mathbf{a}. \tag{11.11}$$

For each variable u_i, the equilibrium solution can be written as

$$u_i{}^o = b_i - \ln(u_i{}^o), \tag{11.12}$$

where b_i is generated by the i-th row of $-\mathbf{A}^{-1}$ times $\mathbf{a}$. Now while u_i^o increases linearly starting from zero, $-ln(u_i^o)$ decreases monotonically from $+\infty$ to $-\infty$. Thus, equation (11.12) should be satisfied for some strictly positive value of u_i^o.

11.2.2 Stability

Instead of equations (11.3), consider

$$\mathrm{v} = (\mathbf{u} + \ln \mathbf{u} + \mathbf{A}^{-1}\mathbf{a})'(\mathbf{u} + \ln \mathbf{u} + \mathbf{A}^{-1}\mathbf{a}), \tag{11.13}$$

from which one can derive

$$\dot{\mathrm{v}} = 2(\mathbf{u} + \ln \mathbf{u} + \mathbf{A}^{-1}\mathbf{a})'(\mathbf{I} + \hat{\mathbf{u}})^{1/2}[(\mathbf{I} + \hat{\mathbf{u}})^{1/2}\mathbf{A}\,(\mathbf{I} + \hat{\mathbf{u}})^{-1/2}](\mathbf{I} + \hat{\mathbf{u}})^{1/2} (\mathbf{u} + \ln \mathbf{u} + \mathbf{A}^{-1}\mathbf{a}). \tag{11.14}$$

Matrix **A** has undergone a similarity transformation which keeps the eigenvalues unchanged (Allen, 1956, p. 468). Again Hahn's argument quoted at the end of Sect. 11.1.1 can be invoked here, which completes the proof of the fact that a sufficient condition for the mixed LVM to be stable is the negativity of the real parts of **A**'s eigenvalues.

11.3 Application

Model (11.10) has been applied to the relative GDPs of the four classical Dutch macro-regions; Table 11.2 lists the numbers (N = North; S = South; E = East; W = West, the latter region being known as the "Rimcity"); see Fig. 11.1. The numbers are percentages and relate to the years 1988–2000.

SDLS estimates using the numbers in Table 11.2 are obtained by introducing the following equations (discrete versions of equations (11.10); tildes relate to the computed SDLS endogenous variables)

$$\tilde{\mathbf{u}}_t = (\mathbf{I} + \mathbf{A})\,\tilde{\mathbf{u}}_{t-1} + \mathbf{A}\ln \tilde{\mathbf{u}}_{t-1} + \mathbf{a}, \tag{11.15}$$

into a mathematical programming model, minimizing the squared residuals of (11.5) or (11.6). As only the fourth observation (relating to 1991) produces a sum diverging significantly from 100 (see also Figs. 11.1,[1] 11.2, 11.3, and 11.4), no extra constraint was introduced. Furthermore, optimal starting points were computed by optimizing simultaneously over the starting vector of computed SDLS variables.

---

[1] We thank Martijn Smit, Vrije Universiteit Amsterdam (VU University Amsterdam), for furnishing us with the digital map necessary to construct this figure.

Table 11.2 Numbers used in the Dutch application

N	S	E	W
10.80984	20.03057	17.32632	51.83327
10.43814	20.19805	17.24065	52.12317
10.41169	20.4517	17.43168	51.70493
10.60662	20.45458	17.50469	51.43412
10.96084	20.45568	17.46016	51.12332
10.62076	20.5621	17.66965	51.14749
10.58301	20.39573	17.88161	51.13965
10.22753	20.68775	17.87189	51.21283
10.06042	20.96244	17.78718	51.18996
10.31112	20.93548	17.71927	51.03413
10.15061	20.71473	17.69594	51.43872
9.74983	20.93522	17.6603	51.65465
9.33526	21.08442	17.85378	51.72653
9.54633	20.99156	17.73859	51.72352

Fig. 11.1 The four Dutch regions

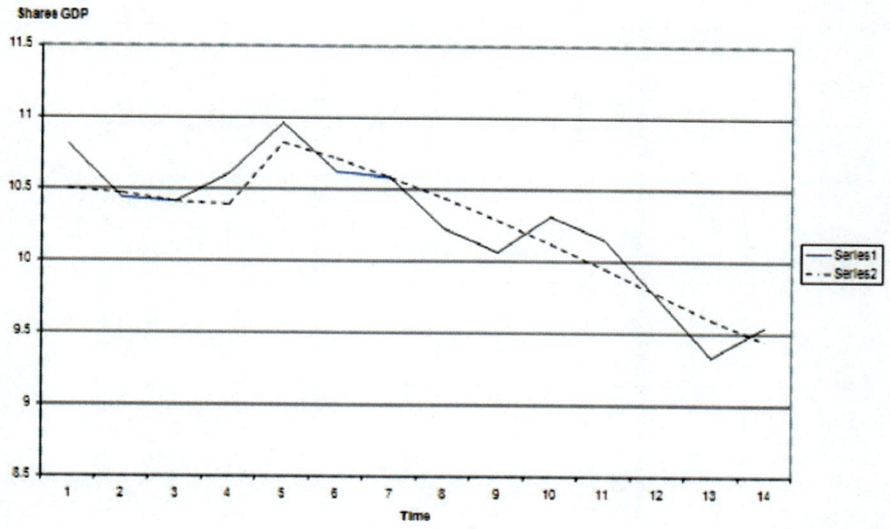

Fig. 11.2 Shares GDP time-series plot: northern Netherlands

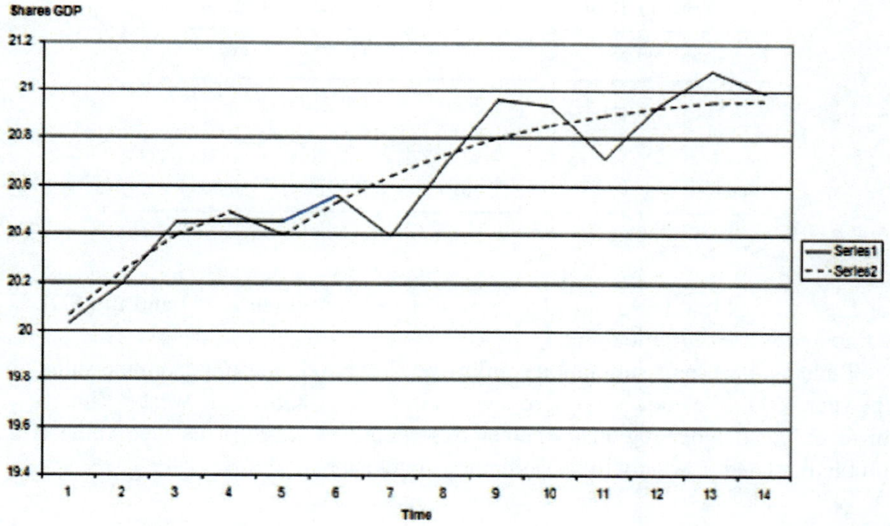

Fig. 11.3 Shares GDP time-series plot: southern Netherlands

Table 11.3 presents the main econometric results.

Every region has been assigned only three parameters: its own influence (a_1), that of the other three regions (a_2), and a constant (a_3).

A *4x4 matrix* can be constructed, dividing each a_2 by *3*. From the trace (*5.5192*) this matrix appears to be non-negative definite (two out of the four a_is are positive), so no mathematical convergence toward the right hand side of equation (11.11) is present.

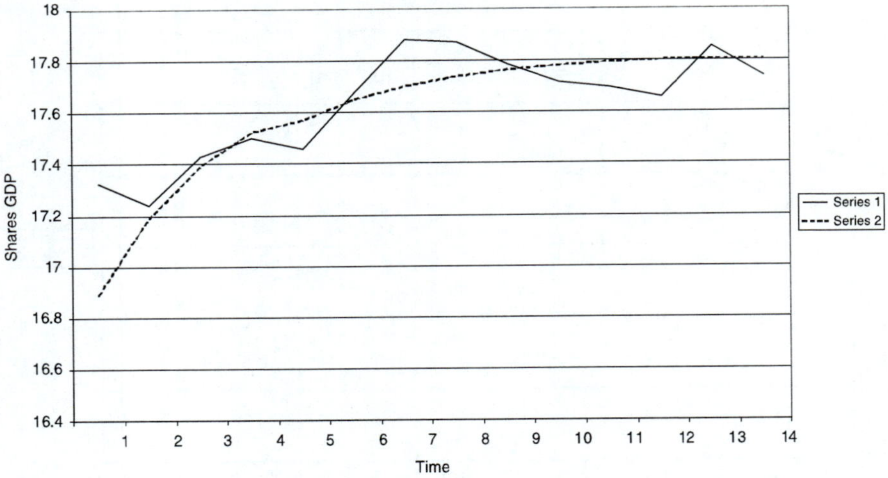

Fig. 11.4 GDP time-series plot: eastern Netherlands

Table 11.3 Econometric results of the Dutch application

Parameters	N	S	E	W
a_1	−1.2368	0.2450	−0.2108	6.7218
a_2	−1.2618	0.4712	0.1072	6.6220
a_3	141.1399	−47.4387	−5.4545	−748.6387
Pseudo-R^2	0.8870	0.8816	0.8074	0.8894

Figs. 11.2, 11.3, 11.4, and 11.5 present the observed (series 1) and the SDLS-computed (series 2) series.

Table 11.4 presents simulation results over 20 periods, starting from the value of the year 2000. One notices a progressive decline in the share of the West ("Rimcity") in favor of all other regions; whether this should be taken at its face value is a problem related to what will be said in the conclusions.

11.4 Conclusion

The method has proven itself to be workable and could be combined with an appropriate estimation method (SDLS); it has moreover the nice property that, if convergence is present, it will lead to economically acceptable (positive) equilibrium values. This means that discrete LVMs, adapted in the way described, could be an ever more useful tool for future research in multiregional dynamics.

Inspection of Figs. 11.2, 11.3, 11.4, and 11.5 shows some local discrepancies between observed and SDLS computed values. Though the specification chosen is

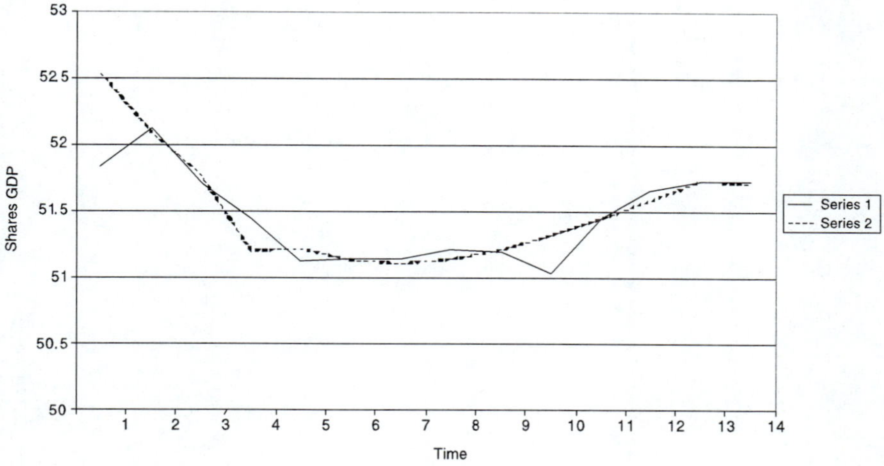

Fig. 11.5 GDP time-series plot: western Netherlands

Table 11.4 Simulation results of the Dutch application

N	S	E	W
9.454351	20.97449	17.81837	51.75279
9.488464	21.08845	17.93343	51.48965
9.513482	21.18737	18.01698	51.28217
9.531703	21.27253	18.07830	51.11747
9.544808	21.34540	18.12382	50.98597
9.554045	21.40749	18.15801	50.88045
9.560352	21.46022	18.18403	50.79539
9.564440	21.50492	18.20409	50.72655
9.566852	21.54275	18.21977	50.67062
9.568005	21.57475	18.23219	50.62505
9.568217	21.60183	18.24217	50.58779
9.567738	21.62474	18.25027	50.55725
9.566758	21.64415	18.25695	50.53215
9.565425	21.66061	18.26250	50.51146
9.563852	21.67460	18.26718	50.49437
9.562128	21.68650	18.27115	50.48022
9.560319	21.69666	18.27455	50.46848
9.558476	21.70534	18.27748	50.45871
9.556637	21.71277	18.28003	50.45056
9.554830	21.71917	18.28226	50.44374
9.553075	21.72468	18.28422	50.43803

already very flexible, it is not the only one. Other candidates (min-algebraic specifications, finite automata; see Chap. 13) are available, and should be tested against the present specification. Tools are also available (see Chap. 12) and will be used in future research.

Chapter 12
Selecting Spatial Regimes by Threshold Analysis

The existence of differential spatial regimes has been revealed on different occasions (see for instance Arbia and Paelinck, 2003a, b; also see Chap. 14). Hence the necessity exists for developing workable specifications to compute possible frontiers or thresholds between those regimes.

The next section describes one possible method. Sects. 12.2 and 12.3 then apply it to two spatial models using Dutch data: an income generating model, and an activity complex model.

12.1 Method

Assume the following model

$$y = ax, \tag{12.1}$$

$$a = a_1 \mid x \leq x^*, \tag{12.2}$$

$$a = a_2 \mid x \geq x^*. \tag{12.3}$$

The model can be respecified as follows, for each observation *i*

$$y_i = a_1\, u_i\, x_i + a_2\,(1 - u_i)\, x_i + \varepsilon_i, \tag{12.4}$$

$$(u_i - \eta)(x_i - x^*) \leq 0, 0 < \eta < 1, \tag{12.5}$$

$$u_i = u_i^2, \tag{12.6}$$

the estimation being, e.g., performed by minimizing $\boldsymbol{\varepsilon}'\boldsymbol{\varepsilon}$. This model was first tested on the following data, appearing in Table 12.1.

The values for a_1 and a_2 are respectively *2* and *1*, the first four observations being governed by a_1. The value of x^*, which is not unique, should lie between *8* and *11*. Estimation reproduced the values of a_1 and a_2, $x^* = 8{,}1333$, with the u_is correctly

D.A. Griffith, J.H.P. Paelinck, *Non-standard Spatial Statistics and Spatial Econometrics*, Advances in Geographic Information Science 1,
DOI 10.1007/978-3-642-16043-1_12,

Table 12.1 Test data for the model defined by Eqs. (12.4)–(12.6)

Variable	Values							
y_i	4	10	12	16	11	13	14	17
x_i	2	5	6	8	11	13	14	17

partitioning the observations. The objective function adopted the value zero, and all restrictions and optimality conditions were satisfied.

If one wants to replace Eq. (12.2) or (12.3) by a strict inequality, the following specification could be used

$$(u_i - \eta)(x_i - x^* - \theta) \geq 0 \tag{12.7}$$

where η is defined as in Eq. (12.5), and θ is an appropriately chosen small positive number. If $x_i - x^* > \theta$, $u_i = 1$; if $x_i - x^* = \theta$, u_i could be equal to either *0* or *1*, whichever value would give the best estimation result; if $x_i - x^* < \theta$, $u_i = 0$. In both Eqs. (12.5) and (12.6) the specification (parameter η) prevents u_i from being zero if the second factor of Eq. (12.7) is non-zero (positive or negative, depending on the case).

12.2 Spatial Income Generating Model

This model was initially developed in Paelinck and Klaassen (1979, pp. 21–23). Its aim is to measure the spatial interdependence between regional incomes or products.

Let **y** be the column vector of regional incomes; then the model in its simplest form is specified as

$$\mathbf{y} = \mathbf{A}\mathbf{y} + \mathbf{b}, \tag{12.8}$$

where matrix **A** integrates some spatial interaction operator. In the present case, a first-order contiguity matrix, $\mathbf{C_1}$, has been selected for matrix **A** to compute total incomes over neighboring regions.

As can be seen from Fig. 12.1, in the Netherlands all regions, except for the provinces of Flevoland and Utrecht, are peripheral, with three purely maritime (Friesland, Noord- and Zuid-Holland), so five "pseudo-border" correcting (additive) parameters have been introduced into the model (Paelinck, 1996b, pp. 4–8). Moreover, the "reaction" parameters *a* and *b* have been split, according to Eq. (12.4) together with constraints (see Appendix).

Table 12.2 presents the data for regional (provincial) products (1987, Dfl 10^6; source: van Gastel and Paelinck, 1995, p. 152). Table 12.3 presents the spatial contiguity structure (degrees of contiguity) for the Netherlands (same source).

Fig. 12.1 Map of the Dutch provinces

Table 12.2 Provincial products in the Netherlands, 1987

Province	Abbreviation	Number	Product
Groningen	Gr	1	22,675
Friesland	Fr	2	13,229
Drente	Dr	3	11,187
Overijssel	Ov	4	25,448
Flevoland	Fp	5	3703
Gelderland	Gl	6	43,861
Utrecht	Ut	7	27,801
Noord-Holland	NH	8	77,997
Zuid-Holland	ZH	9	102,864
Zeeland	Zl	10	10,908
Noord-Brabant	NB	11	59,242
Limburg	Lb	12	29,036

Table 12.3 Contiguity structure of the Netherlands

Province	Gr	Fr	Dr	Ov	Fp	Gl	Ut	NH	ZH	Zl	NB	Lb
Gr	0	1	1	2	2	3	3	3	4	5	4	4
Fr	1	0	1	1	1	2	2	2	3	4	3	3
Dr	1	1	0	1	2	2	3	3	3	4	3	3
Ov	2	1	1	0	1	1	2	2	2	3	2	2
Fp	2	1	2	1	0	1	1	1	2	3	2	2
Gl	3	2	2	1	1	0	1	2	1	2	1	1
Ut	3	2	3	2	1	1	0	1	1	2	2	2
NH	3	2	3	2	1	2	1	0	1	2	2	3
ZN	4	3	3	2	2	1	1	1	0	1	1	2
Zl	5	4	4	3	3	2	2	2	1	0	1	2
NB	4	3	3	2	2	1	2	2	1	1	0	1
Lb	4	3	3	2	2	1	2	3	2	2	1	0

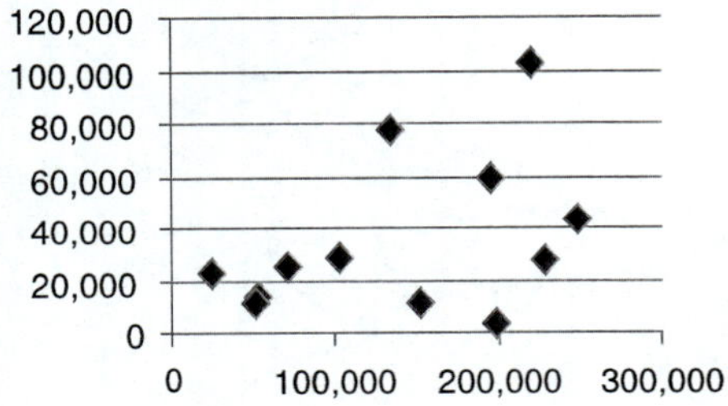

Fig. 12.2 Specifying condition (12.5): the $x_r - y_r$ relation

A typical equation for a region *r* now may be written as follows

$$y_r = a_1 u_r x_r + a_2(1 - u_r)x_r + b_1 z_r + b_2(1 - z_r) + c_r + \varepsilon_r, \tag{12.9}$$

where y_r denotes a regional product and x_r the sum of products in neighboring regions (here divided by *10*, for reasons of magnitude similarity), u_r and z_r are binary variables, and c_r denotes the "pseudo-border" coefficients previously mentioned.

Figure 12.2 hereafter has served to specify the condition referred to in Eq. (12.5). It suggests replacing x_i of Eq. (12.5) by the ratio x_r / y_r.

Table 12.4 presents the results of the estimation procedure (derived by means of a constrained gradient method; see Fylstrom et al., 1998). Model (12.8) including interdependent endogenous variables, an SDLS estimation procedure (see Chap. 11, Sect. 11.1.3) was used, whose optimization principle, as said there, is the minimization of the sum of squared differences between the observed and the endogenously computed y_r variables.

Rather than indicating two regimes, all four combinations of the reaction parameters are present: a_1–b_1 (four times), a_2–b_1 (five times), a_1–b_2 (twice) and a_2–b_2 (once, the case of Flevoland, a recent small new province). The value of the threshold is *0.3486*. Corrective constants c_r are positive for Noord- and Zuid-Holland (two

Table 12.4 Results of the estimation procedure of model (12.8)

Province	$a_1 = 3.8934$	$a_2 = 0.8328$	$b_1 = 10378$	$b_2 = -8276$	c_r	y_r (est.)
Gr	x		x			24, 145
Fr		x	x		220	10, 774
Dr		x	x			16, 598
Ov	x		x			22, 945
Fp		x		x	−2473	7052
Gl		x	x			41, 250
Ut		x	x		−46	36, 258
NH	x		x		3074	77, 518
ZH	x		x		1552	102, 061
Zl		x	x			10, 208
NB	x			x		57, 831
Lb				x		32, 260

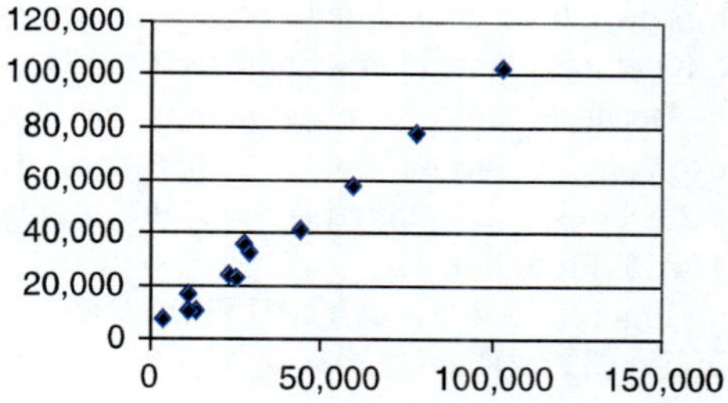

Fig. 12.3 Observed and computed values of y_r

heavily exporting provinces, with important harbors), and negative for Flevoland (see the remark above), the other corrections being negligible.

Figure 12.3 compares the observed and computed values of the endogenous variables; Theil's *U* (Theil, 1961, p. 32) has value *0.0380*, showing the close connection between observed and computed values.

12.3 A Spatial Activity Complex Model

To subject the model to a more terse test, a so-called "attraction model" (first developed by Klaassen; see Paelinck and Klaassen, 1979, pp. 23–30) was evaluated.

Let y_{ir} represent the production level (or value added) of activity sector *i* in region *r*; the model here is specified as

$$y_{ir} = \sum_{j} a_{ijl}\, y^*_{jr} + b_{il}, \qquad 12.10$$

where the y^*_{jr} represent spatially discounted (from Table 12.3) aggregations (potentials) of activity sector production levels (index *j*). The index *l* denotes the relevant regime to which the parameters belong.

Table 12.5 Endogenous activity sectors

1. Industry, public utilities and minerals.
2. Building and construction activities.
3. Trade, catering and repairs.
4. Transport and communication activities.
5. Banking and insurance.
6. Realty and business services.
7. Health care and veterinary services.
8. Cultural, sports and recreational activities.
9. Other services.

Nine endogenous sectors were employed (Table 12.5) for the Netherlands case study.

Appendix 12.5 reports their production (value added) levels. The following sectors have been considered as exogenous: agriculture, forestry and fishery (10), crude oil and gas plants (11), public sector (12).

Because only 12 observations per activity sector are available, aggregated explanatory variables were constructed, whereby contiguous regions have been exogenously discounted at 50%, the aggregation being the following ones: 10, 11; 1, 2; 3 through 9, 12.

Figures 12.4 through 12.7 portray the y_{ir} / y^*_{jr} ratios (inverses of the ratios used in Sect. 12.2) and the $y_{ir} / \sum_r y_{ir}$ one for sector 1; they suggest (at least) three regimes.

Forty-eight binary variables are to be used for only 2 regimes, generating a heavy *0–1* mathematical program; no solution could be reached within 40 hours, so a different strategy was developed.

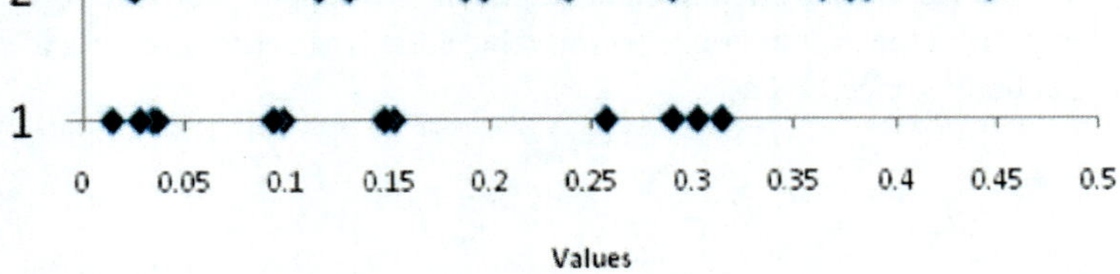

Fig. 12.4 Distribution of xt/yt ratios fot the four Dutch regions

For n observations and k relevant parameters, combining the relative frontiers between the values of Figs. 12.4, 12.5, 12.6, and 12.7, generates $[n(n-1)/2]^k$ simple quadratic programs. In the present exercise the cut-off points have been selected visually, so sub-optimal outcomes might be expected.

Coming back to parameters $a_{ijl,}$ and b_{il} of Eq. (12.10), some outliers might produce reverse results compared to a general trend as revealed, e.g., in Fig. 12.2; therefore an extra condition was introduced, to wit the sign equality of parameters a_{ijl},, $\forall l$. A way to impose that constraint would be

$$\mathrm{abs}\left(\sum_{l} a_{ijl}\right) = \sum_{l} \mathrm{abs}\,(a_{ijl}), \tag{12.11}$$

or, alternatively, binary conditions such as

$$a_{ijl} * a_{ijm} \geq 0, \tag{12.12}$$

which would lead to a quadratic program with non-linear constraints. In practice, and in order to obtain classical results, necessary conditions [like those exposed through Eqs. (12.11) or (12.12)] have been replaced by inspecting all possible sign combinations (i.e., $2^3=8$, as the level parameters, b_{il}, were left free).

Finally, to allow for comparability of results, logarithms are used, yielding non-dimensional elasticities as parameters. Table 12.6 presents the estimation results.

One finding is that only two sectors (1 and 9) satisfy without constraint the equal sign condition; they also show great similarity inside each parameter group, and, as should be expected, the lowest U-values. Sectors with one active constraint number four, sectors with two number two, and only one sector needed all three constraints. The sectors needing two constraints also show the highest U-values (the only ones exceeding 0.05), even higher than the sector needing all three constraints.

At this stage, no SDLS-computations were performed, despite the interdependent specification of the model. Seven out of nine U-values are sufficiently low that no further corrections were deemed necessary (the two high U-values are probably related, to regional accountancy for sector 5, and the requirement of another specification, in terms of explanatory variables, for sector 8).

No significance magnitudes have been shown. However, they could be computed deleting the series of variables corresponding to the zero values.

Moreover no detailed study was made of the different combinations of a_{ijl} and b_{jl} parameters as presented in Table 12.6 (nine tables would be necessary for such a comparison). From that table, however, a remarkable finding can be derived, namely that parameters $a_{2\,l}$ are all non-negative in the solution. As for the other a_{il} parameters, five out of nine are non-positive, but in this exploratory study no detailed analysis has been made of the various cases.

Table 12.6 Estimation results of model (12.10)

S\P	a_{11}	a_{12}	a_{13}	a_{21}	a_{22}	a_{23}	a_{31}	a_{32}	a_{33}	b_1	b_2	b_3	U
1	−2.1819	−2.9368	−3.3316	3.7595	4.4289	4.8247	0.3968	0.4027	0.3721	−17.14	−17.01	−17.12	0.0038
2	1.2198	1.7931	0	3.2543	3.3275	3.4147	−2.6764	−2.7218	2.7692	−7.1934	−10.757	3.8513	0.0432
3	−2.4663	−3.0223	−2.9632	1.1667	3.2278	3.2312	1.4375	0	0.2275	−0.8157	0.1089	−2.3046	0.0248
4	0.2763	0.4163	0.6399	1.2082	1.2566	0	−1.7744	−1,7890	0.0685	9.6793	10.868	10.0858	0.0404
5	−2.7904	−3.4291	0	3.3726	3.3467	0.6926	−6.2102	−0.0158	0	55.907	−1.3444	−0.0898	0.0861
6	−0.0149	0	−0.2519	1.0556	0.1847	0	0	0.9497	1.1796	−3.4719	−3.8712	−1.9225	0.0202
7	1.5117	0	.2852	2.9476	1.1018	1.1558	−2.1952	−0.4717	−0.4853	−11.57	1.3841	−0.4828	0.0073
8	0	1.1709	1.1709	0.0560	4.1141	4.2962	−.0645	0	−3.6705	4.2579	−3.6466	−4.1888	0.1039
9	−0.7343	−0.8851	−0.8655	1.3096	1.4254	1.4840	1.2425	1.4705	1.6052	−14.53	−16.26	−17.75	0.0002

12.4 Conclusion

A workable method to flexibly select parameter regimes has been presented. It is a member of a class of non-standard estimators, many of which will be used in further spatial econometric work. They are indispensable companions of non-standard specifications that will also be required in the field of spatial econometrics.

Recent experience has indeed shown that systems of regions often reveal two regimes. In Coutrot et al. (2009), the introduction of a second regime lifted the R^2 from 0.5156 to 0.9990. Moreover, the regions were behaviorally very different, which separated the main regional activity poles from the minor centers. Another example will be provided in Chap. 14.

In the light of this, still limited, experience, it seems that an appropriate specification-cum-estimation strategy is to systematically test for the presence of multiple regimes.

12.5 Appendix

Activity production levels

R\S	1	2	3	4	5	6	7	8	9	10	11	12
1	1700	408	951	649	18	1307	605	104	451	294	3691	1067
2	1507	481	1051	418	96	1320	458	159	373	598	515	879
3	1085	316	814	245	33	1031	396	95	293	335	625	664
4	3644	981	2273	818	39	2505	889	187	657	599	99	1663
5	345	129	570	93	2	609	138	45	212	352	0	294
6	5399	1519	4316	1310	291	5034	1631	397	1451	922	29	3318
7	2491	890	3473	1127	473	4138	1202	281	1211	274	4	2032
8	6928	1864	7454	4423	502	8594	2346	957	2154	834	84	4483
9	12,323	2846	8782	5058	881	10,947	2930	830	2914	2121	182	6547
10	1841	301	802	408	9	770	264	64	239	244	0	594
11	10,093	2040	5866	1814	178	6842	1815	496	1481	1119	46	3411
12	4225	786	2308	960	177	2729	1100	237	726	580	5	1632

Source: CBS, StatLine; numbers relate to 1993 and are expressed in 10^6 EUROS

Chapter 13
Finite Automata

In Paelinck (2002), attention was drawn to a special algebra—called a min-algebra—that might rule quite a few spatial econometrics specifications; hereafter, applications of this idea are be presented in the form of finite automata.

A finite automaton specification (for a formal definition, see Linz, 1996, p. 2) can be viewed as an "if"-specification; in symbolic terms

$$y\text{: if}(\alpha x_i + \beta < \gamma z_i + \delta; \alpha x_i + \beta; \gamma zi + \delta), \tag{13.1}$$

which reads as follows: if $\alpha x_i + \beta < \gamma z_i + \delta$, then $\alpha x_i + \beta$, else $\gamma z_i + \delta$.

Hereafter, a two-region example of a dynamic finite automaton is outlined.

13.1 A Finite Automaton Bi-regional Dynamic Model

Consider the following numerical case, for which the variables are defined as follows

x_{it}: joint location factors (i = 1, 2);

y_{it}: production levels (i = 1, 2).

The model is specified as follows

$$x_{1,t+1}\text{: if}(y_{1t}/x_{1t} < y_{2t}/x_{2t}; 3; 1), \tag{13.2}$$

$$y_{1,t+1}\text{: if}(y_{1t}/x_{1t} < y_{2t}/x_{2t}; 1.05\ y_{1t}; 1.02\ y_{1t}), \tag{13.3}$$

$$x_{2,t+1}\text{: if}(y_{2t}/x_{2t} < y_{1t}/x_{1t}; 1.5; 1.2) \tag{13.4}$$

$$y_{2,t+1}\text{: if}(y_{2t}/x_{2t} < y_{1t}/x_{1t}; 1.05\ y_{2t}; 1.02\ y_{2t}) \tag{13.5}$$

The logic of the model is as follows: as long as one region has its joint location factors "undercharged" compared to those of the other one—in the sense that they are still attractive for further activity locations—, their joint value stays at a higher level, the growth rate there also being higher, and vice versa.

D.A. Griffith, J.H.P. Paelinck, *Non-standard Spatial Statistics and Spatial Econometrics*, Advances in Geographic Information Science 1,
DOI 10.1007/978-3-642-16043-1_13,

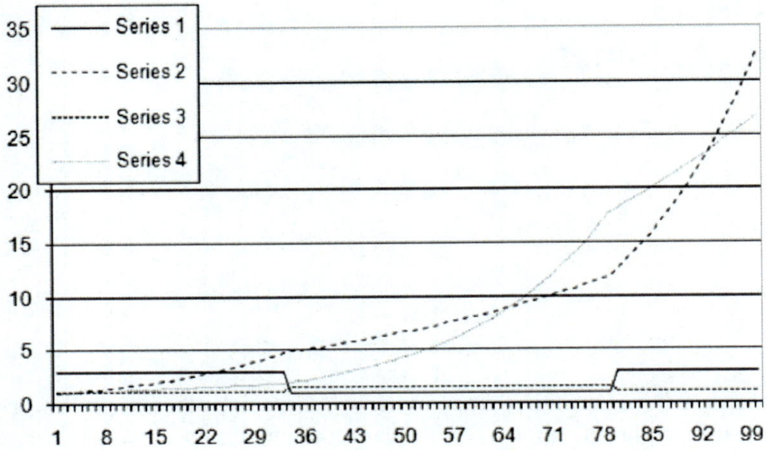

Fig. 13.1 Simulation of a dynamic finite automaton

With the initializing vector

$$(x_{10}, y_{10}, x_{20}, y_{20}) = (3, 1, 1.2, 1) \tag{13.6}$$

the resulting simulation is that presented by Fig. 13.1.

This figure portrays a dynamics in which regions at times lose their competitive edge in attracting activities.

The following problem is that of estimating such a model, possibly using an *if*-condition. Eight parameters have to be computed, to wit for each region the two levels of the location factors, x_i and x_i^*, and the growth rates, ρ_i and ρ_i^*, the lower levels being denoted by asterisks.

An example of an *if*-constraint is the following

$$\text{if}(y_{1t}/\xi_{1t} < y_{2t}/\xi_{2t};\ 1;\ 0) \tag{13.7}$$

where ξ_{it} is one of the values of x_{it}, x_{it}^* determined according to conditions (13.2) and (13.4). The specification of conditions (13.7) implies the introduction of a norm for the x_{it}, x_{it}^* values. Expression (13.7) was combined with the following term appearing in the objective function φ—sum of some function of those terms, as will be seen subsequently—for minimization purposes

$$r_{1t} - \lambda_{1t}\rho_1 - (1 - \lambda_{1t})\rho_1^*, \tag{13.8}$$

where, for region *1* and time *t*, r_{1t} is the observed growth rate, and λ_{1t} is a binary variable, the value of which is determined by condition (13.7) for the previous period, which reproduces condition (13.3) above.

Table 13.1 presents the data used; mark the split of the r_{it}-values in four opposite groups (e.g., for $t=1,\ldots 4$, $r_{1t}<r_{2t}$, then the reverse occurs).

Table 13.1 Data for estimating finite automaton parameters

t	r_{1t}	y_{1t}	r_{2t}	y_{2t}
1	0.05	1.00	0.02	1.00
2	0.06	1.05	0.01	1.02
3	0.04	1.11	0.03	1.03
4	0.01	1.14	0.06	1.06
5	0.03	1.15	0.05	1.12
6	0.02	1.18	0.04	1.18
7	0.06	1.20	0.01	1.22
8	0.05	1.27	0.03	1.23
9	0.04	1.33	0.02	1.27
10	0.02	1.38	0.04	1.30
11	0.03	1.41	0.06	1.35
12	–	1.45	–	1.43

Table 13.2 Parameter values of the estimated finite automaton

ρ_1	ρ_1^*	x_1	x_1^*	ρ_2	ρ_2^*	x_2	x_2^*	φ
0.05	0.022	0.8911	1.0022	0.05	0.02	1.0526	1.0541	0.00148

Table 13.2 presents the estimated parameters, φ being the obtained minimum of the objective function, in which expression (13.8) was squared; use was made of a reduced gradient method (Fylstrom et al., 1998)

The split into four periods mentioned earlier is correctly pictured by the binary variables λ_{it}; to improve the estimates of the parameters, the exercise was repeated keeping the binary variables stable and endogenizing the y_{it}-values, as is explained in the ensuing discussion.

Figure 13.2 portrays results of a simulation of the estimated model over the first twelve time periods. The four-period pattern just mentioned is fully reproduced, but the values, especially in region 1, deviate from the observed ones.

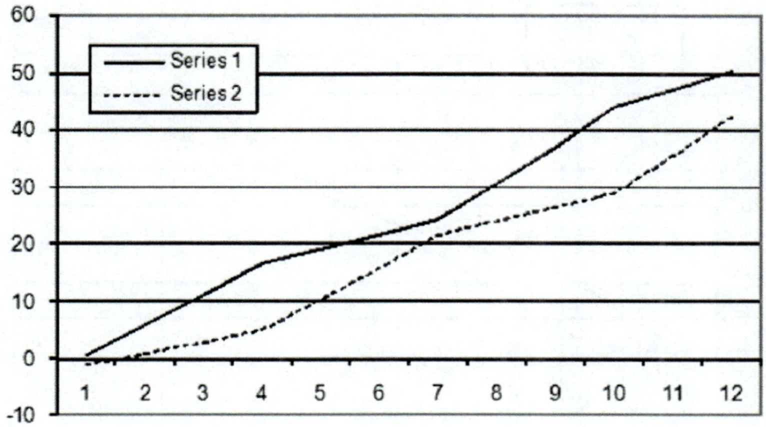

Fig. 13.2 Simulation of the estimated finite automaton

A possible alternative to objective function (13.8) is to reformulate the specification in terms of the production levels y_{it}, and use SDLS (see Sect. 11.1.3) with the computation of an optimal starting point for an endogenous simulation (Paelinck, 1990b). This method minimizes some difference between the observed values and the *endogenously simulated* values of the y_{it}s, those simulated values being at the same time generated within the estimating procedure. This approach is described next.

Of note is that condition (13.7) is expressed in terms of strict inequalities, which implies that if an equality is present, the higher x_i and ρ_i levels automatically are assigned to the other region. An alternative is furnished by

$$(\lambda_{1t} - \omega)(y_{1t}/\xi_{1t} - y_{2t}/\xi_{2t}) \leq 0, \tag{13.9}$$

with $0 < \omega < 1$, which adds another degree of freedom (to be taken up by the estimation procedure) when the second factor in expression (13.9) is zero. This procedure has been applied to the y_{it} data of Table 13.1, again with a quadratic objective function, and subsequently keeping binary variables—which correctly split the overall period into four components—stable.

Table 13.3 presents the obtained parameters; φ_ψ is the value of the objective function, φ_ρ that of φ in Table 13.2.

Figure 13.3 presents the simulated values over the first 12 time periods, again showing that the four characteristic groups mentioned earlier are still correctly

Table 13.3 Parameter values introducing conditions (13.9)

ρ_1	ρ_1^*	x_1	x_1^*	ρ_2	ρ_2^*	x_2	x_2^*	φ_ψ	φ_ρ
0.0483	0.0215	0.9704	1.0134	0.0442	0.0196	.1.0022	1.0139	15.3791	0.0015

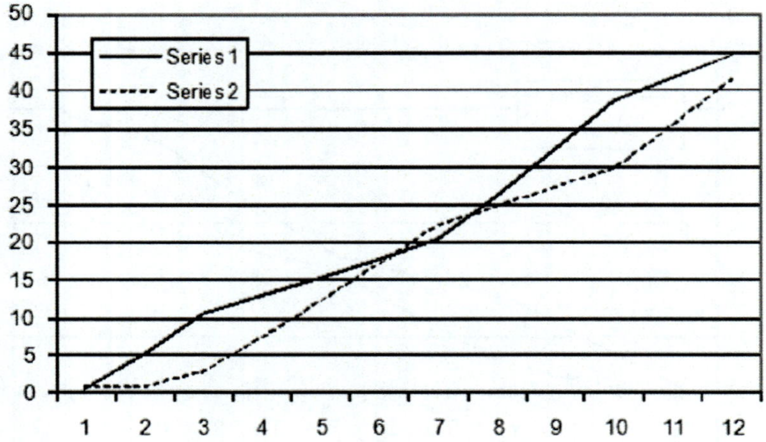

Fig. 13.3 Second simulation of the estimated finite automaton, parameters from Table 13.3

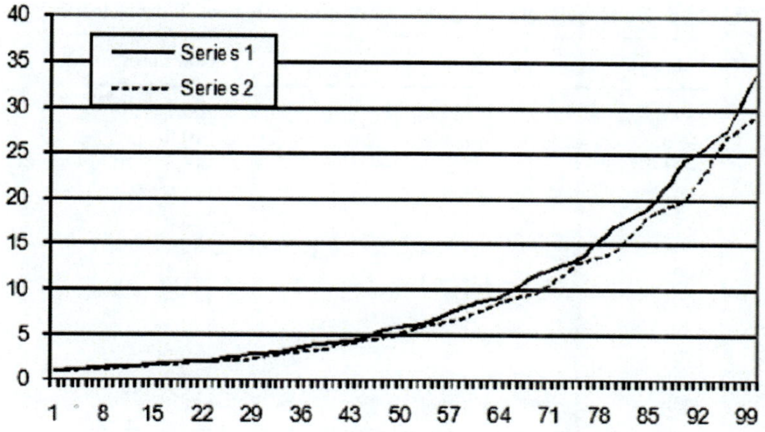

Fig. 13.4 Figure 13.3 results extended over 100 periods, parameters from Table 13.3

represented, as well as that the fit is considerably improved; Fig. 13.4 extends the projection.

To conclude, in the estimation procedure, dependence on initial conditions and multiple solutions do seem to be crucial. The latter especially holds for the x_{it} and $x_{it}{}^{*}$ values, which is due to the multiplicities that can satisfy conditions like (13.7) or (13.9). These points are left for further investigation.

13.2 An Empirical Application

To subject the model developed above to an empirical test in a well-documented case, gross regional product numbers for the Netherlands have been investigated. They were divided in two macro-regional sets, one for the western provinces (Noord-Holland, Zuid-Holland, Utrecht, the so-called "Rimcity"), the other one comprising the data for the remaining provinces (source: CBS, 2003; 10^9 EUROS). Given the low inflation rate (in the order of 1% annually) no price correction was applied. Table 13.4 presents the data.

The data were analyzed with the methodology discussed in Sect. 13.1, using a quadratic objective function for the product levels, and, alternatively, a binary and a fuzzy version. Table 13.5 summarizes the results. The λ_t parameters are generated by constraints (13.9), and are relative to the Rimcity, their values for the other provinces being the complements to *1*.

The curious finding, at first sight, is the respective values of the growth rates for the non-Rimcity provinces: whatever the state of their location factors' attractiveness, they follow the ups and downs of the Rimcity growth rates. This result is completely in line with the Rimcity indeed being the "motor" of the Dutch economy (Paelinck, 1973, pp. 25–40, especially pp. 37–40), imposing its evolutionary rhythm on the other regions, which corresponds to a sort of *Fick* diffusion in thermodynamics (Braun, 1975, pp, 645 a.f.; Philibert, 2005, pp. 2–3).

Table 13.4 Gross regional product data for two macro-regions in the Netherlands

Years	Rimcity	Other provinces
1998	111.0	103.7
1990	116.5	110.0
1991	122.4	116.6
1992	127.3	121.6
1993	131.3	125.5
1994	136.9	130.3
1995	142.8	136.1
1996	147.3	141.3
1997	156.5	147.3
1998	166.8	156.5
1999	176.5	164.7
2000	189.8	177.1

Table 13.5 Parameter values of the model defined by Eqs. (13.2), (13.3), (13.4), and (13.5) with data from Table 13.4

Parameters	Values, binary case	Values, fuzzy case
ρ_1	0.0638	0.0794
$\rho_1{}^*$	0.0385	0.0386
x_1	1,8797	1.1099
$x_1{}^*$	00.9001	0.9675
ρ_2	0.0409	0.0394
$\rho_2{}^*$	0.0538	0.0705
x_2	1.4010	0.9278
$x_2{}^*$	0.6084	0.9564
λ_1	1	0.5976
λ_2	0	0.1
λ_3	0	0.1
λ_4	0	0
λ_5	0	0
λ_6	0	0
λ_7	0	0
λ_8	1	0.4112
λ_9	1	0.6291
λ_{10}	1	0.5021
λ_{11}	1	1
φ_y	16.3326	9.4769

From a technical point of view, the constraint parameter estimates are consistent with one another, which hints at the adequacy of the binary estimation; the lower φ_y-value is in line with the *Le Châtelier*-principle (Samuelson, 1955, pp. 36 a.f.).

Figures 13.5 and 13.6 further down portray the binary and fuzzy cases, respectively.

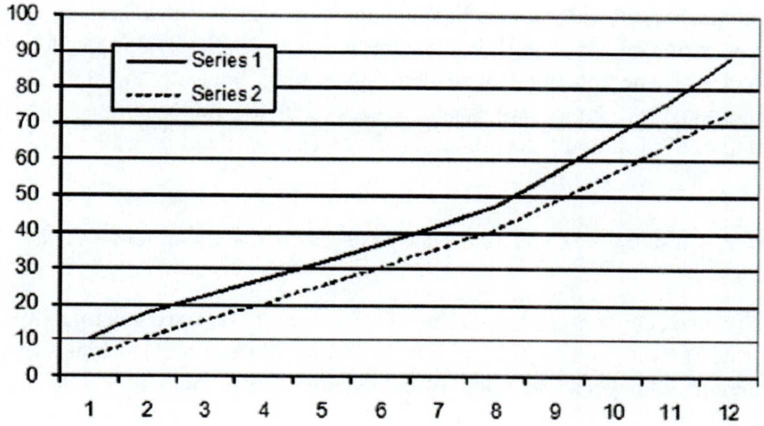

Fig. 13.5 Dutch model, binary case

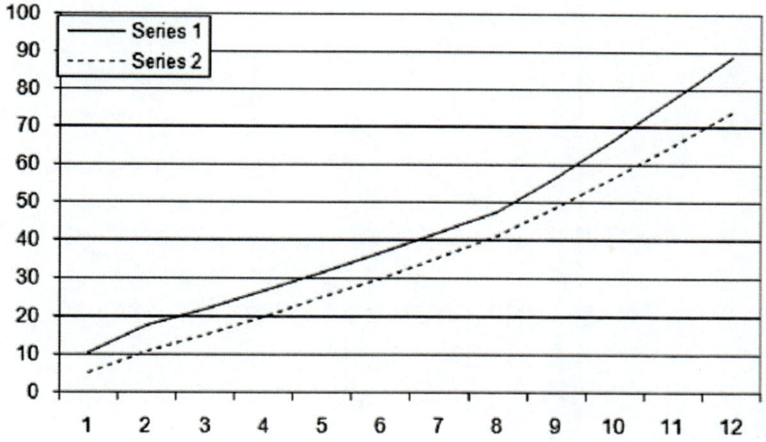

Fig. 13.6 Dutch model, fuzzy case

13.3 Conclusion

It has been shown that finite automata models can be an appropriate specification for multiregional models; the reason is that the development logic of a multiregional system needs for its modeling a special algebra, and a corresponding set-up of the corresponding estimation procedure.

Three more points should still be made.

First, starting an exercise in spatial econometric modeling with a complexity analysis of the data is advisable; obvious candidates for simple exogenous variables are their space-time coordinates. An example can be found in Getis and Paelinck (2004), where regional product data for the Netherlands were analyzed. A model specification implies the choice of exogenous variables, and possibly endogenous ones—in interdependent models—or lagged endogenous variables—in dynamic models—, therefore, they too should be implied in a complexity approach.

Second, the specifications presented can readily be generalized to three or more alternatives (regions, test specifications). For the finite automaton version, for example, the following specification shows how *and* and *or* statements can be added

$$y_i: \text{if}\,((cz_i + d < ax_i + b) \text{ and } (eu_i + f < ax_i + b)\,;\,(cz_i + d) \text{ or } (eu_i + f)\,;\,ax_i + b) \tag{13.10}$$

Finally, one cannot escape from the fact that hypothesis testing is fundamentally theory-laden (Aznar Grasa, 1989, p. 10). Accordingly, theoretical spatial economics will remain an indispensable guide to spatial econometric modeling.

Chapter 14
Learning from Residuals

Residuals often are considered as a troublesome noise in spatial—or, for that matter—non-spatial econometric models. Current practice in spatial econometrics is to set up a spatial error model, more often than not with an exogenous W spatial weight matrix, in order to improve the efficiency of the estimators.

Looking closely into the residuals is less common practice. And still, residuals can represent extremely precious building blocks for further work, as other disciplines have shown. Around 1850 the British chemists, Mansfield and Perkin, had the—for that era of chemistry—strange idea to analyze the composition of tar, until then exclusively used to improve coverage of roads (John London McAdam had his name attached to that technique, tarmacadam); the result of the British chemists' investigation was the roaring development of a whole branch of (industrial) chemistry: carbochemistry.

In the next section, a simple spatial econometric example will be treated, after which further analysis and more results will be presented.

14.1 Residuals

Tables 14.1 and 14.2 present the degrees of contiguity for Belgian regional units, BRU, (the maximum degree being 3) and their gross regional products (1995, 10^5 Euros of 2000); the entries of the two tables follow the same order.

Figure 14.1 reproduces the map of those regions.

The regions are the following. From West to East, northern slice: West-Flanders, East-Flanders, Antwerp, Limburg; same, southern slice: Hainaut, Namur, Luxembourg, Liège (slightly upwards); right in the middle, from north to south: Flemish Brabant and Walloon Brabant, with the Brussels Capital region sticking out.

First the products of 1995 were analyzed. The idea was to investigate the effects of (average) products for different degrees of contiguity (1, 2, 3) on a given GRP, y_i. Hence the equation

$$y_i = ay_{1i} + by_{2i} + cy_{3i} + d + \varepsilon_i, \tag{14.1}$$

D.A. Griffith, J.H.P. Paelinck, *Non-standard Spatial Statistics and Spatial Econometrics*, Advances in Geographic Information Science 1,
DOI 10.1007/978-3-642-16043-1_14, © Springer-Verlag Berlin Heidelberg 2011

Table 14.1 Degrees of contiguity between Belgian regions

BRU	A	BW	VB	OV	WV	LIM	H	N	LU	LIE	BC
A	0	2	1	1	2	1	2	3	3	2	2
BW	2	0	1	2	2	2	1	1	2	1	2
VB	1	1	0	1	2	1	1	2	2	1	1
OV	1	2	1	0	1	2	1	2	3	2	2
WV	2	2	2	1	0	3	1	2	3	3	3
LIM	1	2	1	2	3	0	2	2	2	1	2
H	2	1	1	1	1	2	0	1	2	2	2
N	3	1	2	2	2	2	1	0	1	1	3
LU	3	2	2	3	3	2	2	1	0	1	3
LIE	2	1	1	2	3	1	2	1	1	0	2
BC	2	2	1	2	3	2	2	3	3	2	0

Same obvious of province abbreviations given also in Table 14.2

Table 14.2 Gross regional products for the Belgian units, 1995

Regional units	A	BW	VB	OV	WV	LIM	H	N	LU	LIE	BC
Values	416028	62919	211584	255118	226222	143460	191433	64851	37976	173063	424381

Fig. 14.1 Regional map of Belgium

Table 14.3 First results for model (14.1)

Parameters	Values	t- or F-values	Probability
a	1.8813	1.8900	0.1007
b	1.4596	1.4349	0.1945
c	–0.0378	–0,1198	0.9080
d	–411581	–1.0873	0.3129
R^2	0.4806	2.1590	0.1810

where y_{1i}, y_{2i} and y_{3i} are the average products for different degrees of contiguity.

Table 14.3 presents the OLS estimation results.

Obviously the results are far from being satisfactory. The residual spatial correlation coefficients, $r_c{}^2$ (*c=1, 2, 3*, the observed degrees of contiguity) are respectively *–0.2619, –0.1161* and *–0.2426*. They are not significant, but show that there is no completely random field in the residuals.

Accordingly, further analysis is in order.

14.2 Multiple Regimes

The first column of Table 14.4 shows the residuals of the exercise, and compares them (columns 2 and 3) with the growth rates (averages over 1995–2004) and the GRP levels.

The following scatter plots (Figs. 14.2 and 14.3) picture the partial relations.

The *Kendall-τ* (Kendall, 1955) between residuals (+ or –) and growth rates (above or below the average, 0.0217) is near zero (exactly, 0.0910), and between residuals and GRPs it is 0.4546, but further investigation is still required.

To prepare the latter, a complexity index has been computed (Getis and Paelinck, 2004), derived from a fourth degree polynomial

Table 14.4 Comparing residuals

Regional units	Residuals	Growth rates	GRP
A	132145	0.0202	416028
BW	−159733	0.0335	62919
VB	15140	0.0277	211584
OV	14406	0.0239	255118
WV	−50515	0.0207	226222
LIM	−103617	0.0194	143460
H	−54674	0.0144	191433
N	−31766	0.0245	64951
LU	15883	0.0190	37976
LIE	82491	0.0135	173063
BC	140241	0.0221	424381

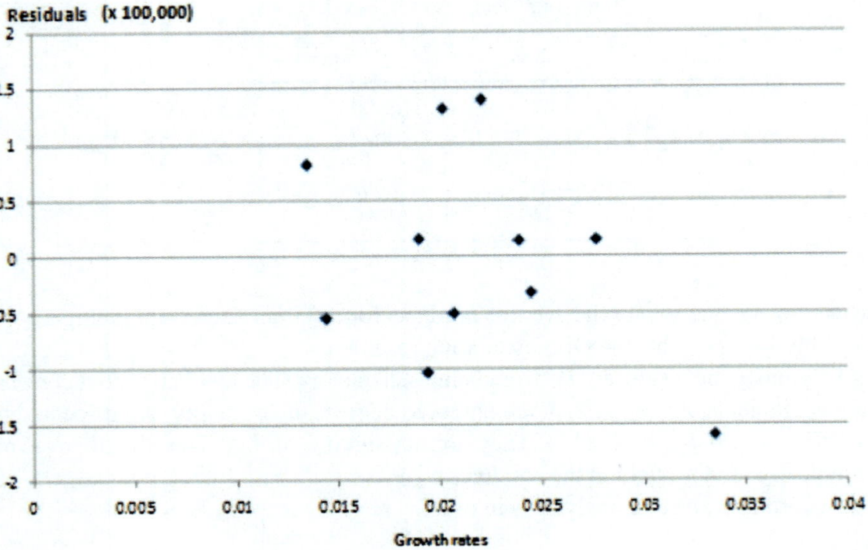

Fig. 14.2 Residuals and growth rates from Table 14.4

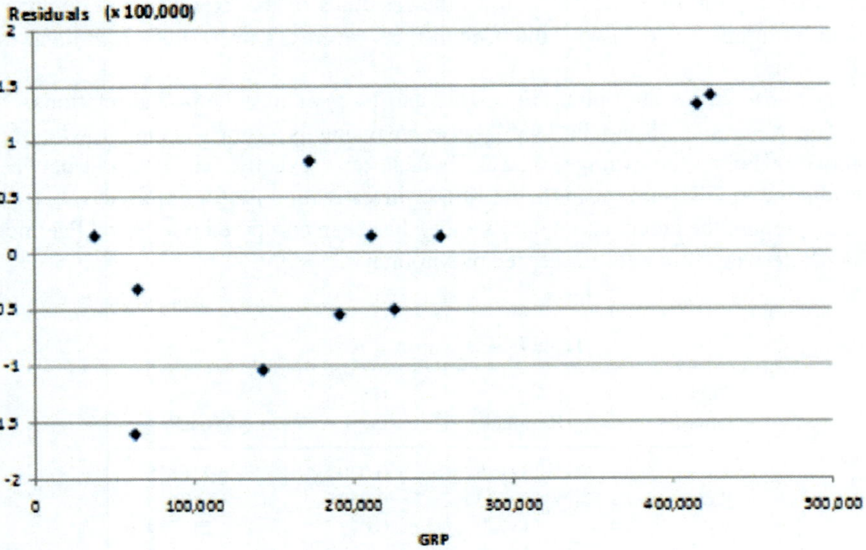

Fig. 14.3 Residuals and GRP from Table 14.4

Table 14.5 Polynomial coefficients from Eq. (14.2)

Coefficients	a	b	c	d	e	f	g	h	i	j	k
Values	1740457	−2.7E+08	55.7706	1.54E+10	−0.00025	−5903.13	−3.01E+11	−1.45E-11	148311	0.02579	−0.62572

$$e_i = a + b^*r_i + c^*y_i + d^*r_i^2 + e^*y_i^2 + f^*r_iy_i + g^*r_i^3 + h^*y_i^3 + i^*r_i^2y_i + j^*r_iy_i^2 + k^*r_i^2y_i^2, \quad (14.2)$$

in which the y_is are again the GRPs, and the r_is the growth rates. Table 14.5 presents the interpolated coefficients of Eq. (14.2).

Coefficients *b, d,* and *g* are extremely high, but they relate to growth rates that are small numbers. Excluding the relatively small coefficients (smaller than one), the complexity coefficient can be computed as

$$C = (v - 1)/(n - 1) = 0.6, \quad (14.3)$$

where *v* is the number of maintained coefficients, and *n* their maximal number (i.e., the number of observations).

Given the rule followed, this is a relatively high value ($0 \leq c \leq 1$), and invites rethinking the model generating the observed residuals, as this model is a very simple one.

The revealed complexity suggests the need for a possible correction by r_i and y_i, but the first correction would not be complete, as shown above, and using y_i would be trivial. A plausible alternative would be to introduce two separate regimes, leading to the following specification (see Chap. 12):

$$y_i = \lambda_i(a^*y_{1i} + b^*y_{2i} + c^*y_{3i} + d) + (1 - \lambda_i)(\alpha^*y_{1i} + \beta^*y_{2i} + \gamma^*y_{3i} + \delta) + \varepsilon_i, \quad (14.4)$$

where the λ_is are the binary variables qualifying the spatial regimes.

This produced the results of Table 14.6 hereafter.

Table 14.6 Results with two regimes, Eq. (14.4)

Parameters	Values 1995	Values 2004
a	0.6726	0.5894
b	1.6484	1.5462
c	−0.7094	−0.7416
d	−69490	−763
α	8.5571	2.9617
β	4.7365	1.0460
γ	−0.1971	1.2467
δ	−2345031	−710212
R^2	0.9853[a]	0.9774[a]

[a]Computed for 6 df

Table 14.7 Regime allocators for two distant years

Regional units	1995	2004
A	1	1
BW	0	0
VB	0	0
OV	1	1
WV	1	1
LI	1	1
H	0	0
N	1	1
LU	1	0
LIE	1	1
BC	0	1

The residual spatial correlation coefficients for contiguities 1, 2 and 3, respectively, are –0.1793, *–0.8029* and 0.2938, showing that there is still some specific spatial autocorrelation, especially of order 2 (coefficient significant at the 0.995 level). Some changes occurred over nine years, as the third column of Table 14.6 shows; this also was the case for the r_Cs (–0.6702, –0.8223, 0.6428, all significant at the 0.995 level), the second order spatial autocorrelation still dominating. But the overall fit is satisfactory, and OLS can be replaced by other estimation methods (see Sect. 11.1.3).

To test the general properties of the residual fields, two statistics have been computed:

- a generalized τ-statistic between all residuals (i.e., 55 cross-products are involved); for 1995 and 2004 they amount to non-significant values –0.1542 and –0.1523, respectively, excluding any general correlation; and,
- the C-statistics of Eq. (14.3); in both cases they equal 0.9, showing a high degree of Chaitin-Wolfram complexity (as independent variables, the numbers from 1 through 12 were used in the test polynomial).

The problem now is: though overall randomness seems to be present, spatial r_Cs show specific dependency, so some further investigation is in order.

Table 14.7 presents the two vectors of λ_I estimates.

The pattern is remarkably stable: most regions (7) belong to the same regime, only the center-south deviating from this.

14.3 Spatial Interpolation

Because there is quite some variability in the coefficients reported in Table 14.6, the question arises asking whether computing local coefficients could give more insight in this phenomenon. One possibility is to interpolate the parameters from

Table 14.8 Results from spatial interpolation to compute the parameters of Eq. (14.1)

Parameter regional	a	b	c	d	a	b	c	d
A	2.8428	3.2064	–0.0712	816849	2.9034	3.0462	–0.8971	–983912
BW	1.1572	–0.4494	0.0776	0	1.0961	–0.4499	0.0747	0
VB	9.0711	5.0851	–0.3824	–2.5058	8.2277	4.6305	–1.2307	–2.6980
OV	3.5057	3.6279	–1.1336	–1.0209	3.3062	3.2851	–1.1589	–1.1109
WV	2.1252	1.0591	–0.1991	–410515	1.6838	0.5410	–0.0943	–282235
LI	1.5086	2.6832	–0.5342	–441860	1.2176	2.2869	–0.6389	–380635
H	8.5517	4.7330	1.0080	–2.3433	6.0211	2,8355	0.7076	–1.8282
N	6.0664	0.1237	0.3975	–833866	4.0873	0.1847	0.2981	–673717
LU	–1.6767	1.0559	–0.4168	214303	–1.7254	1.1119	–0.4436	254356
Lie	.6890	1.3824	–0.5545	–71376	0.6151	1.2632	–0.4947	–76625
BC	9.0717	5.0851	–0,3822	–2.5058	8.2277	4.6305	–1.2307	–2.8980
v=σ/μ	0.9106	0.7594	–2.6792	–1.5842	0.9346	0.7818	–1.3398	–1.7597

groups of— possibly neighboring—spatial units. If more than four regions present themselves as candidates, nearest neighbors—in terms of distances and/or political/linguistic proximity—have been selected.

Table 14.8 presents the results. The parameters are those of Eq. (14.1).

The remarkable finding, again, is that the orders of magnitude are the same for the two years, with some exceptions for the constant *d*. But the variability is large between regions (as shown by the coefficients of variation in the last row of Table 14.8), which suggests the need for further analysis of the available data to complete the picture.

14.4 Composite Parameters

Because time series from 1995 through 2004 are available, composite parameters can be computed (Ancot et al., 1978); for instance, parameter a in Eq. (14.1) can be expanded as

$$a = \hat{a} + a_r + a_t, \tag{14.5}$$

Table 14.9 Generic and region-specific coefficients according to model (14.5)

Region Parameter	A	BW	VB	OV	WV	Lim	H	N	Lu	Liè	BC
a(a$_r$)	–1.26	1.113	0.6643	1.715	1.238	1.249	–0.8	2.055	–.0486	–0.230	1.298
b(b$_r$)	3.132	–2.79	0.1305	–2.39	–2.39	–2.88	–0.93	–3.27	–3.59	–2.40	–1.92
c(c$_r$)	–0.065	–	–	–67.8	0.5279	0.4333	–	–0.493	1.086	0.4754	2.51
d(d$_r$)	132	–97	–116	–67	6.342	–212	41	7.877	–22[a]	0.0326	–0.337

[a]To be multiplied by 10^3

where a is the generic, a_r is the region-specific parameter, and a_t is the time-specific parameter. For reasons of identifiability, one spatial unit should be selected as a kernel (not affected by a_r or a_t); the first region, Antwerp, was picked for this purpose, but any other region would have done.

Table 14.9 hereafter presents the coefficients; the first row, as said above, contains the generic ones, the following rows the region-specific ones. Sometimes a region-specific coefficient c_r is absent, due to the absence of a third order spatial lag.

The entries of Table 14.9 display a large region-specific parameter variability. No measure hereof has been computed this time, but a comparison with Table 14.8 confirms this variability.

Table 14.10 presents the time-specific parameter estimates for 1996 through 2004.

The coefficients are of a much smaller order of magnitude, which confirms a previous remark about the relative constancy of the parameters through time, as opposed to their interregional variability.

Table 14.10 Time-specific parameters for model (14.5)

Parameter FUTL[a]	a$_t$	b$_t$	c$_t$
1	–0.0026	0.0094	–0.0153
2	–0.0016	0.0053	–0.0016
3	–0.0097	0.0206	0.0028
4	0.0044	0.0188	–0.0205
5	0.0050	0.0209	–0.0317
6	0.0028	0.0304	–0.0365
7	0.0122	0.0220	–0.0489
8	0.0153	0.0218	–0.0513
9	0.0176	0.0199	–0.0588

[a]Forward Unit Time Lag, from 1995 on

Table 14.11 Pseudo R^2s

Regions	Pseudo-R^2s
A	0.9538
BW	0.9905
VB	0.9907
OV	0.9650
WV	0.9993
LI	0.9926
H	0.8705
N	0.9861
LU	0.9394
LIE	0.9104
BC	0.9809
Global	0.9735

Finally Table 14.11 presents the partial and global pseudo-R^2-values, pseudo- because the parameters have been computed by least absolute discrepancies to avoid outliers.

The result is remarkably high for 60 df, with a local exception for Hainaut.

14.5 Conclusion

The doggy-bag principle ("never throw away your leftovers") has given insight into a possibly appropriate specification of the spatial econometric models investigated. This is in line with the clear warning that has been given off for time series analysis (G. Mizon, A Note to Autocorrelation Correctors: Don't, Journal of Econometrics, 1995, 69, pp. 267–288).

More research is in order, especially for very large models. But considering residuals as informative should transcend the usual practice of trying to neutralize them. Meanwhile, pure spatial "randomness" also could be interpreted as spatial complexity, and might encourage continued analysis rather than finishing it by discussing "ideal" parameter properties.

In the Belgian case, this has lead to deeper insights in spatio-temporal properties of a static model. Indeed, it appears that each spatial unit possesses its own reaction coefficients with a great stability over time. Problem however is to find out how much of that interregional divergence is due to system heterogeneity, and how much to spatial aggregation. The latter problem is taken up in Chap. 17.

Chapter 15
Verhulst and Poisson Distributions

The logistic curve (or Verhulst sigmoid curve) sometimes is used in spatial econometrics (see, e.g., Domencich and McFadden, 1975; Paelinck and Klaassen, 1979, pp. 68–72, 156–168). Two examples will be given hereafter, one for estimation in the binary case, the other for a dynamic specification. A related Poisson distribution problem is then treated; the latter distribution is less frequently used, because count data have to be available for econometric treatment.

15.1 Robust Estimation in the Binary Case: A Linear Logistic Estimator (LLE)

For a binary variable z = 1, let

$$d_{1i} \triangleq 1 - (1 + \exp(\mathbf{a'x_i} + b_i))^{-1}, \tag{15.1}$$

which, for a variable with subscript *i*,. is the natural distance between *1* and the logistic curve; equally, for *z* = *0*, let for a variable with subscript *j*

$$d_{0j} \triangleq (1 + \exp(\mathbf{a'x_j} + b_j)\big)^{-1} \tag{15.2}$$

which is that distance of the logistic curve from *0*.

Furthermore let

$$\mathbf{a'x_{1i}} + b_i = -\ln\left(d_{i1}{}^{-1} - 1\right) \triangleq \delta_{1i}, \tag{15.3}$$

and

$$-\mathbf{a'x_{0j}} - b_j = -\ln\left(d_{0j}{}^{-1} - 1\right) \triangleq \delta_{0j}, \tag{15.4}$$

D.A. Griffith, J.H.P. Paelinck, *Non-standard Spatial Statistics and Spatial Econometrics*, Advances in Geographic Information Science 1,
DOI 10.1007/978-3-642-16043-1_15,

In both cases, $\partial\delta/\partial d > 0$, and if $d_i = d_j$, then $\delta_i = \delta_j$.

Minimizing $\sum_i \delta_{1i} + \sum_j \delta_{0j}$, and normalizing the vector $\boldsymbol{\delta}$, $\mathbf{i}$ being the unit column vector, namely

$$\min \mathbf{i}'\delta - \lambda/2(\boldsymbol{\delta}'\boldsymbol{\delta} - c), \tag{15.5}$$

yields

$$\mathbf{i} = \lambda\boldsymbol{\delta} = \mathbf{X}^*\mathbf{a}, \tag{15.6}$$

with

$$\mathbf{X}^* = \begin{bmatrix} \mathbf{X_1} \\ -\mathbf{X_0} \end{bmatrix} \tag{15.7}$$

and $\boldsymbol{a}$ including the constant. Because $\lambda < 0$, switching the sign of $\mathbf{X_0}$ in Eq. (15.7), replaces $\mathbf{i}$ by

$$\mathbf{i}^* = \begin{bmatrix} -\mathbf{i} \\ \mathbf{i} \end{bmatrix}$$

A linear estimator of $\boldsymbol{a}$ is given by

$$\mathbf{a} = (\mathbf{X'X})^{-1}\mathbf{X'i}, \tag{15.8}$$

with λ conveniently set equal to -1.

Therefore, for unit λ:

$$\mathbf{V(a)} = (\mathbf{X'X})^{-1}, \tag{15.9}$$

and hence pseudo-t values can be computed as follows

$$t_k = a_k/\sqrt{x_{kk}}, \tag{15.10}$$

The method was applied to the following (unique) explanatory variable: 0.4, 0.5, 0.6, 0.92, 0.95, 0.98; Table 15.1 lists the results (pseudo-t values in parentheses).

The results are graphically presented in Fig. 15.1.

Table 15.1 Estimation results for model (15.8)

Parameters	Values
Slope	−2.2169 (−40.7040)
Constant	1.1158 (30.3286)
R^2	0.9988

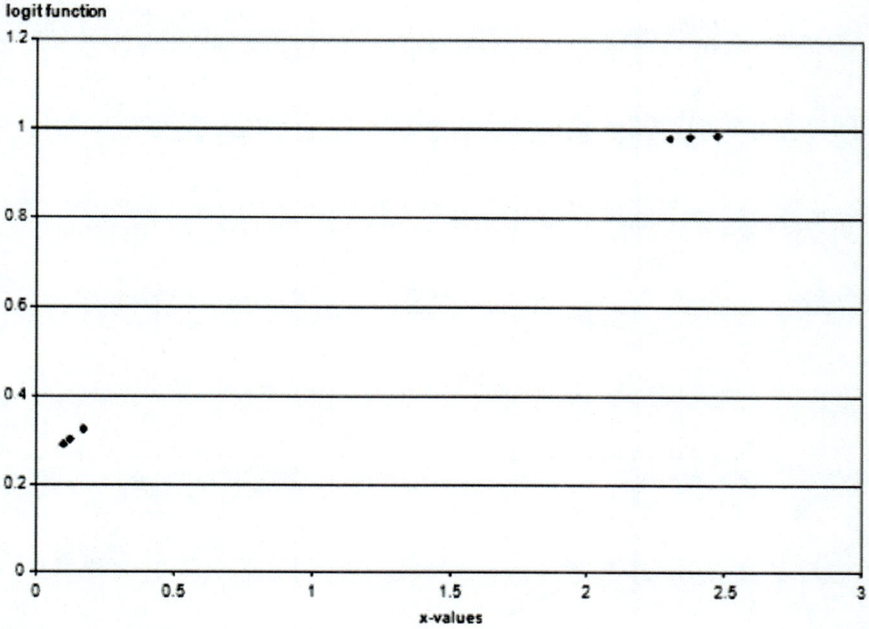

Fig. 15.1 The logistic resulting from Table 15.1

15.2 A Logistic Dynamic Share Model

Let $0 < a_{ij} < 1$ be the share of sector i in region j; $\sum_i a_{ij} = 1, \forall j$.

Let the model be specified as follows

$$a_{ijt} = \left[1 + \exp\left(\sum{}_{ij}\alpha_{ij,t-1} + \beta^{ij}\right)\right]^{-1} \tag{15.11}$$

a generalized logistic function. The superscripts refer to the subscripts of the left hand member.

From Eq. (15.11) one can derive

$$\ln(a_{ijt}^{-1} - 1) = \sum{}_{ij}\alpha_{ij}^{ij} a_{ij,,t-1} + \beta^{ij}, \tag{15.12}$$

In equilibrium, $a_{ijt} = a_{ij,t-1}$, $\forall i,j$. Thus,

$$\ln(a_{ijt}{}^{-1} - 1) = \sum_{ij}\alpha^{ij}{}_{ij}a_{ijt} + \beta^{ij} \tag{15.13}$$

$$= \alpha^{ij}{}_{ij}a_{ijt} + r^{ij}, \tag{15.14}$$

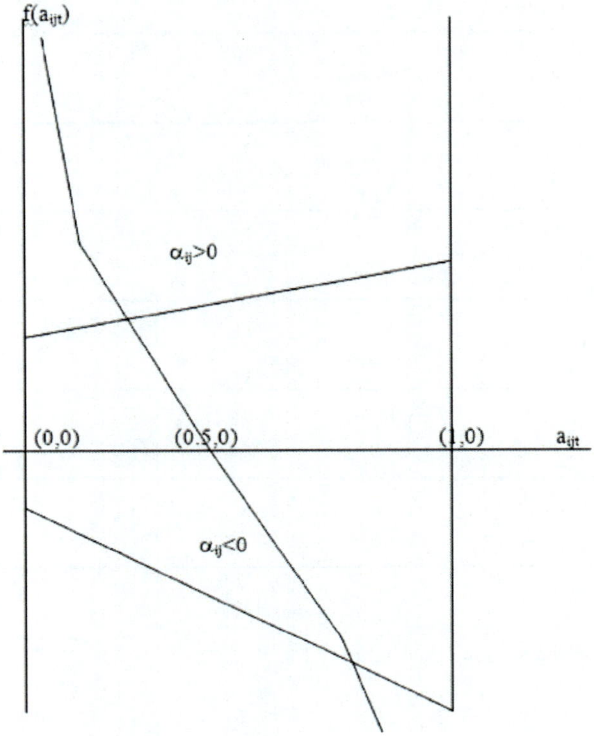

Fig. 15.2 Equilibrium solutions for (15.13) and (15.14)

where r denotes the remaining terms. There are two possibilities according to the sign of α_{ij}. Figure 15.2 shows how possible solutions look (recall that $0 < a_{ij} < 1$).

In terms of stability, the following points can be made:

(1) There exists a confiner defined as follows:

$$\max_{\mathbf{a}} \sum_{ij} \alpha^{ij}{}_{ij} a_{ij} + \beta^{ij} \tag{15.15}$$

$$\min_{\mathbf{a}} \sum_{ij} \alpha^{ij}{}_{ij} a_{ij} + \beta^{ij} \tag{15.16}$$

$$\text{s.t} \sum_{i} a_{ij} = 1, \forall_j \tag{15.17}$$

$$1 \geq a_{ij} \geq 0, \forall i, j \tag{15.18}$$

Table 15.2 Test numbers for model (15.11)

t	a_{11}	a_{21}	a_{12}	a_{22}
1	0.30	0.70	0.65	0.35
2	0.32	0.68	0.64	0.36
3	0.32	0.68	0.64	0.36
4	0.34	0.66	0.62	0.38
5	0.37	0.63	0.62	0.38
6	0.38	0.62	0.61	0.39
7	0.39	0.61	0.59	0.41
8	0.41	0.59	0.57	0.43
9	0.41	0.59	0.57	0.43
10	0.42	0.58	0.56	0.44

(2) One can linearize (Taylor-expansion around 0.5) the left-hand member of (15.12), yielding

$$\mathbf{a}_t \approx -0.25\,\mathbf{A}\mathbf{a}_{t-1} - 0.25(\mathbf{b} - 2\mathbf{i}). \tag{15.19}$$

Convergence depends on $-\mathbf{i} < \lambda(\mathbf{A}) < \mathbf{i}$, whereas divergence is constrained by the confiner. In the case of convergence, the attractor is:

$$\mathbf{a}^o = -0.25(\mathbf{I} + 0.25\mathbf{A})^{-1}(\mathbf{b} - 2\mathbf{i}). \tag{15.20}$$

However, given the approximation, conditions (15.17) are not necessarily satisfied.

The following (fictional) numbers, reported in Table 15.2, have been used to test the model.

Estimation (see Table 15.3) was performed by minimizing the sum of squares between the observed a_{ijt}s and the SDLS endogenously generated ones (see Sect. 11.1.3). The resulting overall R^2 is 0.9989, and, moreover, conditions (15.17) are very closely satisfied in both regions, with erratic divergences not exceeding 2% [see comments about Eq. (15.20)].

Starting from the last observations, a 24-period simulation was performed. Table 15.4 shows the results, which again obey conditions (15.17) very closely. The simulation reveals the inherent dynamics of the model, which could hardly be deduced from the "observed" series; Figs. 15.3 and 15.4 portray this once more.

15.3 A Linear Poisson Distribution Estimator

The Poisson probability mass function is given by

$$p(n) = \exp(-\mu)\mu^n/n!. \tag{15.21}$$

Table 15.3 Estimation results from Table 15.2

Features	α_{11}	α_{21}	α_{12}	α_{22}		
Parameters	–6.16267	11.78416	9.939417	–13.3512		
	4.125329	3.303637	4.513609	–8.34668		
	–6.61019	2.508025	–7.97125	3.446991		
	2.07069	–3.84603	–10.9558	6.180226		
	3.352769	–6.93088	2.237161	6.066311	Conditions (15.17)	
SDLS variables	0.475024	0.505434	0.443078	0.572783	0.980458	1.015861
	0.316985	0.680277	0.63812	0.360616	0.997262	0.998736
	0.324253	0.675817	0.640868	0.357918	1.000071	0.998786
	0.343588	0.656204	0.626998	0.373463	0.999792	1.000461
	0.36097	0.640596	0.616677	0.384327	1.001566	1.001004
	0.379807	0.620659	0.601062	0.401076	1.000466	1.002138
	0.394141	0.608216	0.591866	0.409607	1.002357	1.001473
	0.40885	0.590948	0.575795	0.425779	0.999798	1.001574
	0.414217	0.587613	0.571102	0.427662	1.00183	0.998764
	0.417111	0.57954	0.557563	0.439539	0.996651	0.997101
Conditions (15.12)	1.64E–09	–5.2E–10	1.2E–09	3.81E–10		
	1.27E–10	–1.1E–10	–1E–10	2.49E–11		
	1.1E–10	–1.5E–10	–8.6E–11	3.11E–11		
	1.37E–10	–6.8E–11	–1.5E–10	–1.3E–10		
	3.32E–13	–1E–10	–2.9E–10	–1.5E–10		
	–2.4E–10	1.02E–10	–6.4E–10	–4.1E–10		
	–1.9E–09	1.23E–10	–4.5E–10	5.47E–10		
	–1.2E–09	2.86E–10	–7.5E–10	1.23E–11		
	–3.3E–09	8.85E–11	2.75E–10	2.02E–09		
SDLS minus observed a_{ijt}	0.003015	0.000277	0.00188	0.000616		
	0.004253	0.004183	0.000868	0.002082		
	0.003588	0.003796	0.006998	0.006537		
	0.00903	0.010596	0.003323	0.004327		
	0.000193	0.000659	0.008938	0.011076		
	0.004141	0.001784	0.001866	0.000393		
	0.00115	0.000948	0.005795	0.004221		
	0.004217	0.002387	0.001102	0.002338		
	0.002889	0.00046	0.002437	0.000461		

Its mean μ can be written as a function of various factors, x_k; assume a linear function. Then

$$\ln\left[p(n_i)\right] = -\mu_i + n_i \ln \mu_{iI} - \ln n_i! \qquad (15.22)$$

The first-order maximum likelihood conditions (the second-order ones also are satisfied, as Eq. (15.22) is concave in the parameters) are, for some parameter *a*

$$\sum_i x_{ik} = \sum_i n_i x_{ik}/\mu_i \qquad (15.23)$$

Table 15.4 Simulation results

t	a_{11}	a_{21}	a_{12}	a_{22}	Conditions (15.17)	
1	0.42	0.58	0.65	0.35	1	1
2	0.602225	0.445547	0.496556	0.49466	1.047771	0.991216
3	0.671098	0.357977	0.399227	0.585599	1.029075	0.984826
4	0.663613	0.346834	0.367438	0.615801	1.010447	0.983239
5	0.611098	0.384416	0.383735	0.600675	0.995515	0.98441
6	0.537536	0.447841	0.428239	0.559126	0.985378	0.987366
7	0.457233	0.522775	0.487681	0.503906	0.980008	0.991587
8	0.382394	0.596658	0.551243	0.445049	0.979051	0.996292
9	0.325752	0.655956	0.606947	0.393426	0.981708	1.000373
10	0.295642	0.690855	0.644244	0.358716	0.986496	1.00296
11	0.291214	0.700626	0.659679	0.344255	0.99184	1.003934
12	0.304921	0.691703	0.657195	0.346493	0.996625	1.003688
13	0.328039	0.672193	0.643538	0.359153	1.000231	1.002691
14	0.353605	0.648826	0.624817	0.37652	1.002431	1.001337
15	0.376836	0.626439	0.60558	0.394357	1.003274	0.999937
16	0.394782	0.608214	0.588933	0.409789	1.002995	0.998723
17	0.406029	0.595901	0.576753	0.421086	1.001931	0.997839
18	0.410465	0.589988	0.569829	0.427519	1.000452	0.997348
19	0.409008	0.589897	0.568006	0.42923	0.998905	0.997236
20	0.403273	0.59429	0.570392	0.427041	0.997563	0.997433
21	0.395192	0.601406	0.575628	0.422208	0.996599	0.997836
22	0.386659	0.609425	0.582185	0.416144	0.996083	0.998328
23	0.379219	0.616776	0.588645	0.410161	0.995995	0.998806
24	0.373874	0.622369	0.593931	0.40526	0.996243	0.999191
25	0.371017	0.625685	0.597426	0.402014	0.996702	0.99944

If, on average, $\mu_i = 1$, condition (15.23) also is satisfied on average. Consequently

$$\mu_i/n_i = 1, \forall_i \tag{15.24}$$

from which an OLS estimator can be derived as

$$\mathbf{a} = (\mathbf{X'X})^{-1}\mathbf{X'i} \tag{15.25}$$

where for the constant term elements n_i^{-1} appear in **X,** because each element of the usual unit column vector has to be divided by the counts observed.

The model has been applied to the data reported in Table 15.5.

Table 15.6 presents the results (pseudo-t values in parentheses) for a one-variable (x_i) with the n_i^{-1} terms as required.

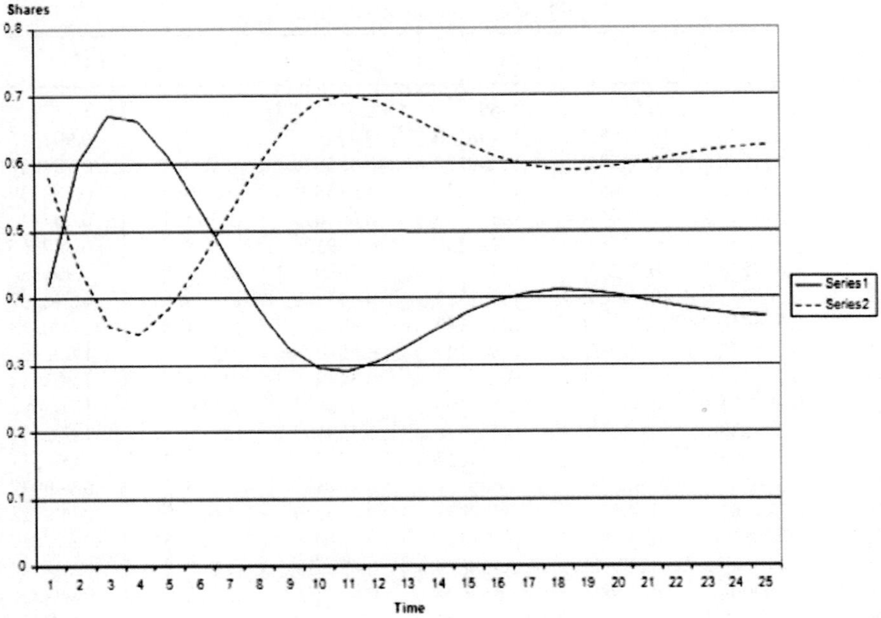

Fig. 15.3 First region simulatioins results graphed

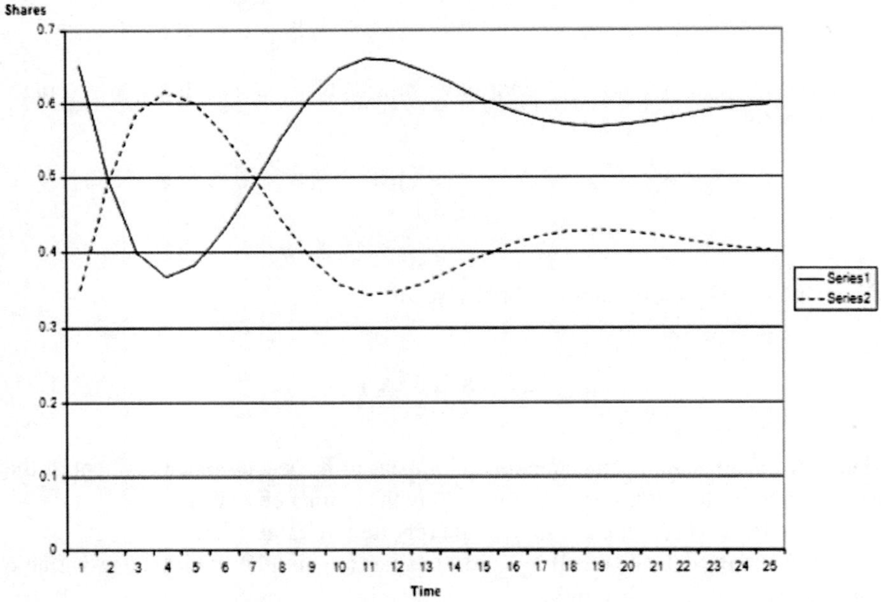

Fig. 15.4 Second region, simulation results graphed

Table 15.5 Data to apply Eq. (15.25)

x_i	n_i
3	1
7	2
8	4
11	5
13	7
17	9

Table 15.6 Estimation results using data from Table 15.5

Parameters	Values
Slope	0.4796 (5.7466)
Constant	–0.4436 (–1.2584)

15.4 Conclusion

Again very robust and simple estimators have been developed for the Verhulst and Poisson curve parameters. Although the processes might be complex, they are readily calibrated.

The examples have shown that the obtained estimation results are readily usable for consistent simulation, which moreover reveals properties that the original series do not show at once. This demonstrates the utility of longer term extrapolations, as the function—in this case, the Verhulst function—does not lead to analyses close to that of classical dynamics (see Chap. 11).

Chapter 16
Qualireg, A Qualitative Regression Method

Circumstances can produce themselves under which no cardinal data are available (see Ancot and Paelinck, 1990). To allow drawing inferences about at least the direction of influence of certain potentially explanatory variables, only available as ordinal data ("rankings"), methods should be developed to treat that problem. The method described here—*QUALIREG*—resulted from work on a qualitative multi-criteria method—*QUALIFLEX*, originated by Paelinck (1976)—which is detailed first, after which the logic of *QUALIREG* will be introduced.

A first application to test the method is then presented, followed by a typical spatial econometric one, to wit estimating first- and second-order contiguity effects.

16.1 Qualiflex

Suppose three objects, O, to be ranked according to three criteria, C, along which they can initially only be separately ranked. Table 16.1 presents such a case.

The relative importance of the criteria is known only in an ordinal manner, i.e., again their ranking (vector **w**).

The optimal ranking of O_1 through O_3 is to be derived out of the $3! = 6$ possible rankings. Those rankings can be classified according to elementary permutations (i.e., permutations of neighboring objects). Table 16.2 presents those rankings, starting from [+++, ++, +], with the rankings being denoted by R_i.

A possible measure of the agreement of two rankings is a so-called rank correlation coefficient, denoted τ, and having values in [–1, 1], much like an ordinary simple correlation coefficient. The easiest choice for attributing values is to divide the *interval* in equal parts, in this case *0.66*, and hence obtain the values shown in the last column of Table 16.2. The values of τ are computed with respect to R_1.

The observed rankings can be laid out in matrix form as Table 16.3 shows for the three rankings of Table 16.1; a descending order is rated *+1*, an ascending one *–1*.

These tables should obviously be skew-symmetric, i.e., the absolute values of symmetric terms are equal, but their signs are opposite. τ-values based on these tables can be computed with respect to R_1 (e.g., the upper-triangular sum of the first ranking of Table 16.3 is *1*, which divided by the sum of all positive scores gives *0.33*,

D.A. Griffith, J.H.P. Paelinck, *Non-standard Spatial Statistics and Spatial Econometrics*, Advances in Geographic Information Science 1,
DOI 10.1007/978-3-642-16043-1_16,

Table 16.1 A qualitative multicriteria table

w (rankings)	C\O	O_1	O_2	O_3
+++	C_1	++	+++	+
++	C_2	++	+	+++
+	C_3	+++	+	++

Table 16.2 Elementary permutations for three elements

R_1	+++	++	+	1
R_2	++	+++	+	0.33
R_3	+++	+	++	0.33
R_4	++	+	+++	–0.33
R_5	+	+++	++	–0.33
R_6	+	++	+++	–1

as in Table 16.2). That part of Table 16.3 also reveals that the maximum correlation coefficient may be observed permuting O_1 and O_2 appears, as the ranking then is identically R_1. This that is the very clue that led to the method that follows.

The idea is to find a ranking that has maximum (possibly weighted) correlation with—or, alternatively, minimum (possibly weighted) so-called Kendall distance (see Paelinck, 1985, pp. 80–98; it is a linear transform of Kendall's rank correlation coefficient; see Sect. 16.2) to—the observed individual rankings This implies constructing a new table—or matrix—from the observed ones by summing them with the appropriate weights, and then—by permuting rows and columns—obtaining a maximum upper-triangular sum. Consider the three parts of Table 16.3, and suppose all weights to be equal. Then the resulting sum table (here the tables may be simply added up) is Table 16.4.

The ranking presented in Table 16.1 is the optimal one. But what if the criteria are only qualitatively ranked? A first possibility is to inspect the weight triangle (or, for higher dimensions, the hyper-triangle). Its endpoints are *(1, 0, 0)*, *(0.5, 0.5, 0)*

Table 16.3 Rankings fot three criteria

Ranking		O_1	O_2	O_3
First	O_1	0	−1	1
	O_2	1	0	1
	O_3	−1	−1	0
Second	O_1	0	1	−1
	O_2	−1	0	1
	O_3	1	−1	0
Third	O_1	0	1	1
	O_2	−1	0	−1
	O_3	−1	1	0

Table 16.4 Sum table from Table 16.3

O_i	O_1	O_2	O_3
O_1	0	1	1
O_2	−1	0	1
O_3	−1	−1	0

and *(0.33, 0.33, 0.33):* see Fig. 16.1. This figure is a two-dimensional cut through the three-dimensional space generated by the weight axes.

For point *(1, 0, 0)* the optimal ranking is (O_2, O_1, O_3), (see Table 16.1). For point *(0.5, 0.5, 0)* the optimal ranking can be calculated (by adding part 1 and 2 of Table 16.3) to be (O_1, O_2, O_3), and this ranking is optimal again in point *(0.33, 0.33, 0.33)* as previously mentioned. The conclusion is that for a relatively high weight attributed to C_1, the optimal ranking would be (O_2, O_1, O_3), with O_2 the "best" (first in rank) object; for lower C_1-weights, (O_1, O_2, O_3) would be optimal with C_1 the "best" object.

In practice, one can randomly scan the weight triangle and find out the zones where certain rankings are optimal. Anyway, in each of those points a matrix permutation is necessary, but it can be shown that this is equivalent to a quadratic assignment problem.

For applications, one can consult Ancot and Paelinck (1982, 1985 and 1986).

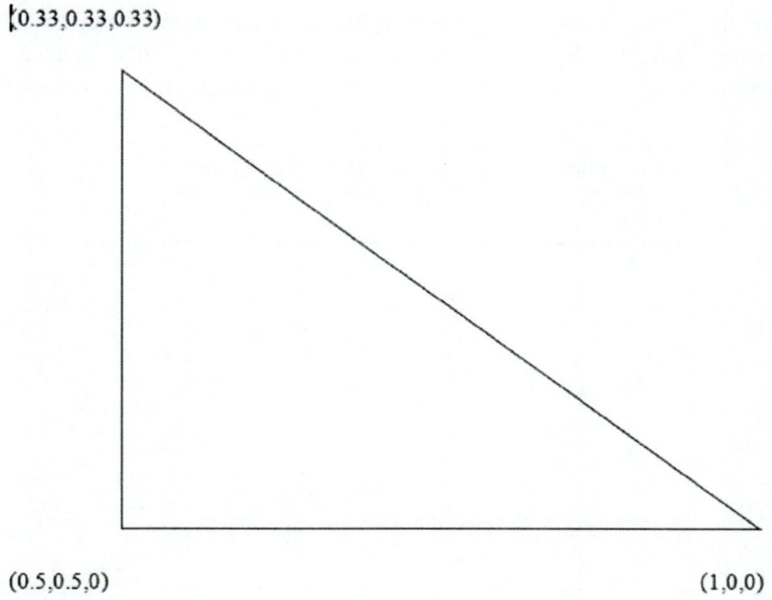

Fig. 16.1 Weight triangle for three criteria

16.2 Qualireg

The problem studied in Sect. 16.1 was to derive an optimal ranking for given rankings and criteria weights. One could ask whether the inverse problem—derive the weights given the final ranking and the initial rankings—has a meaning.

The solution to this optimal ranking problem is equivalent to qualitative regression (for first results, again see Ancot and Paelinck, 1986). Consider the equation

$$y_i = a\,x_i + b\,z_i + c, \tag{16.1}$$

in which parameters a and b have to be estimated. In this case, y_i, x_i and z_i would be elements of three rankings of the respective variables.

Three rank correlation coefficients, $\tau(y, x)$, $\tau(yz)$ and $\tau(xz)$ can be calculated from these rankings (see Kendall, 1955). These correlation coefficients can be used in the classical regression parameter estimation equation, yielding

$$\begin{bmatrix} a \\ b \end{bmatrix} = \begin{bmatrix} 1 & \tau_{xz} \\ \tau_{xz} & 1 \end{bmatrix} \begin{bmatrix} \tau_{yx} \\ \tau_{yz} \end{bmatrix} \tag{16.2}$$

The method has been applied to the data reported in Table 16.5 (source: Plante, 2005).

The data of Table 16.5 have been reduced to their rankings, and the above method applied, rendering the results reported in Table 16.6, which are compared with the original OLS estimates). Their analysis only shows the workings of QUALIREG, with no spatial effects, save distances, having been introduced.

Here the signs of OLS and QUALIREG correspond, but not the relative magnitudes. This result is due, on the one hand to the "data reduction"—"impoverishment"—previously mentioned, and on the other hand to standardization

Table 16.5 Data for a QUALIREG application

Observations	y	x	z
1	2533	53	19
2	962	18	28
3	426	33	35
4	7226	60	3
5	94	20	46
6	411	17	42
7	101	21	61
8	102	27	70
9	27	19	68
10	158	23	69
11	76	24	63
12	269	16	62

Note: y stands for population densities in Northern Virginia counties, x for the share of non-agricultural activity in total activity, and z for distance from Washington

Table 16.6 QUALIREG and OLS estimation results compared

Parameter region	a	b	R^2	t_a	t_b
OLS	0.66	–0.43	0.79	2.50	–2.23
QUALIREG	0.23	–0.66	0.54	0.34	–0.96

of the QUALIREG estimators (i.e., division by the appropriate standard errors, as correlation coefficients have been used). Moreover, for QUALIREG the parameters R^2, t_a and t_b are "pseudo" test criteria, because they can be computed only by analogy. These estimators were obtained as follows.

From general OLS analysis

$$R^2 = \mathbf{y'X'(X'X)^{-1}Xy/y'y}. \tag{16.3}$$

For normalized variables **y** and two τ-variables, this equation translates into

$$\mathbf{R^2} = \left[\tau_{y,x_1};\tau_{y,x_2}\right]\begin{bmatrix} \mathbf{1} & \tau_{x_1,x_2} \\ \tau_{x_1,x_2} & \mathbf{1} \end{bmatrix}\left[\tau_{y,x_1};\tau_{y,x_2}\right] \tag{16.4}$$

and

$$\sigma = (1 - R^2)^{.5}. \tag{16.5}$$

Applying this result to the numbers obtained renders

$R^2 = 0.5388$
$\sigma = 0.6791$
$t_a = 0.3393$
$t_b = -0.9631$

As noted previously, these are all "pseudo-values", because they are simple "analogs" to the theoretical ones. In this case, though R^2 might be significant, a and b are not; this finding is to be expected, as the original data have been "impoverished".

The method finally allows use of even very poor data in order to assess at least the signs (directions) of the partial relations hypothesized.

16.3 Spatial Setting

Next consider spatial interaction effects.

In order to address these effects, a classical spatial specification has been selected, to wit the estimation of first- and second-order contiguity effects. An interdependent linear specification has been selected, which has first been estimated by

Simultaneous Dynamic (here Spatial) Least Squares with endogenously estimated computed values (see Sect. 11.1.3).

The model is specified as follows

$$\mathbf{y} = a\mathbf{C_1y} + b\mathbf{C_2y} + c\mathbf{i}, \tag{16.6}$$

where $\mathbf{y}$ is a vector of regional products, $\mathbf{C_1}$ and $\mathbf{C_2}$ are the first- and second-order contiguity matrices, $\mathbf{i}$ is the unit vector, and a, b and c are parameters to be estimated.

The application concerns the regional products of 11 Belgian regional units, presented in Table 16.7; with their products (in millions of Euros) for 2002, and also with the average contiguity products of order 1 and 2. Only 11 units appear, as the extra-territorial units have been included in the "Brussels Capital" region. The data quantities and map are those of Chap. 14 and, for the contiguity degrees, they are taken from Kaashoek et al. (2004; see Table 16.8).

Table 16.7 Belgian regional units and products, 2002

Number	Unit	Product	C_1	C_2
1	Antwerp	41,483.5	21,307	21,590.1
2	Walloon Brab.	7,639	19,324.2	25,289.1
3	Flem. Brab.	23,232.5	23,908	11,118
4	East Fland.	26,070.5	26,395.8	17,690.7
5	West Fland.	22,766	22,085.9	19,776.8
6	Limburg	14,617.9	27,118.3	17,534
7	Hainaut	18,101.2	17,292.1	23,876.3
8	Namur	6,752.3	11,553.7	21,671.7
9	Luxemburg	3,835.8	11,695.6	15,897.7
10	Liège	16,638.8	11,215.5	32,115.2
11	Bruss. Cap.	42,805.4	23,232.5	23,382.4

Table 16.8 First and second order contiguity degrees for Belgian spatial units

Units	1	2	3	4	5	6	7	8	9	10	11
1		2	1	1	2	1	2			1	2
2	2		1	2	2	2	1	1	2	1	2
3	1	1		1	2	1	1	2	2	1	1
4	1	2	1		1	2	1	2		2	2
5	2	2	2	1			1	2			
6	1	2	1	2			2	2	2	1	2
7	2	1	1	1	1	2		1	2	2	2
8		1	2	2	2	2	1		1	1	
9		2	2			2	2	1		1	
10	2	1	1	2		1	2	1	1		2
11	2	2	1	2		2	2			2	

Table 16.9 Estimation results from Tables 16.7 and 16.8 using model (16.6)

Parameters	a	t_a	B	t_b	R^2
SSLS	0.7963	1.1558	0.3279	0.4464	0.5237
QUALI1	0.9167	1.0501	0.5031	0.5763	0.5128
QUALI2	0.4459	0.4328	0.2209	0.2114	0.1581

Table 16.9 presents three results: the first is the SSLS estimators obtained from the original data (the constant has been omitted, because it cannot appear in the following results); and, the other two are QUALIREG estimators. The first of these latter estimates relates to the average ranks of the contiguity products, whereas the second relates to the ranking of the C_1–C_2 numbers in Table 16.7. The sum of absolute values of residuals has been minimized, rendering very robust estimators.

In this case, not only do the signs correspond, but also the ranking of the values obtained. QUALI1 furnishes the closest fit. Both of these outcomes are remarkable.

Of note is that the SSLS R^2 is not particularly high, despite the low number of df (the four lowest out of the eleven endogenous regional products have been fixed, as they had a tendency to turn negative). This finding corresponds to outcomes reported in Chap. 14, in which each spatial unit (they are the same as in this study) had its own reaction coefficients. Factors are at least twofold: spatial bias (see Paelinck, 2000b; Paelinck et al., 2005, pp. 25–26) and spatial asymmetry proper (due to the differences in economic structure of each of the Belgian spatial units; see Chap. 14).

Nevertheless, the present exercise is concerned with studying only the efficiency of the QUALIREG method.

16.4 Conclusion

It now appears possible to perform spatial econometric exercises with very poor (read: qualitative, ordinal) data, rankings rather than cardinal data. This is of utmost importance when working on multiregional problems in developing countries where the latter sort of information often is not available (for examples, see Ancot and Paelinck, 1990)

Further experience is certainly required with the method proposed, but the improvement with the initial Ancot-Paelinck approach (see Ancot and Paelinck, 1986, referred to above) is very clear.

Chapter 17
Filtering Complexity for Observational Errors and Spatial Bias

In Chap. 12, complexity analysis of spatial data is advocated as a preliminary to spatial econometric regime selection, estimation and testing. That approach assumes that the endogenous variable—only a one-equation model was considered—was measure error free. The present chapter is devoted to controlling for that element of the problem.

Section 17.1 reports on the results obtained earlier; Sect. 17.2 then exposes and applies the method proposed for filtering out assumed observational errors. As spatial bias is an inherent feature of spatial data, an extra correction for this aspect is studied and applied in Sect. 17.3.

17.1 Complexity, Estimation and Testing

The problem studied here arose when spatial data were tested for belonging to one or possibly several possible regimes. In particular a classical linear model and a min-algebraic one were considered.

A first idea of the relevance of one or another specification is to look into the complexity of the problem, by which we mean *computational complexity* of the series to be explained (Chaitin, 1975; Wolfram, 2002, pp. 557–559). This complexity, which we call *conditional complexity*—due to the presence of exogenous variables—can be expressed through the number of parameters necessary to fit the endogenous variable to a polynomial in the exogenous ones. An index whose values are contained in [0, 1] is the earlier mentioned Getis and Paelinck (2004) statistic

$$c = (n_p - 1) / (n_{pm} - 1) \tag{17.1}$$

where n_p is the number of non-zero parameters, and n_{pm} is the maximum number of parameters (equal to the length of the series of endogenous variables, i.e., the sample size. We started by considering especially the endogenous variables as void of measurement errors, because anyhow the observed values are the only ones of which we avail.

D.A. Griffith, J.H.P. Paelinck, *Non-standard Spatial Statistics and Spatial Econometrics*, Advances in Geographic Information Science 1,
DOI 10.1007/978-3-642-16043-1_17, © Springer-Verlag Berlin Heidelberg 2011

Table 17.1 Data for estimation and testing exercise

y_i	2 (1)	4 (8)	5 (5)	7 (6)	6 (5)	8 (20)	10 (15)	9 (12)	11 (11)	12 (12)
x_i	1 (1)	3 (2)	3 (3)	4 (4)	4 (5)	6 (6)	7 (7)	7 (8)	8 (9)	8 (10)
z_i	2 (0)	1 (3)	2 (1)	3 (1)	2 (0)	5 (7)	5 (4)	4 (2)	6 (1)	8 (1)

Table 17.2 Cubic Eq. (17.2) parameters

Variables	1	x	z	x^2	z^2	xz	x^3	z^3	x^2z	xz^2
Values	−15. 3939 (0)	15.5288 (1)	−4.9045 (2)	−1.7288 (0)	−3.8371 (0)	4.5326 (0)	−0.9061 (0)	4.5326 (0)	−5.7258 (0)	2.2432 (0)

The computation has been applied to the following data (Table 17.1, numbers *not* between parentheses).

The analysis resulted in $c = 1$, meaning that 10 parameters are necessary to satisfy the cubic equation:

$$y_i = a_i + \mathbf{a}'\mathbf{u_i} + \mathbf{u_i}'\mathbf{A}\mathbf{u_i} + \mathbf{u_i}'\mathbf{B}'\mathbf{u_i}\mathbf{B}\mathbf{u_i}, \tag{17.2}$$

where $\mathbf{u_i}$ is observation i's vector of exogenous variables. If this test is applied to the numbers between parentheses—generated by $y_i = x_i + 2z_i$—, only two parameters (numbers between parentheses in Table 17.2) are necessary, yielding $c = 0.11$. This gives a clue to a more complex specification for the first series than would be the case for the second one. Table 17.2 presents the values of the parameters that satisfy Eq. (17.2) in both cases.

To test the first series according to this clue, the following model was set up:

$$y_i = \theta\,(ax_i + bz_i + c) + (1 - \theta)\min(\alpha x_i + \beta; \gamma z_i + \delta) + \varepsilon_i, \tag{17.3}$$

with θ binary and for which $\varphi = \min \Sigma_i \varepsilon_i^2$ was chosen as a selection criterion, a minimal variance one (Theil, 1971, pp. 543–545; Aznar, 1989, p. 133). The reason for this choice becomes clear in the ensuing discussion. The second term on the right-hand-side of Eq. (17.3) represents a min-algebraic specification (Paelinck, 2002a).

The computation, involving specification (17.3), resulted in the following parameter values reported in Table 17.3.

Table 17.3 Parameter values for Eq. (17.3)

Parameters	Values
α	1.2801
β	0.9831
γ	5.2934
δ	−1.2935
θ	0
φ	2.8562

Table 17.4 Parameter results for fuzzy cases of Eq. (17.3)

Parameters	a	b	c	α	β	γ	δ	θ	φ
Fuzzy case 1	1.721	−1.7651	1.7398	1.5274	0.921	1.8276	2.1749	0.1749	1.4805
Fuzzy case 2	2.4032	−1.188	0.5634	2.286	0.632	1.2762	0.901	0.3945	≈0

Table 17.5 θ_i*-values for Eq. (17.3)

Observation	Binary case	Fuzzy case 1	Fuzzy case 2
1	1	1	0.9993
2	0	0	0.0260
3	1	1	0.3191
4	1	1	0.5000
5	1	0	0.2172
6	1	1	0.0055
7	1	0	0.1899
8	1	0	0.0589
9	1	1	0.2906
10	1	1	0.1071

Given the complexity analysis performed above, the min-algebraic model is preferable to the classical linear combination one. Obviously the procedure can be generalized to more than two competing model specifications.

The reason model (17.3) was set up, is that it naturally leads to a *fuzzy* generalization, by first relaxing the binary condition on θ to $0 \le \theta \le 1$ (case 1). It also leads to the split between min-regimes (case 2). Further developments of fuzzy spatial econometrics can be found in Paelinck and Klaassen (1979, pp. 136–156). This leads to the values reported in Table 17.4.

Table 17.5 compares the min-constraint ($\alpha x_i + \beta < \gamma z_i + \delta$) parameter values θ_i* in the binary (Table 17.3) and the fuzzy cases (Table 17.4).

As one can observe, although there is no complete correspondence between the three cases, but some of the corresponding cases are striking (see, e.g., observations 1, 2 and 8, the latter at least partially).

Finally, if one applies the method to the "exact" case (numbers between parentheses in Table 17.4, binary case), there results $\theta = 1$ and the exact equation is uncovered, with $\varphi = 1.96 \times 10^{-13}$.

17.2 Filtering for Observational Errors

How would be the shift in the complexity results, if one succeeds in somehow filtering out the stochastic elements from some series of variables? In this section we restrict ourselves to the endogenous variable y_i of Table 17.1.

The model set up has the following specification

$$\min \mathbf{i}'\mathbf{p} \tag{17.4}$$

p, y, a
s.t.:

$$\mathbf{X\hat{p}a} = \mathbf{y} \tag{17.5}$$

$$(\mathbf{y} - \mathbf{y}^*)'(\mathbf{y} - \mathbf{y}^*)/\mathbf{n}\sigma^2(\mathbf{y}^*) = \mathbf{v}^* \tag{17.6}$$

$$\mathbf{i}'(\mathbf{y} - \mathbf{y}^*) = \mathbf{0} \tag{17.7}$$

$$\mathbf{p\hat{p}} = \mathbf{p} \tag{17.8}$$

The various symbols in Eqs. (17.4), (17.5), (17.6), (17.7), and (17.8) are denoted as follows:

- **X** is the matrix of the cubic Eq. (17.3) terms;
- **p** is a column-vector of (binary: see Eq. (17.8)) variables, designed to neutralize certain terms of Eq. (17.3), a cap denoting its transform into a diagonal matrix;
- **a** is a column-vector of complexity coefficients (corresponding to the ones in Table 17.2;
- **y** is the column-vector of filtered $\mathbf{y}^*$ variables, the observed endogenous ones; n and σ^2 are the numbers of observations and the variance of $\mathbf{y}^*$, respectively; and,
- v* is the percentage/100 of the residual variance with respect to the variance of $\mathbf{y}^*$, meant as an indicator of the observational errors present in $\mathbf{y}^*$.

What the model is designed to effectuate is to maximally reduce the complexity of the given series $\boldsymbol{y}^*$ [Eqs. (17.4)], given the cubic relations (17.5) and the observational error indicator (17.6). Equation (17.7) imposes a classical least squares condition on $\boldsymbol{y}$ and $\boldsymbol{y}^*$, and Eq. (17.8) is the binary condition on **p.**

Table 17.6 presents the results of the exercise using the more complex data of Table 17.1; v^* has been set at 0.05. Comparing Tables 17.6 and 17.2, the relevant **a**-coefficients have been halved, and their values are much smaller. The resulting degree of complexity is $c = 0.4444$.

Figure 17.1 hereafter portrays the series **y** and $\mathbf{y}^*$.

To pursue the investigation, vector **y** has been subjected to the test characterized by (Eq. 17.3); the remarkable fact is that the reduction in complexity again does not select the simpler linear model. Table 17.7 reports the results, which should again be compared with those of Table 17.3.

Table 17.6 Observational error filtering results, model (17.4), (17.5), (17.6), (17.7), and (17.8)

Variables and parameters observations	**y**	**a**	**p**
1	0.9603	–2.5371	1
2	3.0623	1.5568	1
3	3.9574	–1.0002	1
4	6.1884	–	0
5	5.4122	–	0
6	9.7864	–	0
7	10.8701	–	0
8	10.5721	–0.0073	1
9	11.8998	–	0
10	11.2913	–0.0057	1

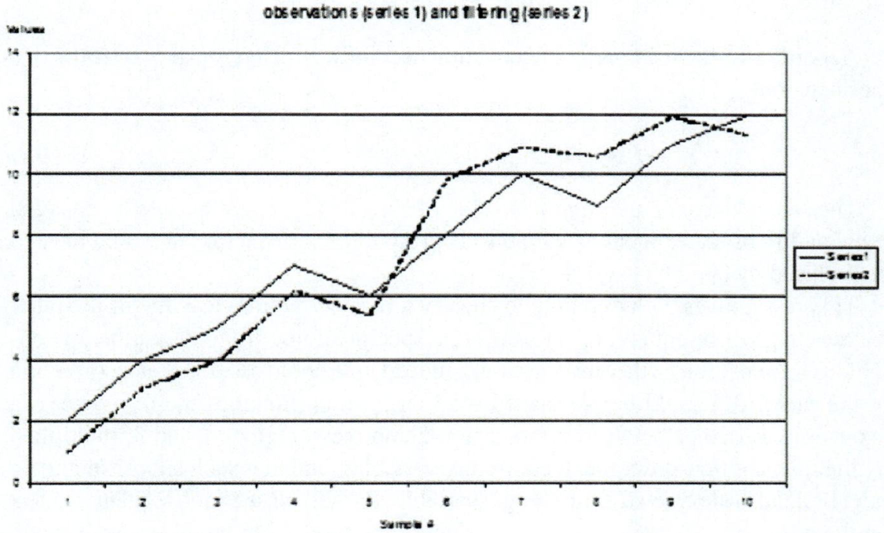

Fig. 17.1 The spatial series **y** and **y***

Table 17.7 Parameters after error correction

Parameters	Values
α	1,5664
β	–0.4709
γ	4.4263
δ	–1.3737
θ	0
φ	1.8955

17.3 Further Filtering for Spatial Bias

The starting point is Paelinck (2000, especially pp. 158–159), but with the simplifying assumption that complete spatial homogeneity is present, the latter being defined by the identity of all reaction parameters and exogenous variables. The following example illustrates this point.

Suppose detailed underlying data—of an additive nature—to those (not between parentheses) of Table 17.1, are located next to each other on a circle or a torus. Assume that only first- and second-order contiguities are relevant. For underlying micro-regions *3* and *4,* the linear equation then is as follows, with only one exogenous variable being taken into account

$$y_3 = ax_3 + b(x_2 + x_4) + c(x_1 + x_5) + d, \text{ and} \tag{17.9a}$$

$$y_4 = ax_4 + b(x_3 + x_5) + c(x_2 + x_6) + d. \tag{17.9b}$$

Accordingly, the meso-regional equation becomes, after aggregation over the two meso-regions

$$y^*_2 = (a + b)x_2^* + (0.5b + c)(x_1^* + x_3^*) + 2d. \tag{17.10}$$

The second term on the right hand side of Eq. (17.10) would have to be changed if inequality of the exogenous variables is present [i.e., the factor 0.5 would have to be replaced by $(x_1 + x_5)/(x_1^* + x_3^*)$].

This result gives a clue to how to specify a bias correction, leaving all the other sources (e.g., different parameters, different specifications, different number of contiguous regions) for further investigation. Indeed, one could assume that a correction of the factor 0.5 could be performed by a term proportional to the differences in exogenous variables between meso-region 2 and meso-regions 1 and 3, multiplied by the relevant first-order meso-contiguity variables, and that such a term should be (algebraically) subtracted from the y_i^*-variable. As will be seen subsequently, a test on the adequacy of the procedure is possible. The model can be generalized easily to more than one exogenous variable, as will be done in the ensuing example.

The preceding suggested correction suggested now is applied to the previous data used, for which the following formal model is suggested

$$\min(\mathbf{y}^c - \mathbf{X}\hat{\mathbf{p}}\mathbf{a})'(\mathbf{y}^c - \mathbf{X}\hat{\mathbf{p}}\mathbf{a}), \tag{17.11}$$

with

$$\mathbf{y}^c \triangleq \mathbf{y} - \alpha\Delta_x\mathbf{C}_1\mathbf{x} - \beta\Delta_z\mathbf{C}_1\mathbf{p} \tag{17.12}$$

s.t.:

$$\hat{\mathbf{p}}\mathbf{p} = \mathbf{p}. \tag{17.13}$$

Table 17.8 Spatial bias correction results

$\mathbf{y^c}$	**a**	**p**
0.1846	–4.0275	1
3.1428	0.0388	1
3.19137	–0.3490	0
6.2020	–0.0081	1
5.4098	–0.0092	1
9.7752	2.3698	1
10.9168	0.9007	1
10.3524	–0.1124	1
11.9934	–0.0972	1
11.4396	0.0522	1

Table 17.9 Linear regression parameters after complete filtering

Parameters	Values	F- and t-values
R^2	0.9968	386.7782
x*-intrareg.	1.6249	16.7149
z*-intrareg.	0.0524	0.5232
x*-contig.	0.3942	4.5931
z*-contig.	–0.4923	–4.6871

Most variables are defined in Sect. 17.2; $\mathbf{y^c}$ is the bias-corrected vector generated in that same section, α and β are the proportionality parameters introduced above, Δ_x and Δ_z are the net deviations between neighboring variables (standardized), and $\mathbf{C_1}$ is the first-order contiguity matrix.

Table 17.8 summarizes the results.

The value of Eq. (17.11) is ≈ 0; the values for α and β are *1.2474* and *–0.7644* respectively. These last two values should have the signs of the corresponding contiguity coefficients of the linear model (Table 17.3), which is indeed the case (Table 17.9). With regard to complexity, only one binary variable is zero, but five other coefficients are smaller than *0.1*. Thus, in practice the degree of complexity is only *0.3333*, which pleads indeed in favor of the simple linear model of which the constant, to be complete, is –1.4242 (–4.5053).

The regression results do not invalidate the assumption that spatial homogeneity is present after correction. However the sample is too small to test that model against an alternative, e.g. a min-algebraic model—which would have 10 parameters—so further investigation with larger samples is in order.

17.4 Conclusions

It appears that filtering for measurement errors followed by spatial bias filtering can reveal an underlying simple interregional model, so the hint is that observational errors and spatial bias jointly are responsible for much of the specification complexity needed to represent the data. Chapter 14 has given an example of the latter case.

Some more points are still in order.

A first point to be made is that we again advise spatial scientists to start an exercise in spatial econometric modeling with a complexity analysis of the data. Obvious candidates for simple exogenous variables are their space-time coordinates. An example can be found in Getis and Paelinck (2004), in which regional product data for the Netherlands are analyzed. A model specification implies the choice of exogenous variables, and possibly endogenous ones—in interdependent models—or lagged endogenous variables— in dynamic models—, so that they too should be implied in a complexity approach.

A second point is that the specifications presented can readily be generalized to three or more alternatives (e.g., regions, test specifications). For a finite automaton version, for example, the following expression shows how *and* and *or* statements can be added:

$$y_i: \text{if}((cz_i + d < ax_i + b) \text{ and } (eu_i + f < ax_i + b); (cz_i + d) \text{ or } (eu_i + f); ax_i + b) \quad (17.14)$$

Finally, spatial economic phenomena should be given very appropriate specifications, as non-linearity is a fundamental principle in the exercise of spatial econometrics.

Chapter 18
General Spatial Econometric Conclusions

What should be clear from the exercises presented is that in most of them, classical "regression" has been combined with mathematical programming to obtain the desired estimators.

Most notable is that multiple regimes have been systematically selected. A case in point appears in Chap. 14, when inspection of residuals leads not only to two regimes, but to as many of them as there are spatial units, this through the use of composite parameters.

However, the exploration is far from being finished (Paelinck, 2004). Many analytical tools, as yet unexplored, lay at our avail, or, better even, wait to be invented. Several of these "inventions" are presented in the preceding discussion—extended Lotka-Volterra models, robust linear estimators for logistic and Poisson distributions, qualitative estimators to treat poor data—but many others, well adapted to spatial *econo*metrics, could be or are formulated, such as PPFDEs, mentioned in Chap. 10 (Introduction: spatial econometrics) to this part (Coutrot et al., 2009).

The loose Latin saying *Contentum sui operis necesse, maximae sunt divitiae* could be the motto of the spatial econometrician: never been satisfied by his work, should be his ultimate riches...

D.A. Griffith, J.H.P. Paelinck, *Non-standard Spatial Statistics and Spatial Econometrics*, Advances in Geographic Information Science 1,
DOI 10.1007/978-3-642-16043-1_18,

Epilogue

The respective contents of Parts I and II might convey to a reader the idea that the focus of spatial statistics and that of spatial econometrics may well be quite separate. The widely cited 1988 books by Griffith (*Advanced Spatial Statistics*, Dordrecht: Martinus Nijhoff) and by Anselin (*Spatial Econometrics: Methods and Models*, Dordrecht: Martinus Nijhoff) may further perpetuate this viewpoint. But this simply is not the case! Rather, these seemingly different foci reflect historical developments, with the first high profile spatial statistics books being penned by Cliff and Ord (*Spatial Autocorrelation*, Pion, 1973) and by Ripley (*Spatial Statistics*, Wiley, 1981), featuring a statistician's point of view, and the first spatial econometrics book being penned by Paelinck and Klaassen (*Spatial Econometrics*, Saxon House, 1979), featuring an econometrician's point of view.

One aspect of both subdisciplines particularly worth stressing is that if mathematical geographers live in a GIS world, spatial econometricians extend a friendly hand to them through metric topology. In reciprocity, mathematical geographers extend a friendly hand to spatial econometricians through GIS functions supporting spatial economics, especially in terms of the new economic geography promoted by Krugman. An article by one of the authors in the *Annals of Operations Research* (Vol. 123, 2003, pp. 371–383) titled "On locations and distances," where a link is established *in both directions*, effectively and convincingly illustrates this two-way cooperative notion.

This perspective illustrates that joint work on spatial analytical problems always will be fruitful, which has been a continual experience of both authors over quite a number of decennia. A reader need only refer to the collection of papers contributed from both subdisciplines that initiated a large amount of this type of interaction—including ones by both authors—appearing in

Griffith and R. MacKinnon (1981), *Dynamic Spatial Models*, Alphen aan den Rijn: Sijthoff and Noordhoff;

Griffith and A. Lea (1983), *Evolving Geographical Structures*, The Hague: Martinus Nijhoff; and,

Griffith and R. Haining (1985), *Transformations Through Space and Time*, The Hague: Martinus-Nijhoff.

D.A. Griffith, J.H.P. Paelinck, *Non-standard Spatial Statistics and Spatial Econometrics*, Advances in Geographic Information Science 1,
DOI 10.1007/978-3-642-16043-1,

These three specific publications were followed by a set of collaborative papers compiled in *Spatial Statistics: Past, Present and Future* (edited by D. Griffith, 1990, Ann Arbor, MI: Institute of Mathematical Geography), and then *Advances in Spatial Modelling and Methodology: Essays in Honor of Jean Paelinck* (edited by D. Griffith, C. Amrhein and J.-M. Huriot, 1998, Dordrecht: Kluwer), again with entries by each of the authors. To this collection of edited volumes can be added the authors' recent study entitled "Specifying a joint space-and-time-lag using a bivariate Poisson distribution" (*Journal of Geographical Systems*, 2009, Vol. 11, No 1, pp. 23–36), which highlights equivalencies between a spatial statistical and a spatial econometric solution to a space-time problem. The authors hope to continue their scientific cooperation over many years to come, and invite others to join them in this specific type of interdisciplinary endeavors.

References

Abramowitz, M., Stegun, I. (eds.). 1964. *Handbook of Mathematical Functions*. Washington, DC: U.S. Department of Commerce, National Bureau of Standards Applied Mathematical Series 55.

Allen, R.G.D. 1956. *Mathematical Economics*. London: Macmillan.

Amrhein, C., Reynolds, H. 1996. Using spatial statistics to assess aggregation effects, *Geographical Systems*, 3: 143–158.

Amrhein, C., Reynolds, H. 1997. Using the Getis statistic to explore aggregation effects in metropolitan Toronto census data, *The Canadian Geographer*, 41: 137–149.

Amrhein, C., Wong, D. 1996. Research on the MAUP: old wine in a new bottle or real breakthrough? *Geographical Systems*, 3: 73–76.

Ancot, J.-P., Chevailler J.-C., Paelinck J.H.P., Smit H., Stijnen, H. 1978. Parametercomponent models in spatial econometrics, *The Econometrics of Panel Data*, *Annales de l'INSEE*, 304/31: 83–98.

Ancot, J.-P., Paelinck, J.H.P. 1982. Recent experiences with the QUALIFLEX multicriteria method. In J.H.P. Paelinck (ed.), *Qualitative and Quantitative Mathematical Economics*. The Hague: Martinus Nijhoff Publishers, pp. 217–266.

Ancot, J.-P., Paelinck, J.H.P. 1983. The spatial econometrics of the European FLEUR-model. In D.A. Griffith, A. Lea (eds.), *Evolving Geographical Structures*. The Hague: Martinus Nijhoff Publishers, pp. 220–246.

Ancot, J.-P., Paelinck, J.H.P. 1985. Ten years of QUALIFLEX, Netherlands Economic Institute, *Series: Occasional Papers and Reprints*, No 2.

Ancot, J.-P., et Paelinck, J.H.P. 1986. Médicométrie régionale et régression qualitative: le modèle QUALIREG. *Journal d'Economie Médicale*, 5: 253–264.

Ancot, J.-P., Paelinck, J.H.P. 1990. *Modèles et choix: une initiation à la modélisation pour pays en développement*. Paris: Economica.

Annals, Association of American Geographers. 2000. 90(issue #3 (September)): 579–606.

Anselin, L. 1988. *Spatial Econometrics: Methods and Models*. Dordrecht: Kluwer.

Anselin, L. 1995. Local indicator of spatial association – LISA, *Geographical Analysis*, 27: 93–115.

Anselin, L. 1998. Exploratory spatial data analysis in a geocomputaional environment. In P. Longley, S. Brooks, R. McDonnell, W. Macmillan (eds.), *Geocomputation: A Primer*. New York, NY: Wiley, pp. 77–94.

Anselin, L., Florax, R. (eds.). 1995. *New Directions in Spatial Econometrics*. Berlin: Springer.

Arbia, G. 2004. Alternative approaches to regional convergence exploiting both spatial and temporal information, *Paper invited for presentation at the First seminar of Spatial Econometrics*, Universidad de Zaragozza, October 22nd and 23rd.

Arbia, G., Paelinck, J.H.P. 2003a. Spatial econometric modeling of regional convergence in continuous time, *International Regional Science Review*, 26(3): 342–362.

Arbia, G., Paelinck, J.H.P. 2003b. Economic convergence or divergence? Modeling the interregional dynamics of EU regions, *Journal of Geographical Systems*, 5(3): 291–314.

Arbia, G., Griffith, D., Haining, R. 1998. Error propagation modelling in raster GIS: overlay operations, *International 1ournal of Geographical Information Systems*, 12: 145–167.

Arbia, G., Griffith, D., Haining, R. 1999. Error propagation modelling in raster GIS: addition and ratioing operations, *Cartography and Geographic Information Systems*, 26: 297–315.

Arbia, G., Griffith, D., Haining, R. 2003. Spatial error propagation when computing linear combinations of spectral bands: the case of vegetation indices, *Environmental and Ecological Statistics*, 10: 375–396; reply to commentary, 399–400.

Armstrong, B. 1992. Confidence intervals for arithmetic means of lognormally distributed exposures, *American Industrial Hygiene Association Journal*, 53: 481–485.

Augustin, N., McNicol, J., Marriott, C. 2004. Using the truncated auto-Poisson model for spatially correlated counts of vegetation, *Technical Report* 04-14. Glasgow, UK: Department of Statistics, University of Glasgow.

Aznar Grasa, A. 1989. *Econometric Model Selection: A New Approach*, Advanced Studies in Theoretical and Applied Econometrics, vol. 16. London: Kluwer.

Bannock, G., Baxter, R., Davis, E. 2003. *Dictionary of Economics*, Third Edition. London: Economist Books.

Barnet, V., Lewis, T. 1978. *Outliers in Statistical Data*. New York, NY: Wiley.

Bartlett, M. 1937. Some examples of statistical methods of research in agriculture and applied biology, *Journal of the Royal Statistical Society*, (Suppl 4): 137–183.

Bartlett, M. 1947. The use of transformations, *Biometrics*, 3: 39–52.

Bartlett, M. 1975. *The Statistical Analysis of Spatial Pattern*. London: Chapman & Hall.

Bergad, L. 1983. *Coffee and the Growth of Agrarian Capitalism in Nineteenth-Century Puerto Rico*. Princeton, NJ: Princeton University Press.

Berry, B. 1961. City size distribution and economic development, *Economic Development and Cultural Change*, 9: 573–588.

Besag, J.E. 1974. Spatial interaction and the statistical analysis of lattice systems, *Journal of the Royal Statistical Society B*, 36: 192–225.

Besag, J., York, J., Mollié, A. 1991. Bayesian image restoration with two applications in spatial statistics, *Annals of the Institute of Statistical Mathematics*, 43: 1–59.

Bivand, R. 1980. A Monte Carlo study of correlation coefficient estimation with spatially autocorrelated observations, *Quaestiones Geographicae*, 6: 5–10.

Blom, G. 1958. *Statistical Estimates and Transformed Beta-Variables*. New York, NY: Wiley.

Boots, B., Tiefelsdorf, M. 2000. Global and local spatial autocorrelation in bounded regular tessellations, *Journal of Geographical Systems*, 2: 319–348.

Borcard, D., Legendre, P. 2002. All-scale spatial analysis of ecological data by means of principal coordinates of neighbour matrices, *Ecological Modelling*, 153: 51–68.

Borcard, D., Legendre, P., Avois-Jacquet, C., Tuomisto, H. 2004. Dissecting the spatial structure of ecological data at multiple scales, *Ecology*, 85: 1826–1832.

Bowers, T., Shifrin, N., Murphy, B. 1996. Statistical approach to meeting soil cleanup goals, *Environmental Science and Technology*, 30: 1437–1444.

Braun, M. 1975. *Differential Equations and Their Applications*. New York, NY: Springer.

Breslow, N., Clayton, D. 1993. Approximate inference in generalized linear mixed models, *Journal of the American Statistical Association*, 88: 9–25.

Brus, D., de Gruijter, J. 1993. Design-based versus model-based estimates of spatial means: theory and application in environmental science, *Environmetrics*, 4: 123–152.

Burmaster, D., Thompson, K. 1997. Estimating exposure point concentrations for surface soils for use in deterministic and probabilistic risk assessments, *Human and Ecological Risk Assessment*, 3: 363–384.

Cameron, A., Trivedi, P. 1998. *Regression Analysis of Count Data*. New York, NY: Cambridge University Press.

Casella, G. 1985. An introduction to empirical Bayes data analysis, *The American Statistician*, 39: 83–87.

Casella, G., George, E. 1992. Explaining the Gibbs sampler, *The American Statistician*, 46: 167–174.
CDC (Centers for Disease Control). 2001. *National Report on Human Exposure to Environmental Chemicals*. Atlanta, GA: CDC.
Chaitin, G.J. 1975. Randomness and mathematical proof, *Scientific American*, 232(5): 47–52.
Chinn, S. 1996. Choosing a transformation, *Journal of Applied Statistics*, 23: 395–404.
Christensen, R. 1991. *Linear Models for Multivariate, Time Series, and Spatial Data*. New York, NY: Springer.
Citro, C. 1998. Model-based small-area estimates: the next major advance for the federal statistical system for the 21st century, *Chance*, 11: 40–41, 50.
Clayton, D., Kaldor, J. 1987. Empirical Bayes estimates of age-standardized relative risks for use in disease mapping, *Biometrics*, 43: 671–681.
Cliff, A., Ord, J. 1973. *Spatial Autocorrelation*. London: Pion.
Cliff, A., Ord, J. 1981. *Spatial Processes*. London: Pion.
Clifford, P., Richardson, S., Hémon, D. 1989. Assessing the significance of the correlation between two spatial processes, *Biometrics*, 45: 123–134.
Cordy, C., Griffith, D. 1993. Efficiency of least squares estimators in the presence of spatial autocorrelation, *Communications in Statistics, Series B*, 22: 1161–1179.
Costanzo, C. 1983. Statistical inference in geography: modern approaches spell better times ahead, The Professional Geographer, 35: 158–165.
Coutrot, B., Paelinck, J.H.P., Sallez, A. 2009. Analyzing the complexity of knowledge-based spatial economic developments, *Région et Développement*, 29: 201–228.
Cowles, M. 2004. Review of WinBUGS 1.4, *The American Statistician*, 58: 330–336.
Cressie, N. 1989. Geostatistics, *The American Statistician*, 43: 197–202.
Cressie, N. 1991. *Statistics for Spatial Data*. New York, NY: Wiley.
Cressie, N., Guo, R. 1987. Mapping variables, in Proceedings of the NCGA Conference, *Computer Graphics '87*. McLean, VA: National Computer Graphics Association, III: 521–530.
Darby, S, Deo, H, Doll, R, Whitley, E. 2001. A parallel analysis of individual and ecological data on residential radon and lung cancer in south-west England, *Journal of the Royal Statistical Society*, 164A: 193–203.
Datta, G.S., Day, B., Basawa, I.V. 1999. Empirical best linear unbiased and empirical Bayes prediction in multivariate small area estimation, *Journal of Statistical Planning and Inference*, 75, 269–279
de Cordova, P. 1968. *Memorias, Geográficas, Históricas, Económicas y Estadísticas de la Isla de Puerto Rico*. San Juan, PR: Edición facsimilar (Editorial Coquí); Instituto de Cultura Puertorriqueña, 1831.
de Jong, P., Sprenger, C., Van Veen, F. 1984. On extreme values of Moran's I and Geary's c, *Geographical Analysis*, 16: 17–24.
Demidenko, E. 2004. *Mixed Models: Theory and Applications*. New York, NY: Wiley.
Dempster, A., Laird, N., Rubin, D. 1977. maximum likelihood from incomplete data via the EM algorithm, *Journal of the Royal Statistical Society, Series B*, 39: 1–38.
Dendrinos, D.S., Mullaly, H. 1981. Evolutionary patterns of urban populations, *Geographical Analysis*, 13(4): 328–344.
Diggle, P., Lophaven, S. 2004 Bayesian geostatistical design, *Working Paper #42*, Baltimore, Department of Biostatistics, Johns Hopkins U.
Dodge, Y. 1985. *Analysis of Experiments with Missing Data*. New York, NY: Wiley.
Dolan, D., El-Shaarawi, A., Reynoldson, T. 2000. Predicting benthic counts in Lake Huron using spatial statistics and quasi-likelihood, *Environmetrics*, 11: 287–304.
Domencich, J.A., McFadden, D. 1975. *Urban Travel Demand, A Behavioral Analysis*. Amsterdam: North-Holland.
Dutilleul, P. 1993. Modifying the *t* test for assessing the correlation between two spatial processes, *Biometrics*, 49: 305–314.
Dutilleul, P., Legendre, P. 1992. Lack of robustness in two tests of normality against autocorrelation in sample data, *Journal of Statistical Computation and Simulation*, 42: 79–91.

Elliott, P., Wakefield, J. 2000. Bias and confounding in spatial epidemiology. In P. Elliott, J. Wakefield, N. Best, D. Briggs (eds.), *Spatial Epidemiology: Methods and Applications*. New York, NY: Oxford University Press, pp. 68–84.

Florax, R., Rey, S. 1995. The impacts of misspecified spatial interaction in linear regression models. In L. Anselin, R. Florax (eds.), *New Directions in Spatial Econometrics*. Berlin: Springer, pp. 111–135.

Flores, L., Martínez, L., Ferrer, C. 2003. Systematic sample design for estimation of spatial means, *Environmetrics*, 14: 45–61.

Flury, B., Zoppè, A. 2000. Exercises in EM, *The American Statistician*, 54: 207–209.

Freedman, D. 2001. Ecological inference and the ecological fallacy. In N. Smelser, P. Baltes (eds.), *International Encyclopedia of the Social & Behavioral Sciences*, vol. 6. Oxford: Elsevier, pp. 4027–4030.

Fylstrom, D., Lasdon, L., Watson, J., Waren, A. 1998. Design and use of the microsoft excel solver, *Interfaces*, 28(5): 29–55.

Gastel, A.J.J. van, Paelinck, J.H.P. 1995. Computation of box-cox transform parameters: a new method and its application to spatial econometrics. In L. Anselin, R.J.M. Florax (eds.), *New Directions in Spatial Econometrics*. Berlin: Springer, pp. 136–155.

Gandolfo, G. 1996. *Economic Dynamics*, Third Edition. Berlin: Springer.

Geerligs, H. 1912. *The World's Cane Sugar Industry: Past and Present*. London: Normal Rodger.

Gelman, A., Rubin, D. 1992. Inference from iterative simulation using multiple sequences (with discussion), *Statistical Science*, 7: 457–511.

Getis, A. 1990. Screening for spatial dependence in regression analysis, *Papers of the Regional Science Association*, 69: 69–81.

Getis, A. 1995. Spatial filtering in a regression framework: experiments on regional inequality, government expenditures, and urban crime. In L. Anselin, R. Florax (eds.), *New Directions in Spatial Econometrics*. Berlin: Springer, pp. 172–188.

Getis, A., Griffith, D. 2002. Comparative spatial filtering in regression analysis, *Geographical Analysis*, 34: 130–140.

Getis, A., Paelinck, J.H.P. 2004. An analytical description of spatial patterns, *L'Espace Géographique*, 2004/4: 68–79.

Getis, A., Mur, J., Zoller, H.G. 2004. *Spatial Econometrics and Spatial Statistics*. Basingstoke: Palgrave-Macmillan.

Gilbert, R. 1987. *Statistical Methods for Environmental Pollution Monitoring*. New York, NY: van Nostrand Reinhold.

Gilks, R., Richardson, S., Spiegelhalter, J. (eds.). 1996. *Markov Chain Monte Carlo in Practice*. New York, NY: Chapman & Hall.

Ginevan, M., Splitstone, D. 1997. Improving remediation decisions at hazardous waste sites with risk-based geostatistical analysis, *Environmental Science & Technology News*, 31: 92–96.

Gneiting, T., Ševčíková, H., Percival, D., Schlather, M., Jiang Y. 2005. Fast and Exact Simulation of Large Gaussian Lattice Systems in Exploring the Limits, *Technical Report* No. 477. Seattle, WA: Department of Statistics, University of Washington.

Goodchild, M. 1980. Algorithm 9: simulation of autocorrelation for aggregate data, Environment and Planning A, 12: 1073–1081.

Gotway, C., Stroup, W. 1997. A generalized linear model approach to spatial data analysis and prediction. *Journal of Agricultural, Biological, and Environmental Statistics*, 2: 157–178.

Graham, J. 1994. Monte Carlo Markov Chair likelihood ratio test and Wald test for binary spatial lattice data, *Mimeographed paper*, Department of Statistics, North Carolina State University, Raleigh, NC.

Green, M. 1993. Ecological fallacies and the modifiable areal unit problem, *Research Report 27*. Lancaster University: North West Regional Research Laboratory.

Griffith, D. 1980. Towards a theory of spatial statistics, *Geographical Analysis*, 12: 325–339.

Griffith, D. 1988. *Advanced Spatial Statistics*. Dordrecht: Kluwer.

Griffith, D. 1992a. Estimating missing values in spatial urban census data, *The Operational Geographer*, 10: 23–26.

Griffith, D. 1992b. What is spatial autocorrelation? Reflections on the past 25 years of spatial statistics, *l'Espace Geographique*, 21: 265–280.

Griffith, D. 1993a. *Spatial Regression Analysis on the PC: Spatial Statistics Using SAS*. Washington, DC: Association of American Geographers.

Griffith, D. 1993b. Advanced spatial statistics for analyzing and visualizing geo-referenced data, *International Journal of Geographical Information Systems*, 7: 107–123.

Griffith, D. 1999. A methodology for small area estimation, with special reference to a one-number agricultural census and confidentiality: results for selected major crops and states, *NASS Research Report* RD-99-04. Washington, DC: Research Division, National Agricultural Statistics Service, U.S. Department of Agriculture.

Griffith, D. 2000a. A linear regression solution to the spatial autocorrelation problem, *Journal of Geographical Systems*, 2: 141–156.

Griffith, D. 2000b. Small geographic area estimation contributions to federal agricultural data: a Pennsylvania case study, *The Pennsylvania Geographer*, 38(#1): 3–29.

Griffith, D. 2000c. Eigenfunction properties and approximations of selected incidence matrices employed in spatial analyses, *Linear Algebra and Its Applications*, 321: 95–112.

Griffith, D. 2002a. A spatial filtering specification for the auto-Poisson model, *Statistics and Probability Letters*, 58: 245–251.

Griffith, D. 2002b. The geographic distribution of soil-lead concentration: description and concerns, *URISA Journal*, 14: 5–15.

Griffith, D. 2003. *Spatial Autocorrelation and Spatial Filtering: Gaining Understanding Through Theory and Scientific Visualization*. Berlin: Springer.

Griffith, D. 2004a. A spatial filtering specification for the auto-logistic model, *Environment and Planning A*. 36: 1791–1811.

Griffith, D. 2004b. Distributional properties of georeferenced random variables based on the eigenfunction spatial filter, *Journal of Geographical Systems*, 6: 263–288.

Griffith, D. 2005a. Effective geographic sample size in the presence of spatial autocorrelation, *Annals, Association of American Geographers*, 95: 740–760.

Griffith, D. 2005b. A comparison of four analytical disease mapping techniques as applied to West Nile Virus in the coterminous United States, *International Journal of Health Geographics*, 4: 18 (14 pp).

Griffith, D. 2006a. Hidden negative spatial autocorrelation, *Journal of Geographical Systems*, 8: 335–355.

Griffith, D. 2006b. Assessing spatial dependence in count data: winsorized and spatial filter specification alternatives to the auto-Poisson model, *Geographical Analysis*, 38: 160–179.

Griffith, D., Csillag, F. 1993. Exploring relationships between semi-variogram and spatial autoregressive models, *Papers in Regional Science*, 72: 283–295.

Griffith, D., Haining, R. 2006. Beyond mule kicks: The Poisson distribution in geographical analysis, *Geographical Analysis*, 38: 123–139.

Griffith, D., Lagona, F. 1998. On the quality of likelihood-based estimators in spatial autoregressive models when the data dependence structure is misspecified, *Journal of Statistical Planning and Inference*, 69: 153–174.

Griffith, D., Layne, L. 1996. Uncovering relationships between geo-statistical and spatial autoregressive models, *Proceedings of the American Statistical Association*, Chicago, IL

Griffith, D., Layne, L. 1999. *A Casebook for Spatial Statistical Data Analysis: A Compilation of Analyses of Different Thematic Data Sets*. New York, NY: Oxford University Press.

Griffith, D.A., Paelinck, J.H.P. 2009. Specifying a joint space- and time-lag using a bivariate poisson distribution, *Journal of Geographical Systems*, 11: 23–36.

Griffith, D., Peres-Neto, P. 2006. Spatial modeling in ecology: the flexibility of eigenfunction spatial analyses, *Ecology*, 87: 2603–2613.

Griffith, D., Sone, A. 1995. Trade-offs associated with normalizing constant computational simplifications for estimating spatial statistical models, *Journal of Statistical Computation and Simulation*, 51: 165–183.

Griffith, D., Zhang, Z. 1999. Computational simplifications needed for efficient implementation of spatial statistical techniques in a GIS, *Journal of Geographic Information Science*, 5: 97–105.

Griffith, D., Layne, L., Doyle, P. 1996. Further explorations of relationships between semivariogram and spatial autoregressive models. In H. Mowrer, R. Czaplewski, R. Hamre (eds.), *Spatial Accuracy in Natural Resource and Environmental Sciences: Second International Symposium*. Ft. Collins, CO: Rocky Mountain Forest and Range Experiment Station, General Technical Report RM-GTR-277, pp. 147–154.

Griffith, D.A., Paelinck, J.H.P., van Gastel, M.A.J.J. 1998. The box-cox transformation: new computation and interpretation features of the parameters. In D.A. Griffith, C. Amrhein, J.-M. Huriot (eds.), *Econometric Advances in Spatial Modeling and Methodology*. Dordrecht: Kluwer, pp. 45–58.

Griffith, D., Doyle, P., Wheeler, D., Johnson, D. 1998. A Tale of two swaths: urban childhood blood lead levels across Syracuse, NY, *Annals of the Association of American Geographers*, 88: 640–665.

Griffith, D., Millones, M., Vincent, M., Johnson, D., Hunt, A. 2004. An assessment of positional error: digital cadastral parcel map versus TIGER line file support for geocoding of address locations in Syracuse, NY, *Paper presented to the joint meetings of the 15th annual conference of The International Environmetrics Society and The Sixth International Symposium on Spatial Accuracy Assessment in Natural Resources and Environmental Sciences*, Portland, ME, June 28–July 1.

Griffith, D., Millones, M., Vincent, M., Johnson, D., Hunt, A. 2008. Impacts of positional error on spatial regression analysis: a case study of address locations in Syracuse, NY, *Transactions in GIS*, 11: 655–679.

Guyon, X. 1995. *Random Fields on a Network: Modeling, Statistics, and Applications*. Berlin: Springer.

Hahn, W. 1963. *Theory and Applications of Liapunov's Direct Method*. Englewood Cliffs, NJ: Prentice-Hall.

Haining, R. 1990. *Spatial Data Analysis in the Social and Environmental Sciences*. Cambridge: Cambridge University Press.

Haining, R. 1991. Bivariate correlation and spatial data, *Geographical Analysis*, 23: 210–227.

Haining, R. 2003. *Spatial Data Analysis: Theory and Practice*. Cambridge: Cambridge University Press.

Haining, R., Griffith, D., Bennett, R. 1983. Simulating two dimensional autocorrelated surfaces, *Geographical Analysis*, 15: 247–255.

Haining, R., Griffith, D., Bennett, R. 1989. Maximum likelihood estimation with missing spatial data and with an application to remotely sensed data, *Communications in Statistics*, 18: 1875–1894.

Haining, R., Law, J., Griffith, D. 2009. Modelling small area counts in the presence of overdispersion and spatial autocorrelation, *Computational Statistics & Data Analysis*, 53: 2923–2937.

Haining, R., Wise, S., Ma, J. 1998. Exploratory spatial data analysis in a GIS environment, *The Statistician*, 47: 457–469.

Hardin, J., Hilbe, J. 2001. *Generalized Linear Models and Extensions*. College Station, TX: Stata Press.

Heagerty, P, Lele, S. 1998. A composite likelihood approach to binary spatial data, *Journal of the American Statistical Association*, 93: 1099–1111

Hill, E., Allen A., Waller, L. 1999. A comparison of focused score tests and Bayesian hierarchical models for detecting spatial disease clustering, *Journal of the National Institute of Public Health*, 48: 102–112.

Hoaglin, D. 1980. A poissonness plot, The American Statistician, 34: 146–149.

Hoaglin, D., Tukey, J. 1985. Checking the shape of discrete distributions. In D. Hoaglin, F. Mosteller, J. Tukey (eds.), Exploring Data Tables, Trends and Shapes, Chapter 9. New York, NY: Wiley, pp. 345–416.

Holt, D., Steel, D., Tranmer, M., Wrigley, N. 1996. Aggregation and ecological effects in geographically based data, *Geographical Analysis*, 28: 244–261.

Hubbell S., Ahumada, J., Condit, R., Foster, R. 2001. Local neighborhood effects on long-term survival of individual trees in a neotropical forest, *Ecological Research*, 16: 859–875.

Huffer, F., Wu, H. 1998. Markov chain Monte Carlo for autologistic regression models with application to the distribution of plant species, *Biometrics*, 54: 509–524

Isaaks, E., Srivastava, R. 1989. *An Introduction to Applied Geostatistics*. New York, NY: Oxford University Press.

Johnson, K., van Hoef, J., Krivoruchko, K., Lucas, N. 2001. *Using ArcGIS Geostatistical Analyst*. Redlands, CA: ESRI.

Kaashoek, J.F., Paelinck, J.H.P., Zoller, H.G. 2004. On connectropy. In A. Getis, J. Mur, H.G. Zoller (eds.), *Spatial Econometrics and Spatial Statistics*. Basingstoke: Palgrave-Macmillan, pp. 217–231.

Kaiser, M., Cressie, N. 1997. Modeling poisson variables with positive spatial dependence, *Statistics and Probability Letters*, 35: 423–432.

Kass, R., Wasserman, L. 1996. The selection of prior distributions by formal rules, *Journal of the American Statistical Association*, 91: 1343–1370.

Kendall, M.G. 1955. *Rank Correlation Methods*, Second Edition. London: Griffin.

King, G. 1997. *A Solution to the Problem of Ecological Inference*. Princeton, NJ: Princeton University Press.

Kuehl, O. 1994. *Statistical Principles of Research Design and Analysis*. Belmont, CA: Duxbury Press.

La Salle, J., Lefschetz, S. 1961. *Stability by Liapunov's Direct Method*. New York, NY: Academic.

Lahiri, S. 1996. On inconsistency of estimators under infill asymptotics for spatial data, *Sankhya, Series A*, 58: 403–417.

Le Sage, J. 1997. Bayesian estimation of spatial autoregressive models, *International Regional Science Review*, 20: 113–129.

Le Sage, J. 2000. Bayesian estimation of limited dependent variable spatial autoregressive models, *Geographical Analysis*, 32: 19–35.

Le Sage, J., Pace, R., Tiefelsdorf, M. (eds.). 2004. Special issue: methodological developments in spatial econometrics and statistics, *Geographical Analysis*, 36: 87–194.

Lee, S. 2001. Developing a bivariate spatial association measure: an integration of Pearson's r and Moran's I, *Journal of Geographical Systems*, 3: 369–385.

Lee, Y., Nelder, J. 2001. Modelling and analysing correlated non-normal data, *Statistical Modeling*, 1: 3–16.

Linz, P. 1996. *An Introduction to Formal Languages and Automata*, Second Edition. Lexington, MS: D.C. Heath and Company.

Little, R., Rubin, D. 1987. *Statistical Analysis with Missing Data*. New York, NY: Wiley.

Mardia, K., Marshall, R. 1984. Maximum likelihood estimation of models for residual covariance in spatial regression, *Biometrika*, 71: 135–146.

Martin, R. 1984. Exact maximum likelihood for incomplete data from a correlated Gaussian process, *Communications in Statistics*, 13: 1275–1288.

McCullagh, P., Nelder, J. 1983 (2nd ed., 1989). *Generalized Linear Models*, 1st ed. London: Chapman & Hall.

McLachlan, G., Krishnan, T. 1997. *The EM-Algorithm and Extensions*. New York, NY: Wiley.

Meijering, E. 2002. Chronology of interpolation: from ancient astronomy to modern signal and image processing, *IEEE*, 90: 319–342.

Méndez, F. 1957. *Cronicas de Puerto Rico (1809–1955)*. San Juan, PR: Ediciones del Gobierno, Estado Libre Asociado de Puerto Rico.

Meng, X. 1997. The EM algorithm. In S. Kotz, C. Read, D. Banks (eds.), *Encyclopedia of Statistical Sciences*, Update vol. 1.. New York, NY: Wiley, pp. 218–227.

Millard, S., Neerchal, N. 2001. *Environmental Statistics with S-Plus*. Boca Raton, FL: CRC Press.

Moore, D. *The Basic Practice of Statistics*. New York, NY: W. H. Freeman.

Mollie, A. 1996. Bayesian mapping of disease. In R. Gilks, S. Richardson, D. Spiegelhalter (eds.), *Markov Chain Monte Carlo in Practice*. New York, NY: Chapman & Hall, pp. 359–379.

Montgomery, D., Peck, E. 1982. *Introduction to Linear Regression Analysis*. New York, NY: Wiley.

Moriconi-Ebrard, F. 1993. *L'urbanisation du Monde depuis 1950*. Paris: Anthropos.

Nelder J., McCullagh P. 1983 (2nd ed. 1989). *Generalized Linear Models*, First Edition. London: Chapman and Hall.

NRC (National Research Council). 1994. *Ranking Hazardous-waste Sites for Remedial Action*. Washington, DC: National Academy Press.

Okabe, A., Boots, B., Sugihara, K. 1992. *Spatial Tessellations: Concepts and Applications of Voronoi Diagrams*. New York, NY: Wiley.

Ord, J. 1967. Graphical methods for a class of discrete distributions, Journal of the Royal Statistical Society, Series A, 130: 232–238.

Ord, J. 1975. Estimation methods for models of spatial interaction, *Journal of the American Statistical Association*, 70: 120–126.

Paelinck, J.H.P. 1973. *Hoe doelmatig kan regionaal en sectoraal beleid zijn?* (How efficient can regional and sectoral policy be?). Leiden: Stenfert Kroese.

Paelinck, J.H.P. 1976. Qualitative multiple criteria analysis, environmental protection and multiregional development, *Papers of the Regional Science Association*, European Conference, Budapest, August 1975, vol. 36, pp. 59–79.

Paelinck, J.H.P. 1985. *Eléments d'analyse économique spatiale*. Paris: Anthropos.

Paelinck, J.H.P. 1990a. Econométrie spatiale: contributions récentes après vingt ans d'histoire, *Revue européenne des sciences sociales*, Tome XXVIII(88): 5–16.

Paelinck, J.H.P. 1990b. Some new estimators in spatial econometrics. In D.A. Griffith (ed.), *Spatial Statistics: Past, Present and Future*. Ann Arbor, MI: Institute of Mathematical Geography, pp. 163–181.

Paelinck, J.H.P. 1992. De l' économétrie spatiale aux nouvelles dynamiques spatiales. In P.-H. Derycke (ed.), *Espaces et dynamiques territoriales*. Paris: Economica, pp. 137–154.

Paelinck, J.H.P. 1996a. Econométrie urbaine dynamique. In P.-H. Derycke, J.-M. Huriot, D. Pumain (eds.), *Penser la Ville, Théories et modèles*. Paris: Anthropos, pp. 91–106.

Paelinck, J.H.P. 1996b. *Studies in Spatial Econometrics*, Research Paper, George Mason University,The Institute of Public Policy, Center for Regional Analysis, Fairfax, VA USA.

Paelinck, J.H.P. 2000. On aggregation in spatial econometric modelling, *Journal of Geographical Systems*, 2(99): 157–165.

Paelinck, J.H.P. 2002. A multiple gap approach to spatial economics, *Annals of regional Science*, 6: 219–227.

Paelinck, J.H.P. 2003. Min-algebraic and finite automata modelling in spatial econometrics, *Paper presented at the 50th annual conference of the NARSA*, Philadelphia, PA, November.

Paelinck, J.H.P. 2004. Veinte años de econometría espacial, *Proceedings of the Primer Seminario de Econometría Espacial Jean Paelinck*, Universidad de Zaragoza, Departamento de Análisis Económico, Zaragoza, Spain, pp. 1–20.

Paelinck, J.H.P., Klaassen, L.H. 1979. *Spatial Econometrics*. Farnborough: Saxon House.

Paelinck, J.H.P., van Gastel, M.A.J.J. 1995. Computing box-cox transform parameters: a new method and its application to spatial econometrics. In L. Anselin, R.J.G.M. Florax (eds.), *New Dimensions in Spatial Econometrics*. Berlin: Springer, pp. 136–155.

Paelinck, J.H.P. Mur, J., Trivez, F.J. 2005. Spatial econometrics: more lights than shadows. In F.J. Trivez et al. (eds.), *Contributions in Spatial Econometrics*. Zaragoza: Copy Center Digital, pp. 15–38.

Pascutto, C., Wakefield, J.C., Best, N., Richardson, S., Bernardinelli, L., Staines, A., Elliott, P. 2000. Statistical issues in the analysis of disease mapping data, *Statistics in Medicine*, 19: 2493–2519.

Peschel, M., Mende, W. 1986. *The Predator-Prey Model*. New York, NY: Springer.

Philibert, J. 2005. One and a half century of diffusion: Fick, Einstein, before and beyond. *Diffusion Fundamentals*, 2: 1–10.

Plante, K. 2005. QUALIREG, *Term paper*, George Mason University, School of Public Policy, December 2005.

Rao, J. 1999. Some recent advances in model-based small area estimation′, *Survey Methodology*, 25: 175–186.

Rayner, A. 1969. *A first Course in Biometry for Agriculture Students*. Pietermaritzburg, South Africa: University of Natal Press.

Richardson, S. 1990. Some remarks on the testing of association between spatial processes. In D. Griffith (ed.), *Spatial Statistics: Past, Present, and Future*. Ann Arbor, MI: Institute of Mathematical Geography, pp. 277–309.

Richardson, S. 1992. Statistical methods for geographical correlation studies. In P. Elliot, J. Cuzich, D. English, R. Stern (eds.), *Geographical and Environmental Epidemiology: Methods for Small Area Studies*. Oxford: Oxford University Press, pp. 181–204.

Richardson, S., Monfort, C. 2000. Ecological correlation studies. In P. Elliott, J. Wakefield, N. Best, D. Briggs (eds.), *Spatial Epidemiology: Methods and Applications*. New York, NY: Oxford University Press, pp. 205–220.

Ripley, B. 1988. *Statistical Inference for Spatial Processes*. Cambridge: Cambridge University Press.

Ripley, B. 1990. Gibbsian interaction models. In D.A. Griffith (ed.), *Spatial Statistics: Past, Present, and Future*. Ann Arbor, MI: Institute of Mathematical Geography, pp. 3–25.

Robert, C., Casella, G. 1999. *Monte Carlo Statistical Methods*. Berlin: Springer.

Rodríguez, F. 1999. SpatialSpatial and demographic change in nine teenth century San Juan, Puerto Rico, 1800–1868, *Journal of Urban History*, 25: 477–513.

Rubin, D. 1972. A non-iterative algorithm for least squares estimation of missing values in any analysis of variance with missing data, *Applied Statistics*, 21: 136–141.

Samuelson, P.A. 1955. *Foundations of Economic Analysis*. Cambridge, MA: Harvard University Press.

Samuelson, P. 1971. Generalized predator-prey oscillations in ecological and economic equilibrium, *Proceedings of the National Academy of Sciences*, 68: 980–983.

Scarano, F. 1984. *Sugar and Slavery in Puerto Rico: The Plantation Economy of Ponce, 1800–1850*. Madison, WI: University of Winconsin Press.

Snedecor, G., Cochran, W. 1967. *Statistical Methods*, Sixth Edition. Ames, IA: Iowa State U. Press.

Spiegelhalter, D., Best, N., Carlin, B., Van der Linde, A. 2002. Bayesian measures of model complexity and fit (with discussion), *Journal of Royal Statistic Society B*, 64: 583–640.

Stehman, S., Overton, W. 1996. Spatial sampling. In S. Arlinghaus (ed.), *Practical Handbook of Spatial Statistics*. Boca Raton, FL: CRC Press, pp. 31–63.

Stein, M. 1999. *Interpolation of Spatial Data: Some Theory for Kriging*. Berlin: Springer.

Stern, H., Cressie, N. 1999. Inference for extremes in disease mapping. In A. Lawson, A. Biggeri, D. Bohning, E. Lesaffre, J-F. Viel, R. Bertollini (eds.), *Disease Mapping and Risk Assessment for Public Health*. Chichester: Wiley, pp. 63–84.

Theil, H. 1961. *Economic Forecasts and Policy*, Second Edition. Amsterdam: North-Holland.

Theil, H. 1971. *Principles of Econometrics*. Amsterdam: North-Holland.

Thomas, E. 1968. Maps of residuals from regression. In B. Berry, D. Marble (eds.), *Spatial Analysis: A Reader in Statistical Geography*. Englewood Cliffs, NJ: Prentice Hall, pp. 326–52.

Thomas, A., Best, N., Lunn, D., Arnold, R., Spiegelhalter, D. 2004. GeoBUGS User Manual, version 1.2. Accessed at http://www.mrc-bsu.cam.ac.uk/bugs/winbugs/geobugs12manual.pdf on 3/4/2005.

Tiefelsdorf, M., Boots, B. 1995. The exact distribution of Moran′s I, *Environment and Planning A*, 27: 985–999.

Ullman, E., M. Dacey, 1960. The minimum requirements approach to the urban economic base, *Papers and Proceedings, Regional Science Association*, 6: 175–94.

Upton, G., Fingleton, B. 1985. *Spatial Data Analysis by Example*, vol. 1. New York, NY: Wiley.

Upton, G., Fingleton, B. 1989. *Spatial Data Analysis by Example*, vol. 2. New York, NY: Wiley.

van Middeldyk, R. 2004. *The History of Puerto Rico: From the Spanish Discovery to the American Occupation*. Project Gutenberg: EBook #12272.

Wakefield, J. 2003. Sensitivity analyses for ecological regression, *Biometrics*, 59: 9–17.

Wakefield, J., Salway, R. 2001. A statistical framework for ecological and aggregate studies, *Journal of the Royal Statistical Society*, 164A: 119–137.

Waller, L., Gotway, C. 2004. *Applied Spatial Statistics for Public Health Data*. New York, NY: Wiley.

Wartenberg, D. 1985. Multivariate spatial correlation: a method for exploratory geographical analysis, *Geographical Analysis*, 17: 263–283.

Whittle, P. 1954. On stationary processes in the plane, *Biometrika*, 41: 434–449.

Wolfram, S. 2002. *A New Kind of Science*. Champaign, IL: Wolfram Media.

Wrigley, N. 1985 (reprinted in 2002 by Blackburn). *Categorical Data Analysis for Geographers and Environmental Scientists*. Longman: London.

Wrigley, N. 1995. Revisiting the modifiable areal unit problem and the ecological fallacy. In A. Cliff, P. Gould, A. Hoare, N. Thrift (eds.), *Diffusing Geography: Essays Presented to Peter Haggett*. Oxford: Blackwell, pp. 49–71.

Yates, F. 1933. The analysis of replicated experiments when the field results are incomplete, *Empirical Journal of Experimental Agriculture*, 1: 129–142.

Yeo, I.-K., Johnson, R. 2000. A new family of power transformations to improve normality or symmetry, *Biometrika*, 87: 954–959.

Author Index

Subject Index

T

U

V